Communicating the Environment to Save the Planet

Maurizio Abbati

Communicating the Environment to Save the Planet

A Journey into Eco-Communication

 Springer

Maurizio Abbati
Sanremo
Imperia
Italy

ISBN 978-3-319-76016-2   ISBN 978-3-319-76017-9   (eBook)
https://doi.org/10.1007/978-3-319-76017-9

Library of Congress Control Number: 2018960355

This Springer imprint is published by the registered company Springer Nature Switzerland AG
The registered company address is: Gewerbestrasse 11, 6330 Cham, Switzerland

Natura est vita memoriae, magistra vitae est.
Inspired by Marcus Tullius Cicero

To my parents

Foreword I

PRINCE ALBERT II
OF MONACO
FOUNDATION

Monaco, 18th April 2018

"People protect and respect the things they love, and to make them love the sea, arousing their sense of wonder is as necessary as educating them".

These words of Captain Cousteau, who for almost 30 years was director of the Oceanographic Museum of Monaco, sum up perfectly the way in which the question of the environment has to be approached nowadays.

When faced with challenges that call for the participation and involvement of the greatest number of people, when faced with matters that involve the very essence of our ways of life, and when faced with issues that influence every aspect of our societies, it is indeed imperative to work collectively—in other words, altogether, without letting anyone fall by the wayside.

No matter what our resources, our goodwill, and our determination to act, we cannot save our planet unless we take on board this essential requirement.

In this context, communication is therefore much more than an adjunct: it is an integral part of the work of those who, like me, have made it a duty to conserve the environment. Thus, while it is obviously not a replacement for action, it does deserve to be treated as one dimension of action, in its own right.

That is what this book spurs us to do, through approaching these issues in a way that is educational and, at the same time, functional. Showing the fundamental mechanisms of communication, exploring the many links in the chain of which it is composed, and suggesting ways to take ownership of it and to avoid the many pitfalls that lie in wait: what we have here is a vital strategy, and all those involved in defending the environment will benefit from taking ownership, if they want their action to have a real impact on society.

That is why I wanted to support this book and to support above all the methods used by Maurizio Abbati, who approaches these issues with a strategy that is simultaneously clear and precise, comprehensive, and practical.

The objective is to ensure that environmental questions continue to feed debate; that the greatest number of people becomes empowered; and that we can, altogether, bring about change!

H.S.H. Prince Albert II of Monaco

Foreword II

SCUOLA DI GIURISPRUDENZA

Bologna, 22nd November 2017

During the years [...], I have been tutoring and training Maurizio Abbati in his academic career as Transportation, Air and Maritime Law Professor and Dean of the Faculty of Law of Bologna University.

I could appreciate his serious, committed, and capable skills together with a very good relationship and interaction with other people.

His professional career brought him to improve communication skills and enabled him to write essays and articles and hold conferences about the *Environment* and *Sustainability* in different cities, both in Italy and abroad (Venice International University, Krakow *Jagiellonian* University), as well as to successfully complete European projects for some Italian administrations.

At present, after a further specialization in journalism, he is a qualified journalist for magazines in the Principality of Monaco, France, and Italy, without forgetting his environmental commitment being the focus of many of his works. In fact, he is now the author of this book *Communicating the Environment to Save the Planet*, a very innovative and multidisciplinary approach that, in a fluent and harmonious way, makes the reader have a precise idea of how a correct communication can highlight the Environmental and Sustainable issues dealing with different fields (e.g., *Environment* and journalism; *Environment* and green & circular economy, law, architecture, art, cinema, music).

Thanks to this "*eco-communication*" the reader will be more aware of this important topic and he will be ready to use good practices to keep the place he lives healthier and healthier.

To sum up, I praise his commitment and efforts in undertaking such a kind of challenge by writing a book which will be propaedeutic for students and a useful tool for scholars, professionals, and all those who want to know more to preserve the Planet Earth.

I would greatly appreciate if his book, mentioned above, could be taken into due consideration.

Professor Stefano Zunarelli
Transportation, Air and Maritime Lawyer and Law Professor at Bologna University.
Former Dean of the Faculty of Law of the same University.
Technical Adviser at United Nations Commission on International Trade Law
– UNCITRAL; at International Maritime Association – IMO; at European
Commission - EC; at Italian Ministries of Transport and Justice.

Credits

This manual is the result of a 2-year personal dedication and research, but I cannot help thanking all those who contributed to its building up, in particular:

- the national and international professionals I have interviewed;
- the Copyright holders of the visual elements;
- any other stakeholders;
- my family and my friends.

I wish to express my everlasting gratitude to *Prince Albert II of Monaco Foundation* for honouring me with His valuable endorsement and to Professor *Stefano Zunarelli* of Bologna University whom I have known, with high regard, since the Law School.

Finally, my special thanks to *Springer Nature International Publishing* and in particular Ms. *Johanna Schwarz*, Senior Publishing Editor at Springer International Publishing, and Ms. *Dörthe Mennecke-Bühler*, Assistant Editor at Springer International Publishing, for their trust and constructive help during the editing phase.

The simple aphorism: *"Think like a wise man but communicate in the language of the people"*, by the Irish writer, poet, and playwright *William Butler Yeats* (1865–1939), contains within itself all the complexity of an action that actually has always accompanied the history of man allowing him to live with the others: communication. But how to make it in an effective way? Which instruments or Media are to be used? How to adapt the language to the addressee's code? How to verify whether the recipient has correctly understood the message?

Communication is an increasing need, a value, a right and a duty, and a fundamental requisite to survive and to find our way on which our success or failure may depend. But it is also an instrument to make people reflect on vital issues as the survival of the Ecosystem Earth which Man belongs to: that is the *Environment*, the Sustainable Development and the Climate Change.

From these premises, **Communicating the Environment to Save the Planet** focuses deeply on the theme of *"Communicating the Environment"* with an innovative and practical *approach* able to point out the main elements which characterize the *"chain of communication"*, its potential criticalities and possible solutions to communicate the *Environmental* and *Sustainable issues* in a simple, synthetic, objective, clear, and flexible way.

Eco-communication in fact is not an activity taken for granted, but it implies professional training and responsibility. How can the journalist influence his readers raising *Environmental* and *Sustainable awareness*? How can he give credibility to the *Environmental messages* conveyed by the *online Media*? Can the *Social Media* convey *Environmental* and *Sustainable values*? Can the images and any other form of artistic creativity "tell the Environment" and become icons of an "eco-sustainable revolution"? In the world of *Green* and *Circular Economy* how can we protect ourselves from the risk of *green washing*? How can we make the *Environmental message* a public *medium* to communicate at the citizens' service? How can we guarantee a qualified Environmental Communication at the level of multiple event management and legal language?

These are some of the main issues we dealt with in the manual, including some exclusive testimonies in *"The Interviews with Professionals"* and practical *"Case Studies"* parts. A *social community* is composed of university professors, consultants, architects, engineers, legal professionals, journalists, economists, artists, and musicians. To communicate the *Environment* and *Sustainability* in a proper way is a fundamental instrument. It means to convey our own commitment in the *Environmental* and *Sustainable field* through a suitable Communication Plan, to catch the audience's attention by stimulating their emotions, conveying good examples, and make our recipient more responsible, be surprised and part of all this, while increasing his knowledge and awareness on eco-sustainable issues thus implementing an *eco-conscience*.

Communicating the Environment to Save the Planet, then, invites the reader to make a journey to the discovery of the *Environmental Communication*, starting from new hints. A starting point to raise awareness of our new role of *eco-communicators* and to do it even better we need *to network* with qualified subjects.

Maurizio Abbati

Contents

Part I ECO-COMMUNICATION

1 The Environmental Communication Under the Magnifying Lens. . . . 3
    1.1 The Origins of the Environmental Communication. 6
    1.2 The Roots of the Environmental Communication 10
    1.3 The Core of Communicating the Environment 17
    1.4 Some Tools to Communicate the Environment 19
    1.5 Conclusions and Reflections: A Feasible Eco-communicative
         Approach 22
    Bibliography and Web Site List 27

**2 Journalism and Environment: Two Opposed Trends Which
Attract Each Other**. 31
    2.1 The Origins of Eco-information. 34
    2.2 The Environment: A *Green Code* to Be Decoded 37
    2.3 An "Eco-system" of Questions and Answers: A Systemic
         Approach to the Piece of News 40
    2.4 The Social Media: Impact 4.0 42
    2.5 Being Environmental Journalists: A Multi-stage Mission 45
    2.6 Conclusions and Reflections: The Strength of Environmental
         Journalism Be with You! The "Star Wars" of News. 48
    Bibliography and Web Site List 52

**3 Which Shades of "Green" Are the New Nets of
Communication Disseminating?** 55
    3.1 The Environmental Message in the *Net* Achieves the
         Four Dimensions 58
    3.2 The Quality of the Environmental Message Through the Net. 60
    3.3 What Really Links the New Media of Communication to the
         Environment?. 61

3.4   How Much Is the Environmental Information Through
      the Net Reliable?   62
3.5   Can We Express Environmental Values Through the *Net*?   66
3.6   The Environmental Messages, Are They Always Understood by
      Their Addressees in the Variegated and Immaterial "Ecosystem"
      of the *Net*?   69
3.7   Communicating on the *Net* Is a Question of Image............   71
3.8   Conclusions and Reflections: The Eco-communication
      on the *Net* Thinks Increasingly "Bigger"...................   79
Bibliography and Web Site List   82

4   **Communicating the Environment Artfully. Ciak, Action!**   85
4.1   Eco-strength of the Images   88
4.2   The Environmental Push of the Visual Arts..................   98
4.3   The New Environmental Boundaries of Bio Architecture
      and Design 105
4.4   Cinema, Music and the Environmental Message............... 118
4.5   Conclusions and Reflections: When Superman Dresses
      Up in "Green" 136
Bibliography and Web Site List 138

5   **Communicating the Environment in the *Green***
    and Circular Economy...................................... 143
5.1   *Green* and *Circular Economy* and the Environmental Message ... 146
5.2   Communicating the Environment Through Products
      and Services.. 149
5.3   *Pure Green* or *Green Washing*? Making the
      Message Credible: This Is the Question 152
5.4   When Does the Advertising Message Inform on the
      Environment and Sustainability?............................ 157
5.5   Conclusions and Reflections: The *Guerrilla Marketing*
      Raises the "Green" Flag!.................................... 164
Bibliography and Web Site List 168

6   **Communicating the Environment Is a "Public Right and Duty"**.... 169
6.1   The Strength of Public Environmental Communication:
      A Multi-voiced Journey...................................... 172
6.2   How Can the *Environmental Matter* Be Better Spread Out
      at the Institutional Level?................................. 175
6.3   The Public *Environmental* Communication: A Citizen's Right
      at the Service of the Community 178
6.4   Let Us Explore the Key Elements of an Effective
      Public Environmental Communication Plan 182
6.5   Cooperating to Communicate the Environment "In Public" 185
6.6   Conclusions and Reflections: The "Biodiversity"
      of the Public Eco-communication 189
Bibliography and Web Site List 191

7   Eco-communication: Environmental *Management*, Big Events
     and Legal *Eco*-language 193
     7.1   Local Agenda 21 and Environmental Accounting: The Shared
           Choice to Communicate the Environment.................... 196
     7.2   When the Environmental Management Systems (EMS)
           Are Turning "Green" 201
     7.3   The Big Events: The Multi-faced Challenge to Communicate the
           Environment.. 210
     7.4   Legal Language and Environmental Communication:
           Utopia or Reality? 217
     7.5   Conclusions and Reflections: The Eco-diversity of Environmental
           Communication, a Future Target or a Present Need? 222
     Bibliography and Web Site List 224

Part II   INTERVIEWS WITH PROFESSIONALS (IWP)

IWP 1   Interview No. 1: Mounir Bouchenaki 229

IWP 2   Interview No. 2: Mariaelena Camerini..................... 237

IWP 3   Interview No. 3: Francesca Carminati 241

IWP 4   Interview No. 4: Cristina Carretero González.............. 247

IWP 5   Interview No. 5: Alice Comble............................ 251

IWP 6   Interview No. 6: Edoardo Croci........................... 255

IWP 7   Interview No. 7: Barbara Frateschi Moreno 263

IWP 8   Interview No. 8: Maurizio Giani 269

IWP 9   Interview No. 9: Rhodri Jones............................ 275

IWP 10   Interview No. 10: Daniela Luise 279

IWP 11   Interview No. 11: Elisabetta Martinelli 283

IWP 12   Interview No. 12: Giulia Meloncelli 287

IWP 13   Interview No. 13: Paola Poggipollini...................... 293

IWP 14   Interview No. 14: Carlo Ratti 297

IWP 15   Interview No. 15: Niccolò Ronchi 301

IWP 16   Interview No. 16: Antonio Salinari 307

IWP 17   Interview No. 17: Omero Soliman......................... 313

IWP 18   Interview No. 18: Jeremy Tamanini 319

IWP 19   Interview No. 19: Joaquim Tarrasó Climent 323

IWP 20   **Interview No. 20: Paolo Taticchi** 329

IWP 21   **Interview No. 21: Paula Cristina Cayolla Morais Trindade** 335

IWP 22   **Interview No. 22: Eleonora Vallone** 339

APPENDIX – INTERVIEW TRANSLATIONS (IT) 347
    IT 1   Translation of Interview No. 1: Mounir Bouchenaki 347
            IT 1.1   Premise to Questions 347
            IT 1.2   Questions 347
            IT 1.3   Answers in a Nutshell 348
    IT 2   Translation of Interview No. 2: Mariaelena Camerini 350
            IT 2.1   Premise to Questions 350
            IT 2.2   Questions 351
            IT 2.3   Answers in a Nutshell 351
    IT 3   Translation of Interview No. 3: Francesca Carminati 352
            IT 3.1   Premise to Questions 352
            IT 3.2   Questions 353
            IT 3.3   Answers in a Nutshell 353
    IT 4   Translation of Interview No. 4: Cristina Carretero González 355
            IT 4.1   Premise to Questions 355
            IT 4.2   Questions 355
            IT 4.3   Answers in a Nutshell 355
    IT 5   Translation of Interview No. 5: Alice Comble 356
            IT 5.1   Questions 356
            IT 5.2   Answers in a Nutshell 357
    IT 6   Translation of Interview No. 6: Edoardo Croci 357
            IT 6.1   Premise to Questions 357
            IT 6.2   Questions 358
            IT 6.3   Answers in a Nutshell 358
    IT 7   Translation of Interview No. 7: Barbara Frateschi Moreno 361
            IT 7.1   Premise to Questions 361
            IT 7.2   Questions 361
            IT 7.3   Answers in a Nutshell 361
    IT 8   Translation of Interview No. 8: Maurizio Giani 362
            IT 8.1   Premise to Questions 362
            IT 8.2   Questions 363
            IT 8.3   Answers in a Nutshell 363
    IT 9   Translation of Interview No. 10: Daniela Luise 365
            IT 9.1   Premise to Questions 365
            IT 9.2   Questions 365
            IT 9.3   Answers in a Nutshell 366
    IT 10 Translation of Interview No. 11: Elisabetta Martinelli 367
            IT 10.1 Questions 367
            IT 10.2 Answers in a Nutshell 367

IT 11 Translation of Interview No. 12: Giulia Meloncelli 369
       IT 11.1 Premise to Questions 369
       IT 11.2 Questions 369
       IT 11.3 Answers in a Nutshell 370
IT 12 Translation of Interview No. 13: Paola Poggipollini 372
       IT 12.1 Questions 372
       IT 12.2 Answers in a Nutshell 372
IT 13 Translation of Interview No. 14: Carlo Ratti 373
       IT 13.1 Premise to Questions 373
       IT 13.2 Questions 374
       IT 13.3 Answers in a Nutshell 374
IT 14 Translation of Interview No. 15: Niccolò Ronchi 375
       IT 14.1 Premise to Questions 375
       IT 14.2 Questions 375
       IT 14.3 Premise to Questions 375
       IT 14.4 Questions 376
       IT 14.5 Answers in a Nutshell 376
IT 15 Translation of Interview No. 16: Antonio Salinari. 378
       IT 15.1 Questions 378
       IT 15.2 Answers in a Nutshell 379
IT 16 Translation of Interview No. 17: Omero Soliman 380
       IT 16.1 Questions 380
       IT 16.2 Answers in a Nutshell 381
IT 17 Translation of Interview No. 19: Joaquim Tarrasó Climent. 382
       IT 17.1 Premise to Questions 382
       IT 17.2 Questions 383
       IT 17.3 Premise to Questions 383
       IT 17.4 Questions 383
       IT 17.5 Answers in a Nutshell 383
IT 18 Translation of Interview No. 20: Paolo Taticchi 385
       IT 18.1 Questions 385
       IT 18.2 Answers in a Nutshell 386
IT 19 Translation of Interview No. 22: Eleonora Vallone 387
       IT 19.1 Premise to Questions 387
       IT 19.2 Questions 387
       IT 19.3 Premise to Questions 388
       IT 19.4 Questions 388
       IT 19.5 Answers in a Nutshell 388

Part III   CASE STUDIES (CS)

**CS 1   Case Study 1: The Buddhas of Bâmiyân—An
         Unsustainable Loss**. 393

**CS 2   Case Study 2: An Adriatic Sea Project That Communicates
         the Idea of Sustainability**. 397

CS 3   Case Study 3: When Communicating the *Environment*:
        The Common Sense Makes the Difference 401

CS 4   Case Study 4: The Communicative and Participatory
        Strength to Make Milan More Sustainable 403

CS 5   Case Study 5: The Sustainability Communicated by the
        Young People—Dribbling the Sustainability "Skateboard" 407

CS 6   Case Study 6: The Tools of the Urban Planning Give
        Voice to *Environment* 411
        CS 6.1 *Metrominuto* Ferrara/Ferrara Metre-Minute 411
        CS 6.2 A Social Wood for the City of Ferrara (Italy) 412
        CS 6.3 ECOWASTE4FOOD Project 413

CS 7   Case Study 7: When Objects Communicate the *Environment* 415
        CS 7.1 Computer Bag/Rucksack (Fig. CS 7.1.1). 416
        CS 7.2 Billiard Cloth Hanger (Fig. CS 7.2.1) 416
        CS 7.3 Half-Light Lamp (Fig. CS 7.3.1) 417

CS 8   Case Study 8: *Treepedia*, the Project That Makes Trees
        the Eco-communicators 419

CS 9   Case Study 9: The "Soul" of Architecture Communicates
        the Respect of the *Environment* Through the Five Senses 423
        CS 9.1 The Case Study of the *Hotel Residence Rossi* of
               Sirmione (Garda Lake) 424

CS 10   Case Study 10: A New Visual Communication Tool to
        Facilitate Any *Decision-Maker* Task—The *State of Green
        Transition Index* 427

CS 11   Case Study 11: Communicating the *Environment* Through the
        Architectural Linguistic Code. 431
        CS 11.1 Architecture and Design Are Evolving into
                Living Organisms. 431
        CS 11.2 The Nature "Embraces" the Architecture to
                Eco-communicate 432
        CS 11.3 The Square Communicates the Importance of the
                Water Resource 433

CS 12   Case Study 12: Communicating the Sustainability as a *Driver*
        of Innovation: Three Italian Eco-Virtuous Examples. 435

About the Author. 439

PART I

ECO-COMMUNICATION

The Environmental Communication Under the Magnifying Lens

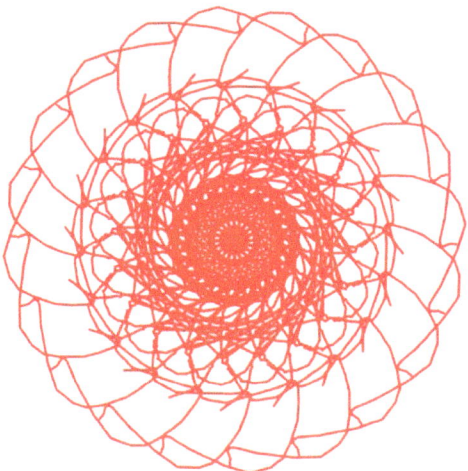

Abstract

Communicating the *Environment* is the result of a thought evolution, launched at the dawn of the *Nineteenth Century* by some researchers in the field of science and social science. Initially criticised and hindered by the most conservative scholars, *Environmental Communication* spread worldwide debating on the key matters on how to preserve the Earth ecosystem. A complex communication process which is influenced by many factors and threatened by different kind of "noise" able to jeopardise the mutual understanding of *Environmental messages*, while moving from the sender to the addressee. Hence the need to prevent in advance most of the potential misunderstandings, caring about every single detail, both lexical and expressive through a smooth, honest, clear, responsible, objective and flexible communication strategy.

© Springer Nature Switzerland AG 2019
M. Abbati, *Communicating the Environment to Save the Planet*,
https://doi.org/10.1007/978-3-319-76017-9_1

"Omnia vivunt, omnia inter se connexa"
"Everything is alive, everything is interrelated"
Marc Tully Cicero
Roman lawyer, politician, writer, orator and philosopher.
(106 B.C. – 43 B.C.)

(**a–c**) > Developer: © Plant-for-the-Planet—Environmental Awareness-raising Campaign aimed at reducing CO_2 emissions through tree transplanting compensation measures. Communication Agency: © *Leagas Delaney Global* (Copyright Holder)—Artist: *Lorenzo Duran*. www.plant-for-the-planet.org, www.lorenzomanuelduran.es, www.leagasdelaney.co.uk

Reading Proposal: In Between Words and Images

The **Project** *Plant-for-the-planet,* an evolution from *Felix Finkbeiner*'s creative idea, a German boy of nine who, being strongly fond of the climate change and very concerned about the health effects of the photosynthesis in the process of absorption of CO_2, started, in 2007, a **personal awareness campaign** with its ambitious project: to plant one million trees in Germany. Following his first success (150,000 trees were planted in Germany only, in 2007) the young Felix, a year after, was able to introduce his project to the UN, during the **International Conference for the Environment Programme of UN** (UNEP) held in Norway. A Foundation was soon established and spread all over the world. At present, more than 15 billion trees have been planted in 193 Countries under the guidance of Plant-for-the-planet and UNEP. On the 9th March 2018, Felix Finkbeiner signed in the Principality of Monaco the Trillion Tree Declaration together with H.S.H. Prince Albert II of Monaco, the World Wildlife Fund (WWF), the Wildlife Conservation Society (WCS) and BirdLife International. The aim is to plant at least a trillion trees by 2050, giving everyone the opportunity to donate one or more trees through Internet. www.trilliontreecampaign.org.

Communication Campaign: the main theme of the plantation of new trees relives in *Lorenzo Duran*'s works of art, the Spanish artist from *Guadalajara* who has chosen the most fragile and sophisticated element in nature to realize his masterpieces: fresh leaves. *Lorenzo*, by making precise cuts as a surgeon on the delicate leaf which acts as a "solar panel" for the plant, creates figures inspired by nature, human world, into geometrical shapes of great visual and communicative impact.

Form of Art: his art takes its inspiration from the *papel picado* (literally: perforated paper) an ability which transforms paper, cardboard or cloth into very elaborate objects of design, traditionally used in *Mexico* in special occasions or during religious festivities. A technique whose deep origins rely upon other countries in the world such as China where, on the contrary, paper is cut with scissors and not carved, following the tradition of *jianzhi*.

Interpretation of figures on the cover: in figure (a) it is self-evident the neat contrast between the plant area, symbol of a complex, powerful, sophisticate Nature but, in the same time, fragile and the silhouette representation of a highly industrial manmade area. Several industrial chimneys are captured while vomiting into the air their dangerous smokes gathering soon in artificial clouds. A situation which is often associated to the atmosphere pollution producing a strong impact on the eco-system and represents one of the casual factor of the growing of CO_2 concentration in the atmosphere. In figure (b) a concentric incision reinforces the environmental message: a never-ending line of vehicles, queuing and emanating gas into the atmosphere. We can reflect then on the uses of biofuel or hybrid systems that, with the same performance, produce a reduced impact on the *Environment*. The same reflection can be applied to figure (c) where the leaf gives birth to the aeroplane, symbol of technological progress as the car. Nevertheless, here the polluting factor is emphasized, that is the emission of gas engine generated into the atmosphere, among them carbon dioxide (but also nitrogen oxide, hydrocarbon, carbon monoxide, etc.). So as in these pieces of Art, the Communication we are going to deal with in the *"Chapter 1—The Environmental Communication under the magnifying lens"*, is the result of a long journey in which any detail has its own role.

1.1   The Origins of the Environmental Communication

Historical *excursus*: communicating the *Environment*, the result of a long thought evolution.

"There is something infinitely healing in the repeated refrains of nature—the assurance that dawn comes after night, and spring after winter" reminds us **Rachel Louise Carson** (Springdale, 1907–Silver Spring, 1964), the American biologist and zoologist, author of numerous books among them her bestseller *Silent Spring* giving birth to what will be defined later, all over the world, the *environmental movement*. Years and years of studies led the researcher to explore the connections between *Environment* and the use of new pesticides and particularly the well-known DDT (*Dichloro Diphenyl Trichloroethane*)—chemical substances widely used by farmers to fight against the storms of insects.

Provided that *"in nature nothing exists alone"* (*R. Carson*) the author published a bleak picture underlying the dangerous risks linked to the disproportionate use of insecticides which inevitably affected the food chain of other animals (e.g.: birds) and man, himself. A position that, in spite of criticisms and pressures, brought *Carson*'s *Environmental message* to raise and influence public awareness to the extent that DDT was banned some years later her death (1972) to avoid the world would lose its Springtime, rebirth for the entire Earth eco-system, generally accompanied by the usual festive chirping of many species of birds.

It is not by chance that *Carson*'s ecological commitment and her literary production is traditionally associated to the birth of the *Environmental Communication* meant as subject of study. The analysis of the *Environment* problems and the search of methods to improve the very quality of the *Environment* came to light in the Anglo-Saxon literature of the years Sixties and Seventies as a consequence of the "green impulse" given by the American writer. A trend that will witness, during the Years Eighties, the birth of the first university courses and faculties specialized in the art of communicating the *Environment*, established originally in the scientific and medical areas. Up to the time of the first Years Nineties when real National and International Associations were set up to focus the communicative aspects linked to *Environment* meant to avoid conflicts, to solve problems and to sustain the proactive exchange of ideas (e.g.: *International Environmental Communication Association, IECA*).

From these preconditions today's reality originates and the *Environmental Communication* is a subject of studies, researches, theories, conferences, conventions, placing itself even more as an interdisciplinary subject which unites the economic-scientific area to the humanities fields and social sciences. A journey of "green" communication which pushes some researchers to analyse it from a critical or analytical point of view or others to suggest tools and methods in order to improve the efficacy, addressed to the *mass communication* or to a selected audience. Finally, there are those who study it in relation to the indefinite *Environment* subject areas: climate change, pollution, endangered species, nuclear energy, acid rains, just to name some.

But the spread of *Environmental Communication* as a subject was so sharp as to influence rapidly, within a decade (between Nineties and Years 2000), even the corporate sector by introducing the concept of the *triple bottom line* (*van Marrewijk 2003; Elkington 1999*). At the base of the *company sustainability* this principle drives the productive sector (e.g.: companies, multinational corporations, enterprises, etc.) to achieve, at organizational level, a balance among three factors: profit, human resources and the respect of the Planet Earth (Profit, People, Planet). The traditional economic aspects integrate, for the first time, into the *Environment* and social ones. And here the business management feels the necessity to communicate the effects of production chain onto the surrounding *Environment* both indoor (addresses: managers, employees, etc.) and outdoor (addresses: suppliers, distributors, sellers, contractors, etc.).

An irreversible trend to *communication* of the *Environment* commitment which engages other commercial and advertising channels of distribution of goods and services. Nowadays, it is a *surplus value*, more and more requested by the consumers and meant to increase the reputation, that is a marking credibility, provided that we rely upon scientifically proved data and implement good *Environmental* practices as we will be dealing later. For example, through the highlighting of annual reports, journal articles or information about the websites on the production phases respecting the *Environment* (*Green signalling*); or underlying the *Environment* performances of a given article (*Green advertising*).

The *Green Communication*, a *must* in the career training of Public Relations, is not however a synonym of *Environment Communication* either of *Sustainable Development*. From that assumption, we decided to prefer the adjective "*Environmental*" instead of "green". Obviously not because we think it is unsuitable. On the contrary, to be "green" in communicating represents a best practice if they use "those Media" which enable a real reduction in the consumption of resources (e.g. paper) or in the efficient use of energy (PC, laptop, scanner, etc. with energetic certifications, use of energy from renewable sources, etc.). Practically it would be very contradictory if those charged in a green communication campaign would print only ten pages of introduction, using a common non-certified paper (i.e. produced by forests run according to the sustainability *Environmental principles*— as the label FSC—*Forest Stewardship Council* states). A label which has been guaranteeing the traceability of the products for 20 years, from the plantation of the tree to its whole life cycle until its substitution during the phase of cutting down to produce wood-based products. A simplified example which reminds us that the best way to communicate "green" is to be always coherent to what we communicate.

As we will see during our "editorial journey" in the heart of the *Environmental Communication* there are different ways of approaching this topic. The communication of the *Environment* may be treated, for example, under the philosophical aspect by analysing the human beings who created the concept itself of Nature (*Neil Evernden*). A very long evolutionary path whose roots must be found in the ancient Greece to attain the duality of the modern vision, *Man* and *Nature*, conceptualized through culture, means of communication and new expressive arts. We may think about the revolution brought to painting by the introduction of perspective which represented landscapes with *Man* and *Nature* interacting. A pictorial technique which has developed since the end of the Thirteenth Century thanks to the Italian master *Giotto*.

Nowadays, the *Environmental Communication* can be expressed by the efficient use of *Mass Media* traditional tools such as newspapers, radio, television, mail etc. or new ones such as *Internet*, *Social Media* etc. Means that, properly used, can catalyse the attention of thousands and thousands of people on the *Environment* issues and its "derivatives" as the concept of Sustainable Development (*John Muir* 1890), which will be examined in depth later on in the next paragraph.

To **Communicate the *Environment*** is above all a discipline closely linked to the audience's perception of Nature, according to how the communicator of the *Environmental message* presents it (*Robert Cox*). An innovative approach which, investigating a system of interrelated elements, affects our choices, our way of thinking, our everyday life, following a holistic approach. What the American thinker *Neil Postman* describes as "*the ecology of Media*" in the most social sense of the term. A word closer to the etymology of the noun deriving by the union of two Greek words: οἶκος (oikos), home and λόγος (logos), speech.

We cannot forget the **lexical evolution of the Latin term "*ambiens*"** from which the word "ambient" [*Environment*] derives. Its ecological meaning is given for granted, nowadays, but it is the result of a long semantic evolution. The entry word is used in the Italian language during the Middle Ages and it was **associated with the term "air"** to identify the space surrounding an object or a person. During the Nineteenth Century, under the French cultural influence, the "*Environment*" was associated with the semantic area of social, economic and cultural sciences and it was known as one of the causes of the thought change, and, by an inspired accident, as one of the cause of the gene mutation. From the physic sphere, it moves to the historical and social sciences including the behavioural aspects of each individual belonging to a specific community. The concept of "**social and cultural *Environment***" is thus developed.

We would like to point out that the same term "ambient" is being used, in everyday speech, as a synonym of "**room**", i.e. an **enclosed area**. "*Aerate an Environment before staying*" literary translation from the Italian advice frequently used in commercial and advertising language. While in biology ***ambient-Environment* means the set of physical and biological condition of all the living creatures are subjected to**.

The industrial and technological evolution of the Nineties changed the ***Environmental resources***: coal, water, wind, etc. into **energy**, thus generating an "*Environmental change*" and consequently an "*Environmental pollution*" of the eco-system, in order to satisfy the economic profit and exploitation. Hence, the further lexical evolution of "*ambient-environment*" associated to ecology.

Finally, the concept of *"biological Environment"* merges into ***"ecosystem"*** in which the *Environment* is the complex system composed of multiple interacting living creatures ensuring the survival of the Planet Earth.

The idea of a unique system is then established beyond the traditional subdivision into competence areas. Techniques, culture, sensorial perceptions, forms of expression, and more recently, new technologies, personal computers and virtual world, tend to merge together thanks to a large group of thinkers, groups and currents coming from different learning and training paths which, since the Years Seventies, have given birth to Media ecology.

An example, the ***New York School***, symbol of the intellectual fervour specifically present in the *Big Apple,* whose territory is scattered by the most prestigious universities at international level (e.g. Harvard University, New York University NYU, Columbia University etc.). And here the first reflections on the relationship which links the technological innovation, i.e. the technique (in its broader meaning) and the *Environment* called "man", influenced by habits, superstition and trends, find their fertile ground strictly linked to Nature and its resources (*Lewis Mumford*). A link between human and natural that man can communicate with a multiplicity of means: words, signs, symbols, art. Each expressive or creative form can send a message and therefore influence our culture (*Susan Langer*). Hence, the educational function of *communication*, with its capacity to enhance our knowledge of the *Environmental Subject*, helps us to "digest" the contents through the explanation of technical terms and the subdivision into sub-matters (*Neil Postman*).

A capacity able to affect, inform and form those who *communicate the Environment* with responsibility. But *Media* have created a new vision of *Environment* from which new forms of knowledge and understanding originate and, if they should become dominant, they could subject to condition the development of the society modifying its culture (*Harold Innis, Marshall McLuhan, Walter Ong,* ***Toronto School***). A responsibility which becomes crucial for those who exercise a profession such as a journalist. The press is in fact one of the main factor of the socio-cultural changes and, as such, can facilitate the diffusion of a correct *Environmental* culture (*McKenzie, Park, Burgess, Toronto School*). The leading function of the *Environmental Communication* is, in fact, to find the common elements between the human and natural dimension considering that man cannot leave out of consideration the *Environment* in which he lives (*John Dewey, Charles H. Cooley, William I. Thomas,* ***Chicago School***). The human and natural world, on the other hand, coexist in the same system: Planet Earth. They can but interact and be interconnected (*Gregory Bateson,* ***Palo Alto School***).

Green Tweets

@RachelCarson   *#SilentSpring*   *#EnvironmentalCommunication*
#EnvironmentalMovement *#EcologicalCommitment* *#MassCommunication*
#TripleBottomLine *#TBL* *#Profit* *#People* *#Planet* *#GreenSignalling*
#GreenAdvertising *#Ecology* *#FSC* *#ForestStewardshipCouncil* *#Ecosystem*
@NewYorkSchool *@TorontoSchool* *@ChicagoSchool* *@PaloAltoSchool*
#Education *#Responsibility*

1.2   The Roots of the Environmental Communication

Communicating is a dynamic journey, in a continuous evolution and never equal to itself.

"*A man could not feel any pleasure in discovering the beauties of the universe, even in Heaven itself, if he had not a partner to whom he might communicate his own joy*", in this famous quotation by *Marc Tully Cicero* the essence of the unavoidable need of the mankind to communicate is expressed with extraordinary modernity. The etymological studies of the term "to communicate" make it derive from the Latin verb *communicare*, evolution of the adjective *communis* [common] to which a lot of semantic meanings are given: a common thing that belongs to everybody; but it includes also something friendly, nice, sociable, from which its tight link to concepts like: community, nation, group of people derives, if we are referring strictly to the human "world", of course. It is necessary to underline this since each living being of the ecosystem Earth, and recently also the objects thanks to the new *high-tech* devices we will be talking later in *Chap. 3*, communicate.

Accordingly, we understand from the very beginning how a simple word connotes a lot of meanings to which other countless ways of communication can be associated according to **what** is communicated, **to whom** is communicated, **to what** purpose is communicated, **under what** circumstances is communicated. Referring to the human species, then, the evolution of *technology* has ensured the original meaning of communicating. So, the *sharing* of the dynamic concepts, ideas and information, considered as if they were *tangible*, *material*, became soon a more sophisticated and complex message. The information spreading from the **addresser** or **sender** to the **addressee** or **receiver** through a **linguistic code** that turns into a **message** rich of knowledge, emotions, psychological elements, gestures, and any other expressive form able to communicate. Theoretically we can state then that a conveyed message can influence another one (*process of social influence*), without taking into consideration its application field (e.g. linguistics, sociology, psychology, information technologies—IT at the dawn of their diffusion).

A *communication* function that nowadays is given for granted and has taken global dimensions with the birth of the Net or *World Wide Web*—the most popular space of information where documents and other web resources are identified by Uniform Resource Locators (URL). "*What information consumes is rather obvious: it consumes the attention of its recipients. Hence, a wealth of information* creates a poverty of attention" as *Herbert Simon*, Nobel Prize for Economy, stated in 1971, foreseeing a debate that will take place in the engine search age, such as "Google".

But the first "evolutionary theories" on the subject of *communication* will point out also another key aspect of the same. The lasting link between the contents of communication and what derives from. "*Communication either affects conduct or is without any discernible and probable effect*" (*Shannon and Weaver* 1949). A reflection that will influence the whole communication—subdivided according to *Weaver* in three levels of analysis: Level 1, **information accuracy**; Level 2, the transmission of **meaning by symbols**; Level 3, the **inductive behaviour** of the **receiver** (social aspect), eliciting the correct decoding of message by the addressee. Therefore, a ***Plan of Communication*** is effective if it is able to foresee and prevent the effects of any possible "**noise**" which could compromise the right understanding of the message, in its journey from the source of information to the decoding by the recipient.

For this reason, it is necessary that the contents of the message are listened to, understood and remembered. Many are the factors contributing to the realization of this: the codes used, the time, the *Environment* in which the communication takes place, the perception, the memory, the capability of information, the informational tools available. Just as many as the typologies of communication are:

– **verbal**, expressed by words and sounds;
– **non-verbal**, expressed by gestures, body movements, look, the so-called *kinesics*; or through the space, the distance, the so-called *proxemics*; or even through the tone of voice, its rhythm, the vocalizations as laugh or whispers, the so called *paralinguistic*.

Everything makes us understand the axiom "*We cannot not communicate*" by *Paul Watzlawick*, an Austrian psychologist and philosopher, naturalized United States citizen. Starting from the thesis "*communicating is behavioural*" and the antithesis "*non-behavioural is not communicating*" we are reaching the conclusion that any behaviour is a message and, being so, it communicates (synthesis). The attempt of not communicating, in fact, is translated into an action. A conduct that may bring to three different main outcomes:

1. the willingness to refuse communication implies the sending of signals that make us interact with the others. This behaviour can be interrupted according to action number 2.
   – e.g. *shaking your head, keeping quiet, taping your mouth and/or face, etc.*
2. The willingness to accept a conversation: deciding to start a communication even if originally you did not have this intention.
   – e.g. *to defend yourselves, to argue on something you disagree, when you are passionate about a matter, when you are pressed to talk or you cannot stand a situation anymore, etc.*
3. The willingness to discredit the communication: doubting the validity of contents given or received.
   – contradicting oneself, changing speech, not being coherent or intentionally incomplete, misunderstanding, giving closing statements, complaining, arguing, etc.

Not answering an *email,* for instance, is a trend which communicates by itself both in a professional or relational field. Originated as a "***defensive approach***" to the numerous advertising and unwanted invitations (*spam*), that made the management of the mail box difficult, the resulting non-message produces "noises" from the point of view of communication. Nowadays this problem is overcome by the installation of a good *anti-spam* software. Silence in fact might means "no" or "yes" or "perhaps". And inevitably this makes the sender unable to decode the unanswered message and hence spontaneous questions rise: maybe the addressee has not read the message yet? Maybe the addressee wants to ignore me? Maybe the addressee is too busy to answer? Maybe the addressee is tired? Maybe the addressee's mobile or personal computer are not connected to a network? All that makes the time of communication to protract longer, distracting from the real contents of the communication or jeopardizing the correct reception. In many cases, in fact, more reminders are necessary in order to have an answer or even we are obliged to change the means of communication (e.g. telephone, *Social Media* etc.). *Email*s in fact do not give any clue and silence is impossible to decode (*Lucy Kellaway*). This result may have unpredictable consequences when *Environment* and Sustainability are involved.

Certainly, the considerations so far start from the assumption that addresser and addressee share the same linguistic code. The problem could be much more complex in case it (the message) would be false. As it is well emphasized by the visual *rebus* suggested by **Umberto Eco** in "***Kant e l'ornitorinco***" (Kant and the platypus). The interpretation of two concentric circles, source of inspiration for yesterday and today architects, (see Fig. 1.1) changes according to the linguistic code of the group identity. Those who use the Latin alphabet could associate the figure of circles to two

Fig. 1.1 St. Peter's Square, Vatican City, Rome—original project by the architect *Gian Lorenzo Bernini* and some glimpses by *Maurizio Abbati* (Photographer) © 2010 * Other examples of architectural concentric cycles would be: *Plebiscito* Square in Naples (IT); *Zentrum Paul Klee*, Berna (CH) by Italian architect designer *Renzo Piano*; *Calatrava Bridge*, *Reggio Emilia* (IT) by Spanish architect designer *Santiago Calatrava*; Greek Theatre of *Taormina* (IT)

"C", but for a Greek would find easy to associate them to "Ω" (omega). A Russian would associate to two "S" (in Cyrillic alphabet "C" corresponds to the sound "S"). For other cultures would be much more difficult to compares a graphic symbol to an alphabet letter (e.g.: Arabic, Jew, Chinese, Japanese culture etc.). For those who are fond of crossword puzzles and are Italian speakers might invent a *rebus "Se-mi-cerchi non C(i) sono" "If you need me, I am not in to anyone"*—original definition. This short semiotic and semantic "experiment" shows that a communicated message would not be universally understood correctly. Its interpretation is then, on the contrary, the result of many factors combined together: a "hybrid" (*Eco*).

On the matter **Roman Jakobson**, Russian linguist and semiologist, naturalized American, concluded that "*the code is not limited to what the engineers call purely cognitive content of the speech*" but is the summing up of the stratification of more constituent elements influenced by signs, symbols, perceptions. In practice, the way of perceiving things around us varies according to the culture of the society in which we live, passed down by our family. If we think about that, effectively, our knowledge about *Environmental matters* relies merely upon our relationship of communication with friends, colleagues, relatives, businessmen, and, in general, with anybody who gets in touch with us (*R. Cox* 2010).

The anthropologist and linguist *Edward T. Hall* (*Chicago School*) was made to state that people grown up in different cultures live different sensorial worlds, so called *sensorial relativism*. A consideration that is clearer when we find difficult to explain *in words* what goes beyond our "cultural code", like a modern abstract piece of art or a musical avant-garde composition which do not follow the aesthetic canons or classical melodies which we are used to.

Communication therefore is always influenced by a series of factors conditioning its chain—meant as an articulated set, also defined as "net" or "system", including the main activities and their material and informative flows. First of all, the so-called **noises of communication** that, as already said, may alter the correct delivery of the message to the addressee. Herewith some communication obstacles:

1 **physiological aspects**: feel hot, cold, suffer from pain etc.;
2 **technical aspects**: to be in a draughty or heated *Environment* or to undergo the sound effects of a construction or road site, to have the telephone network or Internet disturbed etc.;
3 **socio-psychological aspects**: difference of personality, education, culture etc.;
4 **semantic aspects**: multi-meaning words, double sense, false-friends, foreign words etc.;
5 **relation aspects** concerning *how* to communicate: words, gestures, written messages, signals, looks, pieces of art, photos, pitto-writing, television, radio, newspapers, *Internet*, *Social Media* (e.g.: Facebook, Google+, Twitter, Instagram, WhatsApp, etc.), just to make some examples.

Last but not least, the context in which the interchange of messages and ideas takes place has got a precise role in making the communication more or less efficacious. Each interpersonal exchange of messages to be communicated assumes, in fact, a relationship between two or more subjects who may belong to the same social level (*symmetric relationship*) or to two different levels: superior and inferior (***complementary relation***). Furthermore, they can belong to a huge and heterogeneous social group (***external communication***). Starting from this, a mechanism of competition can originate that leads one of the speakers to prevail on the others or, on the contrary, a complementary relationship which can be more or less productive, according to the willingness to reach agreement or stand on its distant ground. This may happen between two subjects belonging to two different generations such as parents and sons: the so-called *generation gap.*

Communicating then is not a mechanic default operation, equal to itself but a dynamic process, constantly evolving and whose single detail takes on a different meaning thus characterizing the relationship between one single individual and the community he belongs to or *vice versa.* A dimension in which, for example, the novelty effect plays a key role in drawing the attention. Only the message whose content is unexpected by the addressee, represents a real novelty to be decoded. In prearranging a conversation, we must take into consideration also our **"audience" expectations**. An element well known by the professionals of communication, and by the advertising world (e.g.: *catchphrase, slogan*), where we try to "shock" the public opinion by the ever-growing use of multimedia contributions or unexpected messages (e.g.: Figs. 1.2, 1.3, and 1.4).

The image of a she Polar bear, symbol of the global warming, together with her bear cud doing cuckoo from a large cardboard normally used by *the urban homeless,* "upsets" any possible expectation focussing on the main environmental problem universally called into question.

The communication is simple, quick, efficacious, catch the buyer's attention, by matching an object to a different colour from "green", the symbolic colour of ecology, sustainability and natural ecosystem; a relationship between an abstract concept to a material one: a metonymy in the literary language.

An eco-message by an Italian e-commerce company specialised in the mail order of shoppers, packaging materials and disposable foodservice products. The bag, in fact, is the combination of different "green" elements considering that:

Fig. 1.2 Awareness campaign for the phenomenon of Global Warming in occasion of the tenth UN Convention about Climate Change—"*Global warming is leaving many homeless*"—© *EcoEduca*, Chile (Copyright Holder).

Fig. 1.3 Awareness campaign for the reduction of waste linked to the large distribution of the retail trade, encouraging the recycling of the same shopper at each visit to the supermarket—*"This blue shopper is green"* © Carrefour DR (Copyright Holder). www.carrefour.fr

Fig. 1.4 *"Eco Bag Think Green"*, shopper bag made of jute (natural fibre) by © Eurofides (Copyright Holder), an Italian e-commerce company specialised in the mail order of shoppers, packaging materials and disposable foodservice products. www.eurofides.com

1. it is made of natural fibre;
2. the green tree as the symbol of Environment;
3. the catch phrase invitation to think about the Environment and buy sustainable goods.

In the *communication* described so far, what makes the difference is not only the **circular movement by the message**, product of a continuous and active exchange between addresser and addressee, it differs thus from the mere information resulting by a simple *one-way movement* from the addresser to one or more addressees. *Communicating* means to reflect on the quality of contents conveyed. A requisite but also a challenge of any kind of information, more essential if linked to the *Environmental subject*, called in the present book *Environmental System* to underline the complexity of structural elements and mechanisms beyond the *Environmental matter*. One of them is the key concept of Sustainable Development first defined in 1987 by the **Bruntland Report: Our Common Future**, drawn up by the World Commission on Environment and Development (WCED), at the time presided over by *Gro Harlem Brundtland*, Norwegian Prime Minister.

> *"Sustainable Development is development that meets the needs of the present without compromising the ability of future generations to meet their own needs"* (WCDE 1987)

A fundamental document which states a historical step doing the groundwork for a long international process aiming at making compatible the world economic development through the precious *medium* of intergovernmental cooperation but also scientific, cultural and social. Principles that are made into *universal values* starting from the **UN Conference of Rio de Janeiro in 1992**, nicknamed *Earth Summit*. A key event to which delegation coming from all over the world gathered to discuss about the world ecosystem protection—some figures just to have an idea: 172 Governments, 108 heads of States and Governments, 2400 representatives of non-governmental Organizations (ONG) and International Organizations (OI), 17,000 members attending the parallel forum to institutional negotiations.

An important result to fix States' rights and responsibilities to facilitate an economic, juridical, socially sustainable and participated development (*Declaration of Rio on environment and development*) which laid the foundation for the development of **Environmental Communication**. We are referring to the **Local Agenda 21**, a *manifesto* to implement at territorial level, becoming soon for many local administrations a sustainable tool of dialogue and participation to public administration at world level. A strong drive to local capacities to communicate and promoting *Environmental policies* through the dialogue approach and sharing the support to participation of all the companies' representatives summoned up to reflect on the implication of human activities on the *Environment* and resources available. Negotiating tables, conferences, *peer review* conferences, community meetings, shared projects, auditors' agreements, education centres, *Environmental research* and innovation, training courses, awareness campaigns on *Environmental* and *Sustainable themes*. These are only some of the activities still being used in many local administrations, at international level, whose start originated from that document.

Green Tweets
@MarcTullyCicero #Communicare #Communication #Verbal #Nonverbal #InternalCommunication #ExternalCommunication #SymmetricRelationship #ComplementaryRelation #WaysofCommunication #Technology #Adresser #Addressee #Message #LinguisticCode #WorldWideWeb #Internet #CommunicationNoise @PaulWatzlawick #WecannotnotCommunicate @UmbertoEco #LinguisticCodes @RomanJackobson #Signs #Symbols #Perceptions #Noise #CommunicationPlan #SocialMedia @EdwardHall #SensorialCommunication #Dynamic #ProcessofCommunication #BruntlandReport #RioConference @UnitedNations #UN #EarthSummit @LocalAgenda21

1.3   The Core of Communicating the Environment

The deep knowledge of what is being communicated makes the *Environment* more sustainable.

What do we mean by Environmental Communication? Before answering this question, we would like to start from a short introduction. Considering the numerous aspects of what is called "Environmental System", in this Manual, we are going to identify the **Environment** as an **interdisciplinary system** overcoming the traditional subdivision in subjects (e.g.: science, economy, law, social sciences, architecture, art, music, design, marketing, etc.), typical of the Western Culture whose roots are to be found in the ancient Greece and then spread out all over Europe by establishing *scholae*, during the Medieval time. *Communicating the Environment* cannot boil down to a mere transmission of information concerning the huge world of *Environmental issues*. We will realize soon the impossibility of managing them without an organic and rational plan. In the same time, we will have to face different topics leading to the most varied debates: from the climate change to the safeguard of cetaceous or grizzly bear; from renewable resources to acid rains etc.

A wiser approach would suggest then to subdivide, in a logical way, the role of the *Environmental message* according to the *medium* of its transmission and delivery: that is through language, art, photography, scientific publication, a demonstration, a street mob etc. As reminded by *Kenneth Burke* (*Language as Symbolic Action, 1966*), in fact, each element of our language or human action expresses something potentially persuasive.

Environment Communication must be treated as a tool to shape the mankind's knowledge, in order to broaden his awareness of the relationship between his world and the natural one. From that the multi-faceted "soul" of communicating the *Environment* springs out with its purpose to inform, to teach, to persuade, to solve problems, to prevent negative impacts, that is our action effects or our behaviours which destroy the *Environment*. We can think about the awareness campaigns to protect a specific area, or those meant to influence the public's opinion to mobilise against the building of an industrial polluting plant or, on the contrary, the diffusion of *sustainability report* (or *Environmental balance*) meant to underline the *trend* and the social commitment of a certain organization towards the *Environmental* themes and *Sustainable Development*.

On this basis, the **Environmental Communication** can be defined as "***the pragmatic and constitutive vehicle for our understanding of the environment as well as our relationships to the natural world***" (*Robert Cox*). The action of *communicating the Environment* becomes itself a *medium* used to solve *Environmental problems* and to manage any possible debate derived by the public opinion.

But why is it pragmatic? To *communicate* an *Environmental Message* implies always an action whether directed to teaching, to raise awareness, to mobilise etcetera. Let us consider an ad campaign towards awareness to saving paper in order to prevent deforestation of protected areas or a *flash-mob*—literally "flash-crowd" a sudden gathering of people, unknown to each other, who got an appointment somewhere through *Internet*—against the building of a public work or an industrial plant or a mining site impactful to *Environment*. Let us also think, for instance, to a factory plan of action to buy "eco" products and services to make the production chain sustainable; or even to the integration programme between the economic and *Environmental balance*, inside of any organization of both private or public administration; or to the integration and management of natural resources in the city planning.

And what do we mean by constitutive? To *Communicate the Environment* contributes towards creating in the addressee of the *Environmental message* a vision or better to say a representation of Nature and its *Environment* problems, thus triggering a process of knowledge-awareness about some themes unclear to the average citizen. A function of great responsibility for the communicator. Rivers, forests, protected sea areas, can be introduced as precious ecosystems to be preserved or as wild areas to be *cleared*. Natural resources can be described as a resource to be exploited according to the human needs or as a source of life for any living being and, as such, a heritage to be preserved for the very survival of the Planet Earth. From that it derives, as we will be stating later on, the right to guarantee a correct information. In a recent interview, **H.S Pope Francis** strongly has underlined a very basic aspect of communication: "*A very dangerous thing in the means of information is the misinformation, that is [...] to say only one part of the truth and not the whole. This means disinformation. You give only half the truth to the listener or TV viewer and so he cannot judge seriously [...] because the disinformation creates a one-direction opinion omitting the other part of truth [...] but the Media should be very clear, transparent [...]. They (Media) are opinion makers and can build, and make limitless right things*" [source: ANSA Press "In Europe today there is a lack of leaders", 7th December, 2016, Copyright ANSA].

Therefore, the new vision of the "*Environmental System*" is depending on the lexis choice and any other expressive form of communication (images, sounds, signs, symbols, etc.). In order to be efficient, the *Environmental Communication* must be able to influence and, if wrong, be able to modify our visions of *Environment*. In doing this, it must be honest without pursuing partial interests whether commercial, political and economic. Environmental issues are, for their nature, *super partes* since they refer to the common good which we are part of and which our survival depend on: the ecosystem Earth. The *Declaration of the United Nations Conference on the Human Environment* (Stockholm, June 1972) reminds us: "*Both aspects of man's Environment, the natural and the man-made, are essential to his well-being and to the enjoyment of basic human rights, the right to life itself*".

Everybody can implement their *Environmental plan* of action and communication, just in their everyday life, provided that their messages are correct. Our personal approach to *Environment* and *Sustainable Development* is derived, then, greatly from the representation made by the different *Media.*

Green Tweets
#Environment   #SustaianableDevelopment   #InterdisciplinarySystem #Medium #Media #Pragmatic #Constitutive #Flashmob @PopeFrancis #RiskofmMisinformation #Awareness #Information #Truth #HumanRights #RighttoLife #EnvironmentalMessage

1.4   Some Tools to Communicate the Environment

Giving the right voice to *Environment* is a *must* of any communicator.

As specified in the Introduction, the *Environmental Message* can be conveyed not exclusively by "words". Since a long time the *Environment* and sustainable issues "have been part" our way of doing, thinking and expressing ourselves. Acid Rains, Biodegradability, Biodiversity, Biological, Biomass, Biosphere, Carbon Dioxide, Climate Change, Desertification, Energy Efficiency, Ecosphere, Ecosystem, Eutrophication, Fine Dust (PM10), Global Warming, Greenhouse Effect, Greenhouse Gas, "Green" and "Eco" Products and Services, Hole in the Ozone Layer, Organic and Chemical Pollution, Radiations, Recycling, Reuse, these are only a few among the many words associated to *Environment* nowadays, commonly used in the world of information, formation and dissemination.

A *linguistic evolution* made by: writers, scientists, film-makers, composers, poets, artists, citizens, politicians and businessmen. A communicative approach made of verbal and non-verbal messages that the more "public" they are, the more efficient. "Public" in this context does not mean something official, linked to public administration or government. It goes further to identify each tool that gives a voice to the public: Events, Fairs, Exhibitions, Videos, Films, Radios and TV Programmes, Photographic Exhibition, Art Exhibition, Eco-labels and Eco-tags. The communicative offer to the public strengthen the message, being more perceived, and it influences their behaviours making them more eco-virtuous. *"There is nevertheless the need to clearly communicate the ecological issues of sustainable development to a wider audience than is presently achieved in society thus raising the importance of the communicative act"* underlines *Pierre McDonagh*, professor of Marketing at Bath University, UK (1998). A key factor even in the *firm field and retail*.

It is not by chance that *Rio UN Conference 1992* gave birth to the policy paper called *Agenda21* spreading out all over the world a participated approach: nowadays a compulsory path in the decisional process, both at central level (national governments) and peripheral (local public administrations), whether a law to approve, or a political programme, or a simple debate or a day-to-day action. On the other hand, since ancient Greece, cradle of democracy, a great importance was given to *agorà*, a symbolic place devoted to public meetings, exchange of ideas, *evolution* of thoughts, changes. From that customs, the rights to express freely our own thought, both in a public or private sphere, is a basic fundamental principle of any modern democratic form.

The *communicative skill* to debating, exchanging and comparing ideas, taking decisions, is then the main characteristic of the human race who has been able, in some cases, to lay the foundation for innovation and progress. We could ask ourselves what communication has to do with *Nature and the Environmental aspects*, or whether an *Environmental Communication* may exist: key-questions also for our Manual. The same doubts called into by American university professors (*Robert Cox, Phaedra C. Pezzullo*) who achieved interesting conclusions: *"Nature"* and *"Environment"* are not only *words* of a specific language (code) but they express *ideas*. This means that their use determines some reactions in the public that, as already said, depend on many different factors.

The "**wild Nature**" was not always considered as a precious resource to preserve and pass on to the future generations. During the Nineteenth and Twentieth centuries, some animals such as the *wolf*, were often labelled as cruel, wicked, symbol of evil. Campaigns in favour of *wolves' extermination* spread out at governmental level in many American areas, supported by scholars, poets, scientists, politicians and businessmen. These actions were banned soon after, thanks to the evolution of the *Environmental protection* and Sustainable Development. Such an example makes us understand how **the perception of *Environment* matters is extremely subjective** and how to communicate them in a correct way is a precise duty of the communicator. And not only for the communication of an ecological disaster or a specific danger for the *Environmental ecosystem*, but also to enhance Nature which is as such *"ethically and politically silent"* (*Robert Cox*).

"The natural world affects us, but our language and other symbolic action also have the capacity to affect or construct our perceptions of nature itself", reminds us Robert Cox, professor at the University of North Carolina at Chapel Hill. Therefore, what Communication and *Environment* System have in common is the need that communicators become *intermediary* of the natural world to spread out the knowledge, give information, and orientate the addressees to recognize, respect and preserve the ecosystem.

Certainly, in order to "affect" our listeners, we must be skilled and able to do that in a correct way because not all the addressees "feel" the messages in the same way but they interpret them according to their cultural and professional know-how. *Ludwig Wittgenstein*, a well-known, Austrian leading representative of the existentialism of the Nineteenth century, supplies us with a clear vision of the dynamics of communication. The philosopher subdivides human beings in *shapes of life* who live in a particular socio-cultural context and are able to interact among themselves through *linguistic games*. Each human being knows different languages and communicates easily to those who share the same social context.

Language is essential to understand the reality and to make one's own thought. The latter, thus, cannot exist without the former. In order to make a thought we need to expand our knowledge. Each single aspect or problem of our life exists only for the fact that we can communicate it and experiment the consequences. With these premises, *Wittgenstein* concluded that reality does not exist in the absolute but it depends on how it is interpreted. *Water! Go out! Ouch! Help! Beautiful! No!* They are all exclamations but they get more and more different meanings. The first, in fact, expresses an invocation to a natural source: the water. The second corresponds to an order. The third is a complaint. The fourth is an invocation to catch the attention. The fifth reinforces so much the concept of "beautiful" to turn it into "wonderful". The sixth is a strong refusal. Each *Environmental message* is thus perceived by the public opinion according to the language chosen by the communicator.

Really there are other variables to take into account. There are many elements that can "disturb" the *dissemination* of *Environmental messages*, as already mentioned, e.g.: the communication codes used, the time, the *Environment* in which the communication takes place, the perception, the memory, the listening motivation, the communication capacity, the information tools available, etc. All that falls within the so defined **distortion arc of the message**, whose radius grows in proportion to the public's dimension. An aspect to be considered in advance trying to contain the effects. Promoting, for example, "*feed-back*", better if simultaneous, between addresser and addressee.

It is fundamental to give a positive image of what is communicated, avoiding to fall into the "*pornography*" of tragic, emphasizing the catastrophic side of the *Environment*. It is very useful to drive the public to be pro-active and solve problems: having a remarkable ability of listening and knowing the message recipients and verifying their satisfaction level. In fact, the role of communication is to raise interest among the *stakeholders*. In order to get this goal, the *medium* to convey the message should be involving, and represent itself a respectful behaviour of the *Environment*.

The message must the clearest and the most unambiguous one and built in such a way that everybody can understand it easily. In other words, it will help us to be concise and popular in explaining too technical terms. This is to apply for both the written and oral communication. In the latter, any detail is not to be neglected: check the voice intonation, talk calmly, show a serene face expression, spell the words emphasizing the most important information.

Finally, *the Environmental message must be flexible*, i.e. be able to adapt to different social contexts and the public addressed to.

Summing up: simplicity, conciseness, honesty, objectivity, clearness, flexibility are the foundations to be respected and found in any Plan of Communication finalized both to the *Environment* commitment of a specific organization (firm, association, foundation, institution, public administration, etc.); or to disseminate the results recorded in a Sustainability Report, in an *Environmental Declaration* or *Environmental Balance*.

Therefore, to train a class of *Environmental communicators professionally certified* represents a challenge but also a necessity considering the interests involved and the high level of responsibility characterizing the *communication of Environment*.

Green Tweets

#CarbonDyoxde #Biodegradability #Biodiversity #Organic #Biomass #Biosphere #OzonLayer #ClimateChange #Desertification #Ecosphere #Ecosystem #EnergyEfficiency, #Greehouseeffect #Eutrophication #GreenHouseGases #Pollution #AcidRains #FineDust #PM10 #GreenProducts #GreenServices #Recycling #GlobalWarming #Reuse @PierreMcDonagh #CommunicationAct #AbilitytoCommunicate #Nature #Environment #perception #Environmentalissues #Knowhow #culturalbackground @LudwigWittgenstein #Rhino #Existentialism #Lifeforms #WordGames #Noise #DistortionEffects #EnvironmentalMessage #Feedback #Flexibility #Semplicity #Synthesis #Honesty #Objectivity #Clarity #EnvironmentalCommunicator

1.5   Conclusions and Reflections: A Feasible Eco-communicative Approach

How can we communicate the Environment efficaciously?

The debate about the *efficacy of communication* in our technological age is one of the most important key-topic of modern culture. As the evolution of technology goes ahead and the sources of information proliferate, to *communicate* seems to be a "vital element" but not always correctly understood. A paradox that makes us reflect with more attention upon the use of the available Media in order to transfer the *Environmental message* properly. For such a wide subject as the *Environment*, the first step is undoubtedly the individuation of a well-known topic to deal with. What apparently seems to be a tautology in reality is the corner stone of a *public speaking*, as stated by the founder of communication **Marc Tully Cicero** (106 B.C–43 B.C). The orator, reflecting on the best strategies to use words and points of persuasion, underlines, with extreme modernity, the importance of a deep cultural background as a pre-requisite to be able to express the message efficaciously.

After fixing the **object to be communicated**, another delicate and compelled step is to establish the **purpose** of the *Environmental message* that, as already said, aims at facilitating the comprehension of the selected *Environmental themes* by the addressee, considering the ecosystem as a whole according to a holistic approach. Let us imagine to work out a Plan of Communication or an Awareness Campaign about the problem of *greenhouse effect.* Given for granted the high competence on the matter, could be useful to find, from the beginning, the main factors being influencing the public opinion on the issue.

We assume, for example, that we already know the problem: i.e. the concentrations in the atmosphere of some gaseous compounds (CO_2, CH_4, H_2O), that produce the phenomenon of the increasing temperature called "*greenhouse effect*". These concentrations, in the last years, reached very alarming values to guarantee the Earth ecosystem balance as the record of +1,35 °C registered in February 2016 show (anomaly reported by NASA compared to the period 1951–1980). Someone could decide simply not to act; somebody else might think that, living in a big town, or a metropolis, it is impossible to reduce the quantity of climate-changing gasses; others could think the phenomenon is not strictly linked to man's activities since it has existed for millions of years, well before man's polluting activities started. According to other points of view: the phenomenon, being on the world scale, does not affect them; some others, on the contrary, might consider the purchase of fruit and vegetables coming from their own Country farming is enough to reduce the emission of pollutant gasses in the atmosphere.

Each public reaction is the product of many factors, some of them consistent with eco-compatible development, others less if not even opposite. The great challenge of *Environmental Communication* from which its efficacy derives is to implement behaviours compatible with the *Environment.* In order to fulfil this goal in the best way, it is important to draw up a detailed plan aimed at achieving the *target* or at least *hit the edge* of eco-communication; a specific objective which produces practical results whose contents and actions are well known.

A journey *at different stages* easily readable, easily replicable and user friendly as we propose herewith as an exclusive example of ***Environmental Communication Plan***. Each reader would be able to adapt it to his own reality, enlarging or reducing it, with stimulating and unmistakable *Environmental Communication* skills. One of the main task of the *Environmental Communication* is in fact to "translate" great ideas into words, sentences and concepts. This function aims at catching the public's attention; informing and involving the audience and pushing their actions with eco-responsibility. In this *Communication Plan*, all the steps are initially analysed in theory and, then, implemented in a hypothetical *Case Study* regarding the Mediterranean Region, inspired by some real professional experiences of the author as *Sustainable Development Project Manager.*

Green Tweets
#CommunicationEfficacy   #EnvironmentPerception   #Environmental Message #EnvironmentalInformation #PublicReaction #Environmental CommunicationPlan #Target #Actions #Results #Media #CaseStudy #PracticalExample #FromTheorytoPractice

ENVIRONMENTAL COMMUNICATION PLAN:
a possible theoretical approach.

Identifying the matter
(environmental area of competence)

A deep analysis on the matter
(based upon authoritative sources and certified data)

Detecting any real or potential criticality with reference to the matter

Case Study in the same area of expertise

Identifyng the communication plan addressees
(stakeholders)

Taking stock:
primary objectives, recipients, strenghts and weaknesses

Overview:
developing a format, diagrams, set of indicators, matrices

Communicating Creativity and Innovation
in order to inform, train, persuade, make decisions, improve the environment

Involving the public through facts reporting *via* fluent, convincing, updated sentences

Action Plan:
encouraging the audience to act = involvement responsibility, engagement, clear objectives, incentives

Positive Thinking:
higlighting the broader benefits of the Action Plan (e.g,: small actions producing big results)

Monitoring and Evaluating:
project development, outcomes, data comparison

Participating: feedbacks, social indicators, focus groups, surveys, etc..

Key Goal:
Reporting on each step of the Communication Plan with reference to methods, outcomes and problem solving

**ENVIRONMENTAL COMMUNICATION PLAN Mediterranean Region
A possible practical approach.**

Identifying the matter:
Protection of Marine Biodiversity in one specific area

A deep analysis on the matter: examining articles, reports, studies, multimedia documents > Climate Change, endangered species, overfishing, sources of pollution, etc.

Detecting any real or potential criticality:
sea temperature increase, chemical or noise sea-water contamination, out-of-control fishing etc.

Case Study:
Life Project: SMILE (reduction coastal zone); Recovery plan of a specific area cetacean Sanctuary International Agreements: RAMOGE, Accobams, etc.

Adressees:
Administrations, Port Authorities, Maritime Transports Companies, Fishermen Associations, Citizens Associations, Tourism Promotion Authorities, etc.

Taking stock:
High Priority: Cataloguing of endangered species Average Priority: search for partners strenghts: testimonies, qualified teams, key strategies Weaknesses: lack of a single communication plan, low cooperation with local Media, etc.

Overview:
microplastic /
macroplastic
percentage detected
in sea-water; list of
marine species and
their ability of
reproduction, etc.

**Communicating
Creativity and
Innovation:**
leaflets, posters,
gadgets, video
interviews, press
reports, web sites,
Social Media
conferences, call
centers, Info points
etc.

**Involving the
public:**
key
words, slogans,
symbols, logos,
etc.

Action Plan:
distribution of
recycling bins, free
eco-training for
fishermen, best
practices addressed
to citizens, tourists,
industrial platform
employees, etc.

Positive Thinking:
action plan
addressed to boost
separate collection,
recycling, eco flash
mob, beach clean-
up operations,
photography
competions on
marine ecosystem,
green guided tours,
eco-exhibitions etc.

**Monitoring and
Evaluating:**
Periodic paper
reports, Official
statements,
information panels,
touchscreen Info
points, Internet
points, etc. aimed at
reporting on marine
and coastal
pollution; sea water
quality; repopulation
of marine
endangered species,
etc.

Partecipating: town
meetings, focus groups,
field visits, peer reviews,
workshops, Info-points,
etc.

Key Goal:
Final Project Report -Leaflets-
Newsletters -Toolkits -Press
Reports -Official Statements

Bibliography

Paolo E. Balboni, Fabio Caon, *La comunicazione interculturale* (The intercultural communication), Venice (Italy), Marsilio Editori, First Edition, 2015, pages: 176.

Erik Balzaretti, Benedetta Gargiulo, *La comunicazione ambientale: sistemi, scenari e prospettive* (The environmental communication: systems, scenarios and perspectives), Milan (Italy), Franco Angeli Edizioni, 2011, pages: 256.

Gregory Bateson, *Steps to an Ecology of Mind*, University of Chicago Pr (Tx), New Edition, 2000, pages: 553.

Gianfranco Bettetini, Fausto Colombo, *Le nuove tecnologie della comunicazione* (New technologies of knowledge), Milan (Italy), Bompiani/RCS Libri S.p.a., 1996, pages: 358.

Ernest Watson Burgess, Bogue D.J., *Contributions to Urban Sociology*, University of Chicago Press, Chicago, 1967.

Kenneth Burke, Language as Symbolic Action: essays on life, literature and method, University of California, 1968, pages: 532.

V.A.F. Casetti (a cura di), F. Colombo (a cura di), A. Fumagalli (a cura di), *La realtà dell'immaginario; i Media tra semiotica e sociologia; studi in onore di Gianfranco Bettetini* (The reality of the imaginary; the Media between semiotic and sociology; a study in honour of *Gianfranco Bettetini*), Milan (Italy), Vita e Pensiero Editore, 2003, pages: 288.

Marco Tullio Cicerone, L'arte del comunicare, a cura di Paolo Marsich (The art of communicating – by Paolo Marsich), Milan (Italy), Oscar Mondadori, 2014, pages: 64.

Rachel Carson, Edward O. Wilson, *Silent Spring*, Boston (Massachusetts, USA), Mariner Books, 2003, pages: 378.

Charles Horton Cooley, *Human Nature and the Social Order*, Florence (Italy), Nabu Press, 2014, pages: 482.

Robert Cox, Phaedra C. Pezzullo, *Environmental Communication and the Public Sphere*, Los Angeles – London – New Delhi - Singapore – Washington DC – Boston, Sage Publications Ltd, 4th edition, 2015, pages: 422.

John Dewey, Democracy and Education: An introduction to the Philosophy of Education, Perennial Press, 2016, pages: 345.

John Dewey, Esperienza e Educazione (Experience and education), Milan (Italy), Cortina Raffaello Editore, 2014, pages: 85.

Umberto Eco, *Apocalittici ed Intergrati – comunicazioni di massa e teorie della cultura di massa* (Apocalyptic and Integrated – mass communication and theories on mass communication), Bompiani/RCS Libri S.p.a., Milan, 2016, pages: 385.

Umberto Eco, Marc Augé, Georges Didi-Huberman, *La forza delle immagini* (The power of images), Milan (Italy), Franco Angeli, 2015, pages: 89.

Umberto Eco, *Kant e l'ornitorinco* (Kant and the Platypus), Milan, Bompiani, 2013, pages: 470.

John Elkington, *Cannibalis with Forks: The Triple Bottom Line of 21ˢᵗ Century Business*, New York (New York, USA), John Wiley & Sons, 1999, pages: 424.

Neil Evernden, *The social creation of nature*, Baltimore, Baltimore (Maryland, USA), Johns Hopkins University Press, 1992, pages: 200.

Paolo Granata, *Ecologia dei Media; protagonisti, scuole, concetti chiave* (Ecology of the Media; main actors, schools of thought, key concepts), Milan (Italy), Franco Angeli Editore, 2015, pages: 159.

Suzie Guth, Modernité de Robert Ezra Park: Les concepts de l'École de Chicago, Collana "Logiques sociales, Paris (France), L'Harmattan, 2008, pages: 312.

Edward T. Hall, *Beyond Culture*, Independent Publication, 1997.

Harold Innis (1950), *Impero e comunicazione* (Empire and Communication), Rome (Italy), Meltemi 2001.

Harold Innis (1951), Le tendenze della comunicazione (The communication trends), Milan (Italy), SugarCo, 1982.

Marcel van Marrewijk, *Concepts and Definitions of CSR and Corporate Sustainability: Between Agency and Communion*, Journal of Business Ethics, Netherlands Kluwer Academic Publishers, 2003, 44:95-105.

Marshall McLuhan, *Theories of Communication*, New York, Peter Lang Pub Inc, 2011, pages: 253.

Marshall McLuhan (1964), *Understanding Media: the extensions of Man: Critical Edition*, Berkeley (California, USA), Gingko Pr Inc, 2002, pages: 500.

Roderick D. McKenzie, *On Human Ecology*, Chicago (Illinois, USA), University Chicago Press, 1969, pages: 308.

John Muir, *A Thousand-Mile Walk to the Gulf*, e-artnow, Prague (Czech Republic), 2015, pages: 126.

John Muir, *My first summer in the Sierra, Steep Trails and the mountains of California*, Palm Springs (California, USA) Ivory Classics, 2017, pages: 344.

Lewis Mumford, *In the name of Sanity*, Santa Barbara (California, USA), Greenwood Pub Group, 1973a, pages: 244.

Lewis Mumford, *Myth of the machine*, London (UK), Martin Secker & Warburg Ltd, 1967, pages: 352.

Lewis Mumford, *The condition of man*, Boston (Massachusetts, USA) Mariner Books – Book Series: Harvest Book, 1973b, pages: 467.

Walter J. Ong, *An Ong Reader: Challenges for Further Inquiry (Hampton Press Communication Series Media Ecology)*, Hampton Press, 2002.

Walter J. Ong, A. Calanchi, *Oralità e scrittura. Le tecnologie della parola* (Oratory and writing. The technologies of the word), Collana "Le vie della civiltà", Il Mulino, Bologna (Italy), 2014, pages: 286.

Robert Ezra Park, *Human Ecology*, American Journal of Sociology, 1936, XLII, pages: 1-15.

Robert Ezra Park, *La ciudad y otros ensayos de ecología urbana* (The city and other essays on urban ecology), Collana "La Estrella Polar", Ediciones Del Serbal, S.a., 1st Edition (November 1999), pages: 152.

Lucy Kellaway, *L'ultima moda nel mondo degli affari: non rispondere alle e-mail*, Il Sole24 Ore, 2nd February 2015.

Luciano Ponzio, *Roman Jakobson e i fondamenti della semiotica*, Sesto San Giovanni (Milan, Italy), Mimesis Editore, 2015, pages: 274.

Neil Postman, *The Reformed English Curriculum*, High School 1980: The Shape of the Future in American Secondary Education. Ed. Alvin C. Eurich, New York: Pitman, 1970, pages: 160–168.

Graziella Priulla, *La comunicazione delle pubbliche amministrazioni*, Rome (Italy), Edizioni Laterza, 2016, pages 182.

Pio E. Ricci Bitti, Bruna Zani, *La Comunicazione come processo sociale*, Bologna (Italy), Il Mulino, 2002, pages: 276.

Ferruccio Rossi-Landi, *Significato, comunicazione e parlare comune* (Meaning, communication and common speaking), Venice (Italy), Marsilio Editori, 3rd edition, 2001.

C.E. Shannon, W. Weaver, *The Mathematical Theory of Communication*, Illinois (USA), University of Illinois Pr., 1949, pages: 144.

William I. Thomas, *The Polish Peasant in Europe and America; Monograph of an Immigrant Group*, Leopold Classic Library, 2016, pages: 376.

Hans C. von Baeyer, *Informazione. Il nuovo linguaggio della scienza* (Information. The new language of science), Bari (Italy), Edizioni Dedalo, 2005, pages: 289.

Paul Watzlawick, Beavin J.B., Jackson D.D., *Pragmatics of Human Communication: a study of interactional patterns, pathologies and paradoxes*, W.W. Norton & Company, 1st edition, 25th April 2011, pages: 301.

Paul Watzlawick, *An Anthology of Human Communication*, Palo Alto (California, USA): Science and Behaviour Books, 1964.

Web Site List

BirdLife International > www.birdlife.org

Lorenzo Duran – personal web site > www.lorenzomanuelduran.es

United Nations > www.un.org

Plant for the Planet Project > www.plant-for-the-planet.org

RAI Radio Televisione Italiana (Italian Broadcast Television) - Dizionario Lessicale: Ambiente tratto da *Lemma, Navigare nelle parole* di I. Moscato (Lexical Dictionary: Environment drawn from the vocabulary word by I. Moscato), linguistic coordination by T. De Mauro, RAI video database, 1998 www.raiscuola.rai.it

World Commission on Environment and Development (1987). *Our Common Future*. Oxford: Oxford University Press, 1987, pages: 27 – www.un-documents.net (United Nations documents)

Wildlife Conservation Society > www.wcs.org

World Wildlife Fund (WWF) www.worldwildlife.org

Journalism and Environment: Two Opposed Trends Which Attract Each Other

2

Abstract

Two "worlds" only apparently distant, *Journalism*, on one side, and the *Environment*, on the other side, tend even more to meet, being pushed by new sustainability forms of communication, in the *Society of Information*. The *Social Media* and the *New Technology* are able to go beyond any geographic and social boundaries by revolutionizing the concepts of "time" and "space". But the revolution in progress must not prejudice the deontological values upon which the journalist job relies. In fact, his main task is not to influence the network or be subjugated by it. The reporter's commitment is much more responsible and multitasking, aimed at building confidence among the readers. In a world in which journalism has not got the monopoly of the information any more, the journalist's main goal is to guarantee constantly the quality of information to prevent the readers' distrust.

© Springer Nature Switzerland AG 2019

M. Abbati, *Communicating the Environment to Save the Planet*,

https://doi.org/10.1007/978-3-319-76017-9_2

a

b

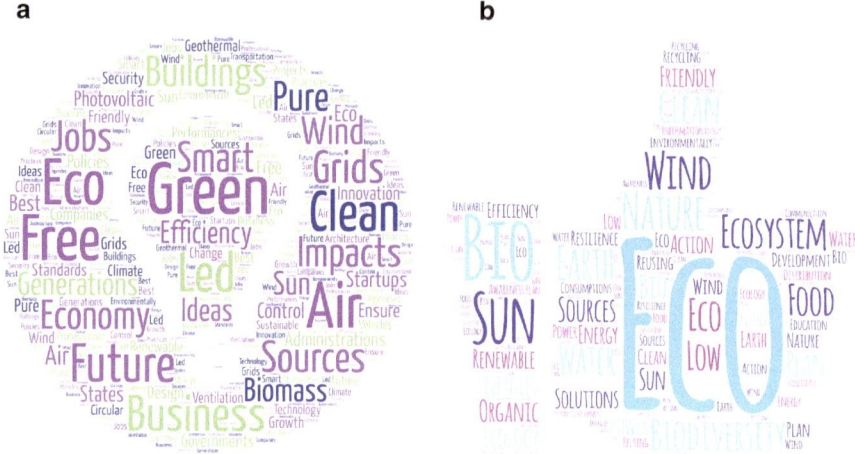

"We must believe in the power and strength of our words.
Our words can change the world."
Malala Yousafzai
Nobel Peace Prize 2014 for her commitement in favour of
civil rights and right of education (Mingora, 12th July 1997)

c

(**a–c**) Word cloud environmentally friendly—sketches by Maurizio Abbati: through WordArt.com (Designer and Copyright Holder) © 2018. (**a**) Key-issue: energy efficiency; (**b**) key-issue: sustainability; (**c**) key-issue: biodiversity/green canopy

Reading Proposal: In Between Words and Images

Thanks to the new Information and Communication Technology—ICT, to convey environmental information to trigger a communication chain has got original forms that show their authors' creativity, becoming thus privileged *Media* because they can make the concepts of *Environment* and *Sustainable Development* easily understandable. Hence, a traditional medium like "writing" is transformed in a synthetic but effective form able to create "mental map" where the key-words are highlighted through a very innovative graphic and the clip art (pre-made images used to illustrate any medium). This term is wildly used to identify any type of graphic project, both for personal or commercial use, that assembles handmade or IT widgets to single words, carefully selected to create a creative narration both visual or written. In some cases, we can find also a real form of art as the Pictographic-writing or Word Cloud, used for educational and recreational purposes. Effectiveness, space rationalisation, and quality of images make these tools perfectly compatible with the mission of the Environmental Communication. In the examples shown on the cover of this Chapter (Figure a–c) words such as "Ecosystem, Recycle, Climate, Danger, Ecology, Sustainability, Life etcetera" help an attentive and informed reader to create by terminology association a mental map about the main environmental to be developed later through insights, idea exchanges, debates and question-time participations and any other participatory form useful to communicate the *Environment*. The positioning of each single word and the chromatic choices tending to green (Colour-symbol of *Environment*) reinforce the environmental and sustainable semantic field of the words they belong to. A green electric bulb (Figure b) symbol of "clean energy"; a green "like" (Figure a), graphic sign used largely in the *Social Media* language to state the approval and so to the "green cause" participation; a tree (Figure c) a further environmental symbol, a visual synecdoche representing forests, main source of oxygen on the Planet Earth, together with the plankton in the oceans. "*Words can inspire but also can destroy*" reminds us *Robin Sharma*, the Canadian writer, author of numerous bestsellers on the development of each individual and his own life. Nevertheless, they are a fundamental element for those working in Journalism as we are going to deal with in "***Chapter 2—Journalism and Environment: two opposed trends that attract each other***".

2.1   The Origins of Eco-information

The right and duty to inform exceed the national boundaries and the knowledge with professional ethic.

In Journalism "*the details tell always the story*" remind us *James McBride*, American modern author of bestseller and musician. Since long time, in the press field, a specialized *green line* has been spreading out and getting more and more important. Flora, Fauna, Biodiversity, Ecosystems, Water resources, Climate Change, Pollutions, Building Exploitations, *Environmental Impacts*, *Environmental Crimes*, are only a few items of the "*green*" topics for reportages.

Environmental Journalism is the ultimate tool for mass communication, a keystone of the "*Information Society*", addressed to a wider and wider and more heterogeneous audience. An heir of the literary production by explorers and naturalists, like: *Henry David Thoreau*, *John Muir*, *Aldo Leopold*, *Rachel Carson*. It has become, without any doubt, one of the most studied subjects at world level. First of all, its importance relies on the capacity to unite worlds apparently different: the political and the scientific; the economic and the cultural and *Environmental*. To realize *reportage* on *Environmental themes* means also to explore simultaneously the local, regional, national and worldwide dimension. Let us think of the *Media* impacts of the International Conferences, where the representatives of the main World Governments and the leading scientific experts agree the suitable measures to prevent, oppose, eliminate and solve one or more *Environmental problems*, the so called "*Environmental criticalities*". An example: *The United Nations Climate Change Conference of Paris*—COP21, 2015—that brought to the ratification of a very important agreement about the Climate Change. Another meaningful event was the following *Marrakech Conference*—COP22, in November 2016—that offered a first balance after 1 year from the Paris agreements.

The **macro-theme "*Environment* and Sustainable Development"** is then a delicate topic to be treated by journalists. They are requested to process information, data, indicators, schemes describing present and future *scenarios* often called into questions by different schools of thought.

- *Which is the cause of the abnormal global heating?*
- *Mankind?*
- *The exceeding concentration of bio dioxide in the atmosphere?*
- *The methane gas produced by intensive cattle-rearing?*
- *The men's choice to feed on reared meat and animal derivatives?*
- *Deforestation?*
- *Vulcan emissions?*
- *The whole of human and natural factors?*

The messages conveyed by modern *Media*, magnified and made almost "immortal" by the diffusion *online*, offer the most different hypothesis concerning their authenticity to be checked at the very source, a deontological requisite for the professional journalist. Those working in journalism, therefore, must follow job behaviours connected to ethics, above all when the information is conveyed by the Net. The freedom of expression and the numerous sources and forms of information offered by the World Wide Web compel the Internet users to choose continuously the informative course to conform to, even though a large use of hypertexts interconnected. A journalist then must be equipped with a "toolbox", that is a principle guide to help him "to navigate in the *Media* ocean" where the information mixes together like in a huge "virtual square" and where it is often difficult to establish which facts are real or imaginary or created *ad hoc* to pursue personal interests. The communicators' and journalists' role is then fundamental.

Substantial Truth; **Information Completeness**; **Correctness** and **Dignity** for human beings and for the whole eco-system are the cornerstones of journalism that, together with the *right of press* freedom and the *duty of responsibility* for what is written, are a real "*handbook of behaviour*". Nevertheless, some authors (*Michele Partipilo*) consider the "***ethics of doubt***" the real model to follow by the journalists' activity, in order to give an answer to a series of questions while seeking facts and circumstances. Journalists always boost a communication process through the news reporting, addressed the audience and encouraging it to communicate the *Environment*. Why should we report on a specific fact? Does it have a social function or a public utility? These should be the first questions needed to be answered.

A **moral commitment** that complies with the **most authentic meaning of "ethics"**, masterfully explained by one of the greatest thinker of the antiquity: ***Aristotle***. Nevertheless, our life style has changed because of the new technologies and the point of view also about the *Environment* has generated a plurality of critical considerations, called ***applied ethics***. The concepts of bio-ethics, *Environmental*, economic and social ethics were born. The ethic of communication and information created an "*Environmental approach*" to the news, thanks to the support of some researchers at the Oxford University. This approach is a new dimension in which man interacts with other organisms, both natural and artificial, able to manage information with a logic and independent process, called ***Info-sphere***. This method is applied to any *medium* of *Environmental Communication* whose purpose is to inform the audience by carrying out a public service. It may be a SMS, a synthetic *tweet*, a *post* on our own *Facebook* diary, a message or an article sent via email or published on a *free press*, a commercial tabloid, a *blog or* an information letter reserved to a small group of subscribers, a journalistic investigation on TV or a picture published on *Instagram* or *Pinterest*. Just to make a few examples.

Further problems may arise when the reporter has to report facts directly caused by the man's impacts on nature. We cannot help considering the catastrophic effect of a flood, of a famine, of an unexpected sea heating or cooling down. The consequences of all this are the displaced people's migrations, escaping from the devastated territories by abnormal forces of nature, the ***climate refugees***.

Hence, the necessity to balance the "objectivity" with the need of sticking to facts, respectful of those who have suffered from human and property losses as well as of the most vulnerable minors' personality. On this purpose, the *Treviso Chart*

(*Carta di Treviso*), signed on the 5th October 1990 by the Italian Journalist Order, by the Italian Press National Federation and by *Telefono Azzurro*,[1] is considered a European key document to protect minors from the Media overexposure.

As already specified, the birth of the *Environmental Journalism* goes back to the early Sixties with the publication of the book *"Silent Spring"* by *Rachel Carson*, about the risks derived from the use of DDT, then spreading out at a world level in the years Eighties. Since its very origins, we have realized that communicating the *Environment*, for the journalist, means examining carefully the *reportage*, understanding the contents, included those techno-scientific, to make the public use them. In deepening the *Environmental issues*, we always look for the right balance between overestimating and underestimating the possible risks for Man and *Environment*. A procedure of great responsibility considering that the *journalistic Media* are the first sources of information for the majority of readers.

Both traditional and new means of communication coexist together the new ones at the service of the mass communication. On the contrary, they are integrated, incorporated, modernized by defying the language, the structure and the contents. Books, newspapers, radio, televisions keep on being side by side the *web* and the new *Social Media*. Consequently, an important *"domino effect"* derives which might affect the government policies, the financial investments, the educational syllabuses, and even the choices for buying products or services. Nowadays, in fact, we can say that the *Environmental matter* dominates the majority of the subjects treated by the press: science, medicine, crime news, judicial cases, agriculture, natural resources, energy, cuisine, wellness, cinema, art, sport, etcetera.

The arrival of new information Technologies, furthermore, has amplified at the global level the capacity of influencing people and the journalists have to act with responsibility not only in the information management but also in foreseeing their effects on the readership exceeding the national borders. This trend has revolutionised journalism that, while maintaining its social function, is increasingly becoming a new communication tool following other logics than the deontological codes: the print run, the audience and the advertising incomes. The doctrine debate is being developed and somebody thinks that the legislative and professional "rules" are the corner stones of a qualified information pluralism (*Sebastiano Malfettone*). In order to defeat the risks of following economic interest journalists must act with a great sense of responsibility when **managing the news** foreseeing also **their effects** on the public.

Words matter and produce consequences! A particular case, in this regard, was the sense of panic in the general population, produced by the false description of a landing of Aliens on the Earth, spread out in 1938 by the famous CBS radio drama

---

[1] *Telefono Azzurro* is a non-profit organization focused on the children's rights protection. It was founded in Bologna (Italy) in 1987. In 1990, the first toll-free telephone helpline for children was set up in the same town. His founder, *Ernesto Caffo* was an associate professor of Child Neuropsychiatry at the University of Modena. Since its inception, the mission has been to give children the possibility to speak and to be heard, a real response to the International Convention on the Right of the Child that was signed by the United Nations in 1989. It currently cooperates with other European organizations operating in the same field, progressed as an information technology platform. http://english.azzurro.it

"*War of the Worlds*", starring *Orson Welles*, taken from the same novel by *Herbert George Wells*. We would like to point out that the listening audience was advised before and after the radio broadcast, Edit. The spread of the *World Wide Web* was still light years far away.

Green Tweets
#Journalism #Journalists #Environment #SustainableDevelopment #Biodiversity #Ecology #ClimateChange #Ecosystem #Fauna #Flora #EnvironmentalImpacts #EnvironmentalProblems #FormsofPollution #WaterResources   #LandSpeculation   @JohannesGutenberg #NewsManagement   #NewsRelations   #AlltheNewsthatfittoprint #20thCentury #SocialMedia #Infotainment #Environmental #Indicators #Impacts   #EnvironmentalManagementSystem   @COP21   @COP22 #Ethics   @Aristotle   #EthicalCode   #FreedomofPressandInformation #UnitedStatesDeclarationofIndipendence   #FirstEmendment #UniversalDeclarationHumanRights#ClimateRefugees#ClimateChange @RachelCarson #SilentSpring @OrsonWelles @HerbertGeorgeWelles #WaroftheWorlds #MassMedia @SocialMedia #WorldWideWeb

---

2.2   The Environment: A *Green Code* to Be Decoded

The Environmental System is revolutionising the concepts of time and space in communicating the piece of news.

The *glocal* (*global* + *local*) nature of the **Environment** System can make it difficult to decode the *Environmental message* by the readers. Sometimes and in some cases, the risks of *Environmental issues* are clear and territorially limited. In some other cases, the negative effects for Man and *Environment* may be distributed globally and they do not occur in the short term. It follows from this a minor attention by the public who may think they are too remote, or "distant", compared to their own daily *routine*. The same *scenario* more or less catastrophic, about the raising sea level destroying megalopolis like New York, is still considered little credible by part of the public opinion and not based on reliable data, but only science-fiction-film material. Finally, the *Environmental themes* are treated exclusively at scientific and technical level with a very difficult language, puzzling most of the people.

Similar communication problems can be found when the pollution effects on the *Environment*, known in economy as "**negative externalities**", appear in the short term and exceed, unexpectedly, the political boundaries of a nation, involving more devastated geographical areas. Consequently, they suffer the implications of the contamination caused by careless or inattention following natural disasters, or caused by the crime activity of terrorists. This is called *freeriding*. Let us give an example by considering the explosion of the nuclear power station of *Černobyl* in Belarus in April 1986, as a consequence of a bad plant management. Another example is the burst of the nuclear plant of *Fukushima* in Japan in March 2011, because of a very strong earthquake hit the country causing a *tsunami*, a seaquake of exceptional proportions. At a journalistic level, minimizing the effects of the eco-system contamination is to be avoided. This is what happened when, in springtime 2015, a huge fire in the woods near *Černobyl* burnt 400 ha of woodland, emitting a new radioactive cloud whose risks were partially minimized by the main newspapers.

Anyway, this sense of a difficult *"eco-communicability"* is generated by the incertitude dominating the scientific world about some ecological issues, as the *climate change*. On one hand, the scientists have reached the unanimous consent in finding a direct link between *anthropic actions* (man-made) and the increase of the *global temperature*, the effects of *global overheating*. In some geographical areas, in fact, their origins are still uncertain. Finally, we must not neglect the different **perception degree** of the piece of news by each reader, according to his cultural, social and economic basis in the Community-Country. Each Country, in fact, brings into being policies and action plans "personalized" in order to achieve the objectives agreed at world level, for instance about the *climate change*.

Furthermore, the **Environmental issues**, very often, are not envisaged as immediate or definitive by journalistic surveys. Journalists must then accept that "*the role of science is to make questions more than answer to them*" (*Michael Keating*). The scientific research can take months if not years to give elements necessary to get a sufficient degree of knowledge on *Environmental* emerging *issues* (e.g.: degree of toxicity of some specific substances and/or waste, causes for the global overheating, causes of the growth of carbon dioxin in the atmosphere).

The role of the journalist is above all "to report the journalistic piece"—idiomatic expression used in *journalese* to refer to an article—with the certainty. Another objective is to show different points of view and theories, always referring to the sources which the information is coming from. Finally, the journalist has to indicate always the evidence: documents, audio registrations, videos, photos etcetera, so that the reader is led to the most logic conclusions on the basis of facts.

An uneasy role, which can lead the journalist to find obstacles by part of the public opinion opposing the *Environmental issues* for ideological or commercial reasons, or simply for stance. For example, the behaviour NYMBY—*Not in my backyard*—identifies those who are protesting against private or public works producing or going to produce negative effects on specific territories. Let us think about largest transport routes, building of quarries, new industrial plants, residential areas, incinerators, waste management, waste deposit of hazardous materials. In such cases, a responsible attitude must prevail to safeguard the freedom of expression, the correctness of information and the legality of action.

Eventually the element *"Time"*, if not adequately managed, risks to put Journalism and *Environment* on two apparently opposite levels. Apart from the melting of glaciers at the Poles and the destructive effects coming from atmospherically extreme phenomena, the majority of *Environmental issues*, as the *global warming*, are very difficult to be perceived daily and be contextualized in a piece of news.

The *"green"* journalist is then pushed to relate with the **concept of *"Space"*** and *"Time"* in a very innovative way thus influencing the communication. Journalism is, for its nature, able to transform anything flowing from his pen into a "public event", contributing to influence the public opinion. This is mainly the social function of a journalist. A big challenge for a "green" journalist is to communicate the *Environmental issues* to a larger public because of the global dimension of the *Environment System*. Facing this *forma mentis* (mind set) the most suitable and coherent approach is that of *"being on top of things"* as some doctrine studies have pointed out (*Henrik Bødker & Irene Neverla*). In a few words: the best way to approach the *Environmental issues* is the critic one. Making questions, deepening all the correlated aspects to a specific *Environmental matter*, also under pressure using oriented reading techniques *skimming*, *scanning* and synthesis of documents.

Other journalists' skills for *Environment* are:

• Building up the case from its "roots".
• Examining the causes, the risks and possible solutions. Understanding of the *"Environmental System"* and the subjects operating with it: Non-Governmental Organizations—NGO, governments, associations, etcetera, interacting among them.
• Simplifying complex concepts, without making them simplistic or source of useless alarmism, but usable by an extended public—in the spirit of *penny press* of the eighteenth century (economic tabloid), without falling into the "literary newspaper" refined in the use of words, sometimes pretentious, and for these reasons addressed to intellectual *élites*.

The multiform and multidisciplinary character of the *Environment System*, in fact, must not be a pretext to dramatize the reality, neither to minimize it. It is a fact that man has compromised his relationship with the natural *Environment*, unbalancing it. It is clear that our age is more and more characterized by the risk, the insecurity, precariousness, loss of any social, economic and *Environmental reference* points.

As already mentioned, the journalist's role is to seek substantial truths, based on facts, without proposing absolute truths or simple points of view. It is not by chance that in the English-speaking world a distinction is made between *news* (from authentic sources) and *views* (subjective news). The final objective for a journalist is then to reinforce his relationship of trust with the audience, constantly maintained to increase credibility on *Environmental Communication*.

In this age of the proliferation of means of communication, furthermore, the journalist's duty is "not making people lose the informative power of information" (*Mihaela Gavrila*). To get a relevant information is the new right of citizenship to be guaranteed for each fully-fledged citizen. *"The good journalist is the one who is able to explain the facts through the use of argumentations rather than exclamations"* states *Philip Meyer*, an American journalist, teacher, and writer, emeritus professor at the Journalism School of North Carolina, USA, and he adds: *"a good journalist nowadays…must […] know many things, but he must know very well a specific topic*". These statements are suitable for the relationship between Journalism and *Environment*, two "worlds" that can coexist, provided that the former shows familiarity with the increasing number of notions and data supplied by the latter.

Green Tweets
#Glocal   #EnvironmentalSystem   #NegativeExternalities   #Freeriding #Perceptions   #PieceofNews   #CircularEconomy   #GreenEconomy #GreenCommunication @MichaelKeating #Notinmybackyard #NYMBY #GlobalWarming   #Space   #Time   @BodkerNeverla   #Skimming #CulturalBackground #Pennypress #News #Views @MihaelaGavrila @PhilipMeyer

2.3   An "Eco-system" of Questions and Answers: A Systemic Approach to the Piece of News

Science and journalism to raise the readers' awareness towards the *Environmental issues*.

How to find the news? How to evaluate and analyse them? How to convey them in the huge and confused voice of Media so that they can reach the recipients? And how to establish and achieve the degree of precision necessary to describe facts?

The journalist facing the "green" subject matters must keep in mind the most important *mission* of the *Environmental Communication* that is to attract interest in the reader, enriching his "green" culture. The reader's involvement in *Environmental issues* will be influencing thus his lifestyle making it more sustainable. The "*Environmental*" journalist, then, must keep his *mission* as a "service of social utility". In doing so some authors suggest to act in two directions:

(a) **the scientific and institutional "world" should transmit the *Environmental messages*** to the newspaper offices **in a more effective way** making the understanding easier and quicker; adopting, for example, synthesis, schemes, tables, sustainability reports, *Environmental balances*;

(b) **a much more positive approach by the journalists leading to focus more on the results** for the *Environmental improvement* than on the catastrophic effects of the human impact on the *Environment*; this means also to be stuck to the real facts, trying to show news of collective interest through people's everyday life contents and new forms in evolution, avoiding *prefixed standards*.

At the base of the binomial "***science-journalism***" lies, in fact, the **quality of the Environmental information**, which the public's interpretation of the *Environment*

System depends on and also it should include the critical guided interpretation of the message and *Environmental data*. It derives that the communication involves both the scientist and journalist. They are then called to behave in a very ethic and professional way abstaining from partisan positions and any misleading information, subjugated to commercial or political interests. *"The management of uncertainty and risk needs clarity and transparency of information, essential prerequisites for an ecology of communication in addition to a different quality of the individual social life"* underlines *Serena Rugiero*, co-ordinator of the *Osservatorio Energia e Innovazione dell'Associazione Bruno Trentin—Ires-Is* (Energy and Innovation Observatory—Bruno Trentin Association—Ires-Is).

For this reason, *what makes the difference in the green journalism is to "visualize the stories"* in a very free, critical and responsible way. The best way is to *go on-the-spot*, as pointed out *Navin Singh* Khadka (@NavinSinghKhadk), Nepalese journalist of the *BBC World, Science and Environment*, who has specialized on *Environmental issues* for many years, with a particular attention to the *climate change* in Nepal and Southern Asia. The reporter's consideration highlights the necessity for the *Environmental* journalists to read up constantly on the evolution of the *Environmental System*, a reality that affects all human beings. In doing that they must deal with the matter from different points of view: economic, political, social, public safety, so that the *"Environment"* theme can *"go on the front-page"* for as long as possible. A separate discourse is deserved by international events as the annual *World Environmental Day*, fostered by United Nations, on the 21st September.

The photo-reporters must be subjected to the same *Nine Rules* of the *Ethic Code of the National Press Photographers Association*, founded in 1947 in the United States and followed all over the world:

1. Be accurate and comprehensive in the representation of subjects.
2. Resist being manipulated by staged photo opportunities.
3. Be complete and provide context when photographing or recording subjects. Avoid stereotyping individuals and groups. Recognize and work to avoid presenting one's own biases in the work.
4. Treat all subjects with respect and dignity. Give special consideration to vulnerable subjects and compassion to victims of crime or tragedy. Intrude on private moments of grief only when the public has an overriding and justifiable need to see.
5. While photographing subjects do not intentionally contribute to alter or seek to alter or influence events.
6. Editing should maintain the integrity of the photographic images' content and context. Do not manipulate images or add or alter sound in any way that can mislead viewers or misrepresent subjects.
7. Do not pay sources or subjects or reward them materially for information or participation.
8. Do not accept gifts, favours, or compensation from those who might seek to influence coverage.
9. Do not intentionally sabotage the efforts of other journalists.

These ethical rules remind us that "*to explain the world through images is information*" (*Pasquale Spinelli*). And how is stated by *John Blewitt*, researcher in *Environmental sustainability* and communication at *Aston Business School* (Coventry, UK) what we see ad say are not properly the same thing. Images and words are both forms of communication, signs and symbols, tools to give voice and see what we listen to, we perceive, we touch, we smell and we think.

The *news report*, then, will enjoy a brilliant style of writing, "personalized" and suitable to the local dimension of readers and their everyday life, accompanied by images or other *cross-media references* like: QR-code, Internet sites link, *social network* etcetera; trying "to plunge oneself" into the facts deepening as much as possible. *Khadka* reminds us that the word "*time*" (atmospheric) is a single word, the "*Climate Change*" is an evolutionary path requiring time and periodical checks made by experts (Nepal Monitor interview, 13th January 2010). In fact, we cannot link each event to the phenomenon of the changing climate, regardless its deepen survey.

Green Tweets
#Media #RaiseAwareness #Science #Journalism #Quality #Information #Communication @SerenaRugiero #NewsReport @NavinSinghKhadk #Environment @BBCWorld #BBC @NPPA (National Press Photographers Association) @UNEP (UN Environment—World Environmental Day) #WorldEnvironmentalDay #NationalPressPhotographersAssociation @PasqualeSpinelli #PressReport #ClimateChange

2.4   The Social Media: Impact 4.0

The revolution of information starts from the "bottom".

"*The seasons are not as good as they used to be, anymore!*" One of the most used commonplace in Italy and so obvious in its contents, may acquire, perhaps, a new meaning if referred to the IT, the Information Technology revolution that, in less than 20 years, has literally transformed the way of publishing, archiving, seeking and using information.

The "huge wave" of the Web Net, as it was defined when *Google* started influencing our lives, in the "far away" 1998, changed definitely our means of communicating: giving information was the birth of *Social Media*: Facebook, Myspace, LinkedIn, Twitter, WhatsApp, and many others that enable people to exchange written texts, videos, blogs, photos or short messages whose main purpose is to collect the highest number of approvals, the so called "*likes*". Let us think about how much effective short sentences are "buttered" with abbreviations, cut words, acronyms and key-words. Furthermore, they are made more functional with *hashtag*: real "*tags*" allowing a very quick access to thousands of information linked to a specific word or concept.

Journalism has shown soon a great interest towards these new tools of communication. Potentially, in fact, they are able to influence a large audience thanks to the possibility to replay indefinitely the message using "*copy and paste*" process. It is not by chance if **Twitter**, the *social* medium whose strong point is the synthesis, is having a rapid diffusion among the main and authoritative newspaper offices at an international level. *Haewoon Kwak*—a scientific researcher specialized in game analysis, journalism, and computational social sciences at the *Qatar Computing Research Institute* QCRI—following up his specific study on the use of *Tweet* went so far as to envisage its "added value" for two fundamental reasons: the great capacity to motivate those who use it and the possibility to get a synthetic answer (140 alphanumeric symbols in its original version) and an immediate *Retweet* to the message transmitted.

Perplexities are not missed. For instance, on one hand the *Social Media* have improved the techniques of research of the *Internet* surfers contributing to increase their knowledge and competence. On the other hand, they have determined a general "personalization" in giving and receiving the information. A **process from the bottom to the top (*bottom-up*)** that allows everybody to give information, even though not always suitable or correct. In 2008, the Report on the *State of Blogosphere*, made by *Technorati*—a highly qualified Californian Company, based in San Francisco, a Consultant in applied Net Technology—highlighted as the *blog* phenomenon had become a milestone of information. Unfortunately, its contents (e.g. multimedia elements, written texts, posts etcetera) visualized in an anti-chronological way (from the most recent back to the most dated) has been creating disorientation between the professional journalism and a simple exchange of ideas, not always qualified. Nowadays many journalists have started doing a qualified *blogging* (*Brian Solis*, "Engage").

The debate is therefore still open. The majority of the public opinion is wondering whether they can rely upon this new form of communication-information, continuously evolving. We are talking about Web 3.0 and Web 4.0 to identify the future devices: *semantic* research engines able to collect metadata independently, without keywords. This question gets more importance if we contextualize it with reference to the *Environmental Journalism*, multidisciplinary and multiform for its nature.

When "**confidence**" is involved, in the world of journalism, we refer to three levels of application. First of all, "*confidence through journalism*", created by the journalists in the social system both of single individuals or organizations. "*Giving guarantee*" to the reader that, nowadays, can be translated into the survival in the market of information, inflated by non-conventional sources. But the "confidence" must be reciprocal. The "confidence in journalism" is the direct consequence of professionalism of any single journalist when reporting the piece of news in a very precise, transparent, unprejudiced way, following the professional deontology. We are going now to consider the third level: "*building trust in journalism*" is the product of the style used by the journalist in reporting facts, once they are communicated by the interviewee, that is: politicians, sport men, scientists, environmentalists, etcetera, as well as by the perception of any single reader about what he reads. In other words, about his more or less propensity to "trust", that is to "believe" the piece of news, depending on his socio-cultural background. All that contributes to create, confirm or remove confidence to the written text. A power of influence on the public opinion that must be managed with great responsibility, not being only the journalist's exclusive prerequisite.

That being said, we would like to know how much the *Social Media* and other technologies of new generation affect the way of "communicating" the *Environmental news*. In order to answer this question, we need to start from a matter of fact. In the past, one of the journalist's main purpose was to look for information necessary to create a story that became then a piece of news. Nowadays, most of the information come from "unknown" sources (e.g.: Facebook post, Tweet, Digg, etc.) that rush ahead the *reporter* who, deprived of his "scoop"—a piece of news given well before the other competitors—is compelled to find out the most original way to tell a story but always very respectful of its truth. Here-hence the newspaper offices' trend to publish the largest number of last minute news, *breaking news*, in order to move with the times and the "non-conventional" sources of information.

Concerning the delicate matter of the *Environmental information*, we underline again the important quality of "green" communication and the necessity of a **qualified "green" journalism**. The *Environmental Journalist*, specialized in *Environment* and Sustainability through a constant, refreshing and updating route such as: post university masters, specific courses, conferences, *webinar*, *peer-review* and guide-tours, is allowed to live "professionally" the *Environmental matter*. The deep knowledge of the *Environmental issues* is an efficient tool to facilitate the day-by-day fully interpretations of facts. The *Social Media*, then, should be used only to spread out information based on reliable sources so that the readers will be able to think of and reflect, trying to elicit answers and solutions. In a world where journalism does not keep the monopoly of information anymore and where everybody can contribute to inform, it is essential to keep a "***critical sense***", analysing and evaluating the facts in a rational way, without stopping to the mere *spot* news.

The *Environmental journalist's* role in view of the *Social Media* age 4.0, seems to be more and more *multitasking*—able to deal with many things at the same time—by using simultaneously more than one "platform" to convey a piece of news: television, radio, writing paper, web etcetera. His responsibility is therefore multiplied regarding writing and printing.

The *reportages* in fact can expand their effects far beyond the "traditional" Net dissemination e.g.: newspaper and magazine on line, newspaper, *Social Media* etcetera. The "participatory spirit" of information has led to create and spread out the "***aggregators***", *web* platforms that put together extracts of news taken from different articles or sites published on the web.

From this point of view, the journalist is supposed to use the *social medium* without being subjugated in order to disseminate ethic values, civilization, *Environment* protection and respect for the eco-system we belong to. The objective should be to make a "public" service, in favour of the "average" reader. This idea is shared by *Geneva Overholser*, a journalist, former professor and director at the School of Journalism *Annenberg* of the Southern California University.

A doubt still remains: *are Social Media really safe?* According to a survey carried on in 2010, on the occasion of Haiti tragic earthquake, among the most well-known international newspapers agencies such as: *BBC UK; Agence France Presse, Singapore and US & Canada; India Today Group Digital; Antara News Agency Indonesia*, the answer was anything but positive, due to the unsatisfactory degree of precision, impossibility to verify the news and recover the sources of information. These are the main risks linked to the use of *Social Media* as it was denounced by the editors in chief. This does not mean to "demonise" these tools. In fact, if they are used correctly they become an *approval rating* for the journalist to test the readers' response, pushing them to "be connected" and informed in real time about facts to investigate.

In view of the *Web 4.0 Age*, these new tools might be the main sources of information. Of course, it is difficult to think that they can work alone, or they may outclass the other *Media* in communicating the *Environmental message*. The twenty-first century journalist cannot isolate himself in an "ivory tower", he can be but a spectator. In the *Information Society*, the journalist must participate, expose himself, express his thought, always ready to answer the questions and to compare himself with a civil way towards those who do not share his opinion. With *Social Media,* all that happens in a few seconds. The journalist must transform himself into an *"orchestra-man"* conducting himself and his way of *"making music"* using in the same time traditional and non-traditional means of communication: a camera, a microphone, a tape recorder, a video camera, a laptop, an iPad, a smartphone, a Bluetooth, a block notes etcetera. Anyway, last but not least, the *corner stones of journalism* do not change: fluent style, correctness, fact check and news based on reliable sources.

Green Tweets
#Web #SocialMedia #SocialNetwork #NewsMedia #Trust @KwakHaewoon #DataJournalism   #Tweet   #Retweet   #GameAnalytics #ComputationalSocialScience   #Blog   #blogsphere   #Communicator #Journalist #Webinar #PeerReview #Multitasking #Environment #Ethics #Civilisation #Ecosystem #Web2.0 #Web4.0 #Laptop #IPad #Smartphone #Bluetooth

2.5   Being Environmental Journalists: A Multi-stage Mission

The new dimension of journalism: being multi-tasking and "curious" in the same time.

Who is an Environmental journalist? A journalist who has an important objective to achieve: to inform the audience on complex topics but in a very accurate and precise way, thus not excluding *a priori* other theories shown. Other points to be considered for a good journalist: balanced, educational, cross Media, always able to keep a fluent style and tell about facts strictly linked to the reality in which the readers live.

Finally, it is essential to refer to the *"further steps"* besides the classical standard of the *"wh-questions"*: *who, what, when, where, why*. Sources highly recommended, certified, institutional and *super partes* are another journalist's must. Furthermore, there is a risk that some documents are following exclusively economical and occupational interests in order to create work places, *bypassing* the issues in favour of the sustainable growth. Nevertheless, this phenomenon is less and less frequent thanks to the spread of "Green Economy", now *Circular Economy* that has changed the main financial companies in reliable sources for the *Media*. For instance, besides the *Gross Domestic Product* (GDP) of a Country, another calculation is added: the *Environmental Impact* i.e. the potential *Environmental danger* registered by the *production chain* starting from the raw materials to the following steps of transportation and energy transformation and final product as far as the removal and disposal. *Stern Report*, an economic, solid, and reasoned analysis drawn in 2006, in order to evaluate the *Environmental* and *macroeconomic impact* of the current climate change pointing out the negative effects on the world GDP, is undoubtedly a good example.

The search of the *absolute impartiality* when talking about journalism is a mere utopia, however. Each press agency follows an editorial policy as regards the topic dealt with, the *Environmental* ones included. On the other hand, the selection itself, made by the journalist when choosing the questions to be treated or not, is a subjective operation.

What is asked the journalist to do? To report "facts separated from opinions", to keep on a precise documentation taken from well-known and truthful sources and supply the "story" with opinions of different trend, faithfully reported and using a quoted text. This behaviour includes also a great sense of *deontological honesty* that allows the journalist to make corrections on what published in the case of new information might modify the version facts. For the *Environmental matter* this means a constant updating of the evolution of the scientific theories, the legislative acts,[2] the data and *Environmental indicators*, and any other hypothesis theorized by experts such as: professors, researchers, non-governmental organizations (NGO), *Environmental journalists*, and so on. In the *Environmental matter*, as in many other sectors, "*the flux of events is continuous*" with the complicity of technological innovation as reminded by *Raymond Williams*, an English critic and historian.

Finally, a key factor is *to know how to report "green" news*. This target can be achieved easily if we think of the meaning of the word *"ecosystem"*. Set of living and non-living organisms interacting among themselves in a specific *Environment*: whether a lake, a river, a forest, a territory or the Planet Earth; what is called also *ecosphere* or *biosphere*. Metaphorically we can say that as in the ecosystem each element is depending on the other, the words reported in an article depend one another, thus creating a balanced structure.

What must the journalist avoid in writing Environmental news? Concepts clearly incorrect, drastic omissions of data so that they influence negatively the process of comprehension of the information.

---

[2] Such as: Laws, Legislative Orders, Law Decrees, Consolidated Law, European directions and decisions, international treaties and agreements etcetera.

Next to the study of documents and interviews, *to experiment* the facts in person is a must: The *Environmental matter* should *be lived* directly.

For example, visiting:

- a plant of natural water purification;
- a deposit of toxic waste;
- a waste-processing plant;
- a sustainably managed forest.

Also, to understand the *entire life cycle* of the products is fundamental:

- the process of production of plastic and glass bottles;
- the production chain which makes aluminium or paper products;
- the process of reusing and recycling of materials;
- the waste disposal at the "end of life" of a manufactured product.

The *de visu* approach (in person) is always the most efficient to understand the real "green" nature of products and services and it is able to denounce *green washing* cases (see Chap. 3).

How can we find the right questions when we face a project, a product or an idea, introduced as Environmental? *Michael Keating*, a journalist, writer, consultant specialized in Sustainable Development suggests some; underlining that sometimes it is not so easy to answer correctly.

Let us imagine to write a *reportage* on a new governmental programme for *Environmental matters*.

1. *Can we define it as "sustainable" in the sense that it can be managed without reducing significantly the natural resources?*
2. *Can we prove it?*
3. *In which measure its implementation foresees the use of non-renewable resources as the fossil fuel i.e. coal, crude oil etc.?*
4. *Has been the employment of renewable resources considered? In a superior way as to their degree of regeneration? In other words, the trees planted to reforest a deforested area are they able to compensate the number of those been cut down, considering the time they take to reach their adulthood?*
5. *The substances we want to use may erode and contaminate the ground? May reduce the resources of clean water available? May compromise the natural resources of food? May pollute the air, the soil, the water? May reduce the stratospheric level of ozone? May enter greenhouse gases into the atmosphere? May produce waste? May interfere with the biodiversity? Are they toxic? What level of toxicity? Which effects on the living beings?*
6. *Which are the maximum levels of concentration of those substances allowed in the legislative system of the Country in which the facts are going to be?*

We leave the readers to deepen the cues that can be elicited from the questioning so far, and keeping in mind that, in the age of "*communication research*" with the outbreak of the communicative system, helped by multimedia technology and Web net, it is impossible to present an exhaustive "approach" to the *Environmental journalism*. As already specified in our dissertation, the multidisciplinary of mass communication means at our disposal always need a careful evaluation of the social, historical and economic context, case by case. Eventually, we cannot help talking about "*journalisms*" and not journalism as reminded by *Angelo Agostini* and *Umberto Eco*. They are "*the work and the product of social institutions that answer the public, the audience, the social groups who find the tool that allows them to make a point of the day, of the week, of the month, of the year in the newspaper [whether made of paper, of words, of image, or multimedia]*". And the five *WH-Questions* cannot do without multiplying themselves in a many-sided system of questions and answers in a continuous evolution.

Green Tweets
*#Whquestions   #WhoWhatWhenWhereWhy   #GreenEconomy
#CircularEconomy   #GrossDomesticProduct   #EnvironmentalImpact
@NicholasStern   #SternReview   #ProductionChain   #EditorialLine
@RaymondWilliams   #GreenNews   #Culture   #IndustrialRevolution
#Sociology #GreenWashing @MichaelKeating #EcoSystem #GreenNews
#LifeCycle   #CommunicationResearch   #EnvironmentalSustainability
#RenewableResources @UmbertoEco #Semiologist #Philosopher #Author
@AngeloAgostini #Media*

2.6   Conclusions and Reflections: The Strength of Environmental Journalism Be with You! The "Star Wars" of News

How can the information be transformed into a *medium* of communication? May the information turn into knowledge?

As highlighted in the previous chapters, finding a united model of journalism dealing with the *Environmental matter* would be an ambitious operation and maybe inappropriate. We have to consider that the stunning increase of information, supported by the new *Mass Media* in addition to the traditional ones, has made ever more complex the *Environmental Communication* System. Therefore, it is better to operate, case by case, bearing in mind that the simple information in itself is not enough to raise awareness in the public opinion about the *Environmental ecosystem* we belong to.

How to be successful to manage the information then, making it a tool of communication? Master Yoda in the film *Stars Wars*, maybe would answer with this maxim: "*A Jedi—defender knight of peace and justice in the Galactic*

Republic—uses the Force for Knowledge and Defence, never for attack". Equally, the *environmental journalist*, effectively, has the duty to "put out" his pen like a "laser sword" to hit his target: the reader.

The history of the eco-journalism has taught us that all that happens when the **information** is transformed into **knowledge**, as the case of the *Environmental matter*, fosters more sustainable lifestyles. This occurred to the pioneers of *Environmental journalism* who showed the strict link between "economy" and "ecology", starting from their very etymology. Both words have the same common root: οἰκος (oikos), an ancient Greek term that means "home", "mansion", thus metaphor of the ecosystem Earth in which both Man and Nature coexist, representing the two-sides of the same coin.

The journalist, then, can exercise a great power: to modify or overturn our point of view on *Environmental issues* through a critic approach able to make the reader fully understand them. The seeking of a substantial truth can turn around the perspective of our *Environmental vision*: from "Element to be exploited for the benefit of Man only" to a "natural resource to be protected and preserved".

On the other hand, if we are walking in a forest we can consider each single tree as a *"source of earning"* for the production of wood or derivate (***anthropocentric vision***) or just appreciate the high landscape value. In this case we are pushed to protect the autochthone flora and fauna and made them known through paths of sustainable tourism (***natural vision***). The journalist's pen can do a lot to draw the readers' attention on considering "the forest ecosystem" as a *simple chain of production* at man's service or as a *patrimony for the humanity*, equal to a monument symbol of civilization.

This attitude does not change into actions of *Environmentalism* or *ecology*, which could induce to a wrong partisan vision of the *Environment*. The journalist is able, on the contrary, to be more effective despite being "soft" by using a strategy that *Joseph Nye*, a political scientist and professor at the well-known Harvard University, defined as *"**soft power**"*. *Nye* opposes the traditional method of the stick (strength) and of the carrot (money) to a third power which is to *"**influence the behaviour of others to get the outcomes one wants**"*. In other words, it means to convince the others "spontaneously" to do what it has been planned. A "seductive" power that aims at a shared consensus.

Unfortunately, this approach is not so easy to take if we consider the trend, by part of the public opinion, to close itself like an oyster when we deal with the *Environment*. A topic which is too often considered "uncomfortable", "problematic", "difficult to understand" or even "not relevant" because apparently it seems not to affect our lives that much. This attitude is strictly linked to our Western *mainstream* culture, based on *consumerism*, that can be defeated with a constant commitment of the journalistic communication.

It follows that, before disseminating a specific *Environmental message*, any journalist should ask himself if the actors involved: politicians, jurists, entrepreneurs, industrialists, citizens, and so on, are seriously committed in favour of the protection and the preservation of the *Environment*. The pioneers of the *Environmental journalism* in India and Asia, have highlighted the necessity of going beyond the piece of news and foreseeing the changes, in case of uncertain issues that might make the public feel sceptical.

When we use terms such as *"Sustainable Development"* or *"Environmental Action"* it is necessary the facts prove the real meaning of *"**Sustainability**"* and *"**Environment**"*, against "confusing" or "unbalanced" messages. The deepening of facts, observations, deductions and objectivity become then precious allies of the *reporter*. This is the opinion of *Carl Zimmer*, a scientific writer, keen supporter of the importance of the fundamental role of the *"**research of information**"* in the *Environmental journalism*. Only one preventative measure: to follow a logical course. Collecting too much information can "sink" any effort. A journalistic *reportage*, like a book dealing with the *Environmental matter*, cannot contain everything. This ultimate ambition is the aim of the encyclopaedias. It would be like *"to make a ship in a bottle"* states *Zimmer*, himself.

In the mid Nineties, under the so called *"Green Revolution"* a new form of intensive agriculture, already experimented in the *Punjab*, was attempted in the *Kutch District*, in the Indian region of *Guyarat*. The promoters of the project emphasized the high *Environmental value*, by stating that, thanks to the use of water and fertilizers, a deserted and unproductive area could have been transformed into its opposite, resulting in a better quality of the local ecosystem. *Could the territory of Kutch be assimilated to Punjab?* The simplest investigative journalistic report would have discovered that the arid region of *Guyarat* could not have been converted in a fertile area without significant unsustainable funding whose result would be uncertain, as some technical documents proved later on. Unfortunately, any *Newspaper Office* dedicated neither headlines or the end credits describing the risks linked to the *Environmental impact* of such a kind of project, afterwards put aside. This "press blackout" should make people think.

Environment should enter the main door of the review *"News from the World"*. *Man* and his *Environment* must be considered as a whole linked by a mutual respect and not by a relationship of supremacy of a "world", the human one, submitting the natural one. Such a message should dominate the *Media* of information, not like the result of a romantic and abstract ideal but as a consequence of a pragmatic and objective approach to the news. This is the true challenge of the *"sustainable" journalism*: rethinking the *Environment System*, its values, its principles and its priorities; taking it out of its dimension merely *economic* and *business oriented*.

Nevertheless, *Communicating Environment* through a qualified journalism means to report the facts in details, contextualize them with a fluent and clear language, whatever the topic: about a *campaign for the forestation* in favour of an Amazon wooden area, or about the *effects of climate change* on the Polar pack or even about the *pollution of the groundwater* in a quarter of our town. What is really important is the way used for communicating: the choice of words and the newspaper column the piece of news is put, avoiding the last pages. Much better integrating it in the political, economic, cultural pages; and why not? In the very *editorial opinion article*, dealing with recent and relevant themes. It would be a mistake, anyway, to catch the average readers' attention by a simple *make-up of the piece of news*, aiming at making it more attractive or "acceptable, easy to digest" thus trivializing it or spreading out too superficial messages.

The journalist should not forget his target: the average reader. His style of writing then should be captivating, able to surprise, make a mark, to excite, "dotted" with some key-definition necessary to understand the concepts and the *Environmental technical nomenclature*.

Let us now remind and summarize the professional values and their must:

- Report facts correctly through an accurate and repeated check also for the most predicted news.
- Act in a transparent manner in building up the *Environmental information* considering all points of view, included the fake sources in order to get a complete idea, starting from the very facts.
- Be faithful to the integrity and seriousness of the professional ethic.
- Spread out the facts by using a responsible communication and the most suitable *medium*: this means not to omit facts only because they are unpleasant for the story or "troublesome" for one or both the parties.
- Feel involved in his own role of journalist, establishing goals and fixing key-paths through which grasp the readers' attention and influence their point of view.
- Try to be neutral and/or independent in comparing to the information channels, dividing what is a piece of news from what is just an opinion.
- Keep an ear out in order to guarantee a shared approach to the piece of news able to leave space to the readers with a view of the maximum "democracy" in communicating the *Environment*.
- Let the journalist conscience guide us.
- Select the huge amount of *Environmental* and *sustainable questions* offered by the *Media on paper and online*, choosing from safe and unsafe sources.
- Prefer complete information, feeling always on the side of the piece of news. This means the journalist's direct participation to the event or to visiting the place where the facts happened possibly interviewing the involved parties, witnesses, or other subjects.

To Sum up:

Chart on "Environmental Journalism" by Maurizio Abbati (Copyright Holder) © 2017

These are the "weapons" available to the journalist of today in order to defeat the superficiality, unjustified scaremongers and any partisan position when speaking about Environment and Sustainability.

And let the strength be with us!

Green Tweets
#Information   #Tool   #EnvironmentalCommunication   #Knowledge
@StarsWars   #EnvironmentalJournalist   #Journalism
#EnvironmentalMatter   #Ecology   #Ecosystem   #Information
#AnthropocentricVision #NaturalVision #Environmentalism #Ecology
#SoftPower   #Influence   #Sustainability   #Environment
#ResearchofInformation   #GreenRevolution   #Punjab
#EnvironmentalCommunication   #EnvironmentalInformation   #Media
#Innovation #Creativity #Multitaskoing

Bibliography

Alejandro Jennifer, *Journalism in the Age of Social Media,* Reuters Institute for the Study of Journalism, University of Oxford, Hilary and Trinity Terms 2010, pages: 47.

Aristotle, *Le tre etiche* (The three ethics), Milan, Bompiani Editore, Il pensiero Occidentale, 2008, pages: 1620.

Attenborought Sir David, Blewitt John, *Media, Ecology and Conservation—using the Media to protect the world's wildlife and ecosystems*, Cambridge (UK), Green Books, 2010, pages: 209.

Bernd Blobaum, *Trust and Journalism in a Digital Environment*, Reuters Institute for the Study of Journalism, University of Oxford, 2014, pages: 66.

Blank-Libra Janet, Overholser Geneva, *Pursuing an Ethic of Empathy in Journalism (Routledge Research in Journalism)*, New York/London, Routledge, Taylor and Francis Group, 2016, pages: 232.

Bodker Henrik, Neveria Irene, *Environmental Journalism*, Journalism Studies, 2012, Vol. 13.

Carson Rachel, *Silent Spring*, Introduction by Linda Lear, Afterword by Edward O. Wilson, Boston—New York, A Mariner Book—Houghton Mifflin Company, 2002, pages: 400.

Eco Umberto - AA.VV.: *Giornali e Giornalismo* in "Il Novecento—Comunicazione, teatro, e cinema" (Newspapers and journalism in "The Twentieth Century—Communication, theatre and cinema"), Milano, EM Publishers s.r.l., 2014, pages: 439.

Gavrila Mihaela, *L'onda anomala dei Media—il rischio ambientale tra realtà e rappresentazione* (The tidal wave of Media—the environmental risk between reality and representation), Book Series: Scienze della Comunicazione (Communication Sciences) by Marino Livolsi and Mario Morcellini, Milan, Franco Angeli Editore s.r.l., 1st Edition 2012, pages: 336.

Keating Michael, *Covering the environment. A handbook on environmental journalism*, Collana "National Round Table Series on Sustainable Development, published in cooperation with The Graduate School of Journalism, The University of Western Ontario, 1993, pages: 164.

McGuigan Jim, Raymond Williams on Culture & Society: Essential Writings, Los Angeles, London, New Delhi, Singapore, Washington DC, SAGE Publications, 2014, pages: 342.

Meyer Philip, *Letters from the Editor: Lessons on Journalism and Life*, Editor, William F. Woo University of Missouri Press, 2007.

Meyer Philip, *Precision Journalism: A Reporter's Introduction to Social Science Methods*, Indiana University Press, Bloomington, 1973, pages: 342.

Meyer Philip, *The Vanishing Newspaper: Saving Journalism in the Information Age*, University of Missouri Press, 2004, pages: 269.

Morresi Enrico, *Etica della notizia* (News Ethic), Collana Saggi, Bellinzona, Edizioni Casagrande, 2003, pages: 282.

Partipilo Michele, *La deontologia del giornalista—Dalle Carte al Testo Unico* (The Journalist ethical code: from Charters to the Single Text), Centro di Documentazione Giornalistica, Ordine dei Giornalisti—Consiglio Nazionale, Rome, January 2017, pages: 209.

Rowe Duncan Graham, *How to communicate Sustainability*, The Guardian, 10th November 2010.

Randerson James, *Science & Environmental Journalism: A 60-Minute Masterclass (60-Minute Masterclasses Book 8)*, The Guardian, 2014, pages: 58.

Rugiero Serena, Salvati Luca, *Territori virtuosi. Temi per una geografia economica e sociale dei rifiuti in Italia* (Virtuous Territories. Issues for a waste geographical economic and social map of Italy), Acireale, Bonanno Editore, 2014, pages: 152.

Solis Brian, *Engage! The complete Guide for Brands and Businesses to Build, Cultivate and Measure Success in the New Web*, New Jersey (USA), John Wiley and Sons Inc. Copyrighted material, 2011, pages: 307.

Spinelli Pasquale: *Tutti sanno fotografare, ma la deontologia?* (Everybody knows how to take photos, what about ethics?), included in V.A.: *I pilastri del giornalismo—Libertà di stampa, etica e deontologia: facile o no?* (The pillars of journalism—freedom of the press, ethics and deontology), Ordine dei Giornalisti dell'Emilia-Romagna—Il Torchio snc, San Giovanni in Persiceto, dicembre 2015, pages: 46.

Stern Sir Nicholas, *The Economics of Climate Change: The Stern Review*, Cambridge University Press, 2011.

Wells Herbert G., *The War of the Worlds*, Collana Collins Classics, William Collins, 2017, pages: 240.

Williams Raymond Williams, *The Country and the City*, The Hogarth Press Ltd, London, 1985.

Williams Raymond Williams, *Culture and Society: 1780-1950*, Spokesman Books, 2013, pages 364.

Web Site List

Condemi Josephine, "Essere è essere interattivi": la rivoluzione dell'infosfera secondo Floridi ("To be is being interactive": the info sphere revolution according to *Floridi*), Il Sole 24 Ore, 28th September 2014 www.ilsole24ore.com

Green Report—magazine on eco-economy www.greenreport.it

National Wildlife Federation: www.nwf.org

Navin Singh Khadka, The Kathmandupost, Ekantipur, kathmandupost.ekantipur.com/author/navin+singh+khadka

World Wildlife Fund (WWF), Environmental Journalism and its challenges www.worldwildlife.org, www.wwf.it

Which Shades of "Green" Are the New Nets of Communication Disseminating?

3

Abstract

The modern society, dominated by the "philosophy" of *being always connected to the Net* allows us to communicate live with thousands of *Internet* users, thus revolutionizing the *concept of time and space. An opportunity or an obstacle to the Environmental Communication? How to guarantee a correct understanding of what is communicated? Which precautions to be taken?* Questions that make the modern communicator reflect on the management of the *Environmental message* when surfing the Net. A process that requires a suitable degree of competence on the environmental matter, as well as a specific training in communicating responsibly. Only in this way, we are able to know, face and foresee the message "noise", which might compromise the contents, creating unjustified alarmism or underestimating real dangers. It is increasingly necessary, to invest in *qualified information*. A target that becomes a *social duty* especially through the *Net* and the use of *Social Media*. The latter have proved to be useful allies to spread out the *Environmental values* with words and visual elements, also in occasion of emergencies.

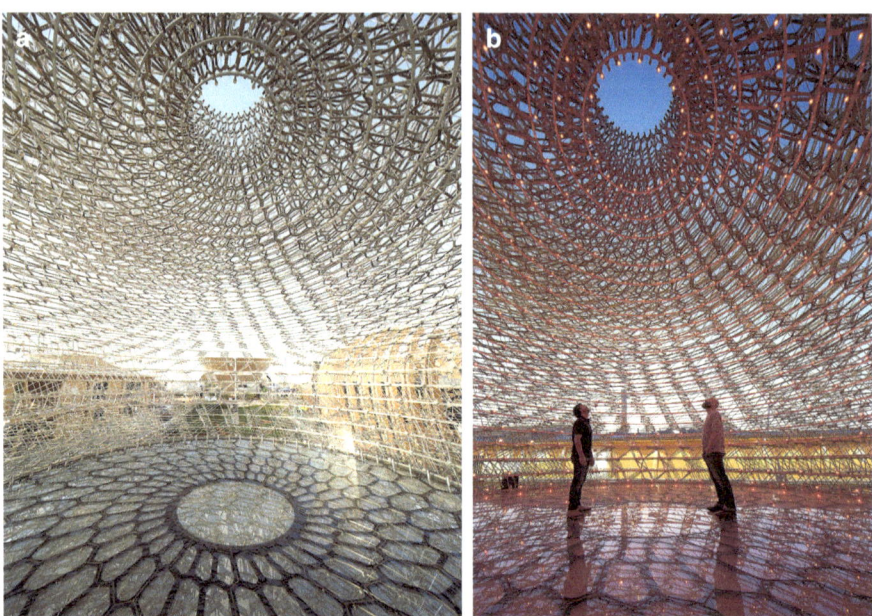

"When one tugs at a single thing in Nature, he finds it attached to the rest of the World"
John Muir
Scottish engineer, naturalist, writer, naturalized as American citizen (1838 – 1914)

Figs.   **a-d**   Great   Britain   Pavilion—Artist: Wolfgang   Buttress—Milano   EXPO 2015—photos by *UK Trade & Investment* (**UKTI**) © @UKTI (Copyright Holder). www.gov.uk/government/organisations/uk-trade-investment, www.wolfgangbuttress.com

Reading Proposal: In Between Words and Pictures

UK Pavilion, named *"Be Hive"*, designed for *Milan EXPO 2015*, by the English artist and creative lead *Wolfgang Buttress* and realized by the art studio *Stage One and Rise*, was awarded the prize by the technical jury as the most related to the theme given for the Universal Exhibition: *"Feeding the Planet, Energy for Life"*, for its interdisciplinary and architectural structure.

Theme: The exhibition premises reproduced faithfully a huge beehive surrounded by a fruit yard and a grass, not by chance, of hexagonal shape—to recall the classical shape of the cells of the beehive where honey and pollen are stocked (Figs. a-d). An evocative environment, at night lit by electrical impulses generated by the movement of the bees' wings, in which the visitors could experiment the world vision from their point of view (the bees). Mini teaching films to discover the hive life and its tight link with the environment we live. The grass field was created taking into account the bees visual field. The faithful reproduction, inside the structure, of vibrations and hums reproduces "the language" used by the bees to communicate among themselves. These tiny insects, in fact, certainly not "spectacular" for their size, have a fundamental role in the ecosystem, allowing many vegetable species to reproduce themselves, through the pollination, and thus guaranteeing the productive cycle of many other animal and vegetable species (e.g. the grass field is the main food for herbivores such as: cows, sheep, goats etc.) at the bottom of the food chain of men (30% of food at the world level).

Architecture: A cubic steel structure, containing a hexagonal sphere made with a knitting framework, 14 m tall, supported by pillars, and made up by 169,300 fragments interconnected to each other and embellished by thousands of LED lights able to change the buzzes and vibrations inside a real hive situated in Nottingham (UK), into visual impulses linked to an integrated audio-video system.

Interpretation: The project focussed on the pun of "be", auxiliary "to be", sharing the similarity of assonance with the word "bee". The insect, symbol of industriousness—but also of work, virtue, hope, according to the traditional heraldry—encourages us to reflect on the role of man in the delicate puzzle of the *ecosystem* in which each "piece" is linked to the other, as *John Muir's* statement reminds us. A "net" system like the one which stands at the basis of the *World Wide Web* spread and *Social Networks* we will be going to deal with in the **"Chapter 3—Which shades of "green" are the new Nets of communication disseminating?"**. The bees' cooperation must be emulated by man with his daily contribution by adopting a lifestyle respectful of the environment and preserving it for the future generations. The solid, but at the same time, "delicate" structure, which looks like dissolving in the air, emphasized all that.

3.1   The Environmental Message in the *Net* Achieves the Four Dimensions

To network and to do it well: these are the first steps to spread out an *Environmental Message online*.

How many of us are surfing the Net to look for information about the most used environmental words by the Media? Climate change, Green Economy, Circular Economy, Agenda 21, Man-Made Environment, Environmental Impacts, Environmental Audit, Benchmarking, Environmental Balance, Bio-indicators, Bio-masses, Rehabilitations, Environmental Certifications, Environmental Accounting, just to list some of them. Nowadays, the main search engines are the greatest virtual "encyclopedia" ever existed, able to offer the *Internet* users a considerable amount of information overlapping one another as to form the *"matryoshka effect"*, triggering a rationalized interchange of ideas and engendering new ones. In fact, we are living the age of *robotics* which does not communicate within the boundaries of a more and more compact and advanced screen but it interacts constantly with its users. It is not secret that technologically and highly **evolved societies** like the: United States, Japan and Europe, **are always on**. *Sherry Turkle*, an American sociologist and technologist underlines bitterly that the technological *Media* are acting as *"stimulators of the human contact taken away by the Net"*. The digital world has become, then, the extraordinary instrument of *Environmental Communication* if used in a correct and efficacious way. We must not forget, in fact, that the new generations learn to interact with technology easily. The new digital grows with electronic toys, mobiles, iPad, laptop, considered as their main source of information. Hence, the need to reflect on the real capacities of the *Net to* communicate correctly the *Environmental messages*.

Consequently, some preventative measures should be taken in order to make the **Information and Communication Technologies—ICT** become really "clever" thus contributing to the "green growth". The "smart" technology to improve the energetic efficiency is already a reality in many sectors: Construction Industry, Transportations, Logistics, Energy Production, Distribution and Consumption. Sophisticated systems of sensors applied to the Net enable the objects to interact with us supplying us with immediate data about the *Environment* status and about future scenarios as consequence of our day-to-day actions. Let us see then the potentiality of the Net takes a three-dimensional shape, at the service of the *Environment* System: **Length**, **Width** and **Depth**. We could better think of a "four-dimension" shape considering also the **Time**, as envisaged by some evolutionary theories of Euclidean geometry, essential element to draw up an *Action Plan* in the *Environmental field*.

"*Making things communicate has a meaning, provided that the Nets of Objects and men are integrated*" underlines *Carlo Ratti*, an architect, engineer and professor at the *Massachusetts Institute of Technology* of Boston, USA, where he is the director of the *MIT Senseable City Lab. A* working party promoting, all over the world, innovative projects to reconsider the urban spaces with a view to sustainability that uses technology in order to protect the *Environment* and the wellness of the citizens.

Accordingly, we can say that the Net outdid itself changing into the *Internet of Things*, an expression coined for the first time in 1999 by *Kevin Ashton*, a researcher at the *MIT Laboratory*, and that clearly gives the idea of the "revolution" we are witnessing now. An example is, for instance, the devices which allow the indoor acclimatization and lighting plants to adapt, hour by hour, to the consumers' needs adjusting the light intensity, the temperature, or the levels of humidity in each single room of a building (MIT Project: redecoration of the *Agnelli* Foundation in Turin). This is the same "revolutionary wave" that enabled *MIT experts' team* to realize the **Supermarket of the Future** inside **Milan EXPO 2015**. A real space where consumers could interact directly with the food products on display on the highly technological *shelving*. This technology informs them, immediately, about the quality and *Environmental impacts* of each single product simply with their physical contact, thanks to a system of "smart" electronic labels.

Furthermore, the *Net* can convey specific *Environmental messages* directed to qualified groups of consumers. The spread *on line* of the technology **Cloud Computing**, thanks to a remote server offered by a provider, makes the users get thousands of information through *software* or *hardware*, mass memory for the data files. This technology, applied to the *Environmental Management Systems—EMS*, allows managers, administrators, employees and suppliers to access easily to the *Environment* information, necessary to obtain or maintain the *Environmental Certification*. Furthermore, it contributes to contact constantly with all the *stakeholders*, responsible for the development of the *EMS*. It has become, then, a qualified *Environmental Communication* that reduces time, expenses and paper waste.

From the Net derives, therefore, a proliferation of new instruments that can guarantee a use more and more personalized of the *Environmental* information, shaped according to the interests and needs of each single user. The *Communication* pattern cannot help including the message sharing between a sender (*addresser*) and one or more receivers (*addressees*) and the further reworking and internalization of its contents. A "journey" during which each "actor" includes his own cultural *background* made of past experiences of real life. *How does all that influence the delicate process of understanding the Environmental "language"?*

Green Tweets
#ClimateChange #GreenEconomy #CircularEconomy #Agenda21 #ManMadeEnvironment   #EnvironmentalImpacts   #EnvironmentalAudits #Benchmarking #EnvironmentalBalance #BioIndicators #Biomass #EnvironmentalRedevelopment   #EnvironmentalCertifications #EnvironmentalAccountability #DigitalConnectivity #Media #Net #InformationandCommunicationTechologies   #ICT   @MIT #MassachussettsIstituteofTechnology #InternetofThings @KevinAshton #SupermarketoftheFuture   #ExpoMilano2015   #CloudComputing #EnvironmentalManagementSystem   #EMS   #Stakeholder #EnvironmentalCommunication #MeansofCommunication

3.2   The Quality of the Environmental Message Through the Net

The *Environmental Communication online* must be qualified, and be able to qualify itself.

With regard to our overview about the **Environmental Communication** some questions arise spontaneously linked to the *quality* of the information conveyed through the Net both whether addressed to the whole public of *Internet* surfers or to small groups. The possibility to interact through the **Net** to anybody who may access to it, increases the risk to get an environmental message distorted by possible **"noises" of communication**. Considering the *social function* of the Net on the *Environmental* issues, it might be translated into unjustified alarmism or commercial boycotts plotted *ad hoc* by criminal organizations, devoted to unbalance the market or the socio-political equilibrium of a specific geographic area.

We would like to make it very clear how much delicate the management of the *Environmental* message can be. The senders should always be checked to verify their reliability and quality. This does not mean an *a priori* lack of confidence of the *Net* as a tool of communication. As shown in many occasions, in fact, the use of *Social Media*, we are going to deal with later on, constituted the *medium* to recruit, for instance, volunteers to rescue populations affected by natural disasters such as flood, hurricanes, typhoons, and earthquakes. For example, during the flood of *Queensland in Australia*, in 2010, the *Baked Relief* Campaign largely used *Facebook* and *Twitter #bakedrelief* to recruit volunteers. No better example of digital "democracy" could be found. That "government by the people", in the strictest etymological sense, who turns itself from virtual into real to help those who are in need.

"*We don't have a choice on whether we do Social Media, the question is how well we do it*", says with absolute conviction *Erick Qualman*, a *digital marketing guru*, author of the bestseller *Socialnomics*. This *like-to-question statement* makes us think about the consequences that the new *Media* have created between the addresser and addressee conveying the *Environmental* message. As numerous studies of semiotics, sociology and hermeneutic—philosophy that, through the interpretation of texts, looks for the essence of man—the role of the receiver of information is completely changed. The addressee of the message is an "*active*" subject who creates his own interpretation of everything conveyed.

In the case of information by the net, *World Wide Web* (**www**), the "disturbing" factors (noise) of an original message multiply and amplify effects. The **Environmental Communication in the Net** interfaces itself with thousands of other themes such as Culture, Literature, Show, Exhibition, Cinema, Art, Science, Economy, Politics etcetera, thus comparing, interconnecting, excluding themselves with *a click*. Favourable and unfavourable opinions influence also our perception of the *Environment*, subjected to dissenting "voices" like the ones of certain critic literature: "**eco-criticism**". The latter is a movement born at the beginning of the Years

Sixties in the United States and still alive in some public opinion. We are facing then a real "*melting pot*" of information with a myriad of direct or indirect references to the most important *Environmental* questions and problems like: news reports, economy news, politics and culture. Therefore, a large wealth of languages has influenced and still influence the public, "plunging the people" in the System "*Environment*", an overused word these days. According to the international press there are five *Environmental macro-theme*s that worry the public opinion greatly: (1) the atmospheric pollution and the climate change; (2) the deforestation; (3) some animal and vegetable extinction; (4) ground deterioration; (5) the overpopulation.

Green Tweets
#Communication   #Environment   #EnvironmentalCommunication   #Net #SocialMedia #Noise #Medium #Media @Erick Qualman #WorldWideWeb #www #EcoCriticism #MeltingPot #Nature

3.3   What Really Links the New Media of Communication to the Environment?

Is there room for the *Environmental Sustainability*, in the Net?

Starting from the axiom that "*you cannot not communicate*" (*Paul Watzlawick*), undoubtedly the innovation of *Internet* has produced epochal changes due to the increasing demand for a large interactivity among the various regions of the world. **Web** is a synonym for *speed* in the message delivery, available in the same instant to a very huge and heterogeneous public, *a priori* unknown. Furthermore, the *Net access* is in *real time*, *personalized* and *available* when and how someone wishes to interact and use its contents for his own needs and interests. This is very useful to disseminate *Environmental messages* as well, contributing to the growth of the global *Environment* and Sustainability awareness. On the other hand, most of the experts agree that the *information* passes from addresser to one or more addressees with the same one-direction path of **communication**—a shared, rational and multi-directional **information**—and they represent "crucial stimuli" in any process of changes in the society.

Communication essence then lies in the interchange of shared information between **addresser** and **addressee** which allows the evolution of the human race. The *Environment* and the world of the new *Media* in the *Net*, defined *as Web 2.0*, cannot help being tightly linked. *In what terms?* The **pair Web-Environment** not always has been successful to deepen the most authentic meaning of *Environment* System and its numerous sub-systems: energy, waste, biodiversity, pollution, sustainable tourism, climate change, sustainable cinema, etcetera.

In order to encourage the ***global change*** towards more ***eco-sustainable behaviours***, a greater commitment is necessary to map and disseminate the *Environmental matter* to a very large public. A constant challenge for those who deal with communication when using the new *Media*. In fact, their attention is always drawn to adequate the instruments to decode the eco-messages and adapt them to their own "public", that is to update them and put them in the right "context" of life of the stakeholders. *Environment* must not be considered an abstract and immaterial identity but an instrument to put in contact every single individual the essence of his territory made of culture, art, nature, traditions and values.

It is not by chance that the recent Doctrine suggests to follow the path of the *Culture of the Environment*, instead of just stopping on the mere definition. Let us talk about one of the many themes: "waste". The *Media* at our disposal: cinema, theatre, video, narrative, art, music, and so on, have contributed to deepen the most "cultural" effects. The interpretations are innumerable, herewith a few examples: *waste* as a metaphor of physical or social death; *waste* as background of a social scenery; *waste* as a result of the consumers' lifestyle; *waste* as workplace or bed refusal for the poorest strata of the society; *waste* as a raw material to create new forms of art or new tools for the day-to-day life. Once we have found the source on the *Net* and organized the information according to logics we can identify the most efficacious *tools of communication*, able to arise emotions with a large ***communicative strength***. Nevertheless, there is an open question about the correct assimilation capacity of the *Environmental message* disseminated by the Net: *which quality*?

Green Tweets
@Paul   Watzlawick   #Web   #Speed   #Sender   #Addresser   #Addressee
#Recipient   #Information   #Communication   #Environment   #Web2.0
#ClimateChange   #GlobalChange   #SustainableHabits   #Culture
#Media   #Ideas   #MeansofCommunication   #EffectiveCommuncation
#QualityofCommunication

3.4   How Much Is the Environmental Information Through the Net Reliable?

To disseminate the *Environmental matter* efficaciously is a question of responsibility.

The communication potentiality and the relationships offered by the *Net* cannot help relying on a synthetic but clear and fluent communication. We must not forget that: "*The awareness is like the sun, when it shines on things, they are transformed*" reminds us *Thích Nhất Hạnh*, a Buddhist monk, a poet and an activist for peace. This maxim reinforces the idea on a crucial element for the

Environmental Communication, the **quality** of its **contents**. The insiders are then compelled to find the most efficient way to transmit complex scientific concepts involving many areas of competence. The *Environment* and the *Sustainable Development*, the economic development compatible with the *Environment* protection are the main concepts of a new science seeking the integration of *Environmental*, economic, social, cultural, institutional aspects. They are pushing the academic world, the institutions, the school and business companies worldwide to promote a coordinated action in order to look for new models of innovative development to redesign the relationship between Planet Earth and its inhabitants, starting from the "local".

To achieve these goals, it is necessary to start from well-founded information. On the matter, two studies on the *e-democracy*, a system explaining the new technologies to promote decisional, transparent and participated processes open to public, help us to find out two prerequisites by the *web*. First of all, it must be "*credible*" and state the sources of inspiration clearly; **explain the facts smoothly** and supply with adequate tools to improve the lifestyle of any single individual on the basis of *Environmental problems* such as: supplying with Internet sites/addresses of "public utility", Citizen Support, Consumer Support, Associations, Chambers of Commerce, Governmental Offices of the European Union (UE), available to the citizens. Secondly, it must be immediately "**recognisable**", while surfing the *Net* through the research engines. It is then very important to create a relationship of confidence with the addressee of our message and also with our co-workers. This means to "**sign**" the *medium* we are using whether a brand mark, a symbol, a logo of an organization or a real digital signature.

We must never forget the principles of **ethics** and **correctness** contributing to create a work *Environment* based on principles of cooperation and reciprocal respect also essential in the *Net*. To *communicate the Environment* and the *Sustainable Development* means **to bear our own responsibilities**, be aware of the strategic role of our action and the effects it can produce, whatever our purpose is: to inform, to teach or simply to disseminate the message.

An ever-growing number of citizens get informed through the *Net* and *Social Media*, by exchanging a lot of *multimedia information* (e.g.: images, written texts, videos, emoticons, gifs, etc.). Very recent statistical data by the Italian CENSIS (Italian Centre for Social Investment Studies) point out that the *Social Media* users are above all *under* 30, because of the general request of information. At this point it is useful to talk about the capacity to approach some little-known words like "biodiversity" if not supported by an educational tool whose start comes from the real-life experience. The places we are around daily and then broadening in the *Net* originate the necessity to interpret the *Environmental message* correctly. Council Centres, Commercial Centres, Libraries, Museums, Places of Worship, Parks, Squares, Train Stations, Medical Centres, Recreational Places, become "primary" tools of communication able to give the gateway to the *Environmental language* available in the *Net*.

Of course, the *Web* must allow an easy "searching". The **clarity**, the **rational organization** of the multimedia contents, the **easy finding**, and the **possibility to deepen the topics** dealt with through *link* and other Sites, are only some of the many elements making the site more accessible for surfing and are the objects of checking and testing its quality. **Web surfing** is one of the major indicators of evaluation used at the Institutional bodies for the *Environment* protection, like the ISPRA's—Italian Higher Institute for *Environmental Protection* and *Research*—Sicaw26 indicator "*Use of communication and web information tools*".

The same path must be followed for the very new communication tools dedicated to *smartphone* and *tablet*, rapidly spreading out. The Environmental **Apps** are already present in the Italian reality where the concept of *smart city* (literally a "clever town") identifying a set of strategies of urban planning (e.g.: urban development, sustainable mobility, waste management etc.) direct to optimise the public services by approaching the citizens to the institutions and other social parts: university, research centres, police, etc.

The **quality of the Environmental Information** must be guaranteed by **quality communicators** found in the *Net* after a very rational information choice. This to counteract the spasmodic search for a "scoop" in the *Media*, linked to a crime story or *gossip*, nowadays heightened by the thousand sources of information competing with each other. A more oriented approach to *business* than to an educational and informative function thus lowering the quality of the contents disseminated by the *Web*. Furthermore, we cannot but notice that the *Environmental matter* is treated, more and more often, by some Sites ran by associations, foundations, and other private organizations that might exploit the *Environment* and *Sustainable Development* issues for political, economic and commercial reasons.

Piero Angela, a journalist, a writer, a TV anchor man, and one of the most successful Italian science communicator reminds us how to deal with any piece of news "*a* **creative presentation***, also* **amusing** *and* **emotional***, but with a sensibility meant to support the* **rationality**". His reflection makes us focus on the importance of the translation of difficult concepts about the *Environmental System* into simple ones: a challenge for everybody who wants to communicate the *Environment*. "*On the other hand, the worst enemy of the cultural communication is the boredom or the difficulty to understand*", *Angela* adds. Hence his innovative idea to disseminate complex knowledge, as the *quantum mechanics*, for instance, through graphics of fantasy, that is the comics of the fancy character *Mister Rossi* animated by *Bruno Bozzetto*, a well-known Italian animator, cartoonist and film maker (see Fig. 3.1). A symbolic case of employment of images to convey *Environmental messages* which was welcomed greatly by the scientific and educational world in Italy, Spain and Germany, in the Seventies.

Evidently, the **sources of information** must be **reliable**. This means to consult good sites and to prefer the information supplied directly by the responsible subject. There is a tendency nowadays to create also, on institutional portals, press offices or virtual *info point* producing documents validated by competent offices or by experts of communication. The **Data bank** offers different kinds of documents, both juridical or scientific, available also in *centres of information on line*, often supplied by Universities or National or International Research Institutions. In such *certified platforms* of cultural dissemination, we can often find bibliographical and site references, lists of experts to whom you can ask about specific themes.

Fig. 3.1 From the documentary film *"Bozzetto ma non Troppo"* (Bozzetto - small sketch - but not Too Much) Film-maker: Marco Bonfanti—Producer: Anna Godano—Co-producer and Distributor: © Istituto Luce Cinecittà (Copyright Holder). www.studiobozzetto.com, www.cinecitta.com

An example is the ***Platform of Knowledge Site***, good practices for the *Environment* and climate. A bilingual platform (Italian and English) realized by the *Italian Ministry of Environment and the Sea Protection* in cooperation with the *European Fund for Regional Development* and the *Italian Agency for the Territorial Cohesion*. In the Home Page, we can find the subdivision of *Environmental themes*: Resources Efficiency, Soil Protection, Atmospheric Pollution, *Environment* and Health, Waste, Climate Change, etcetera, accompanied by visual elements to reinforce the message. The texts are short, but efficacious, and they allow the user to get the key-information of each section with the possibility to deepen the topic by a simple *click*. This flexible approach allows readers or *Internet* users different types of reading:

- superficial reading (*skimming*);
- deepened reading (*scanning*);
- *"qualified"* reading aimed at the consultation of regulations, policy and financial instruments;
- *reading meant to get information* about events and last news;
- *reading for teaching purposes* to understand the main *Environmental themes*.

A *horizontal drop-down index*, well highlighted, enables the users to expand further theme sections (Policy and regulation, Themes, Programmes, Financial Instruments, Geography, Projects, News and Events) subdivided clearly into topics. The *Newsletter*, furthermore, can guarantee a constant updating about the best *Italian Environmental experiences* (*best practices*) establishing with the users a personalized link whose purpose is to inform them in a precise and functional way about the deadlines, thus arousing "emotions".

The *quick registration*, in fact, allows each single *Internet* navigator to be "called by name" when delivering the message. The link to the main *Social Network*: *Facebook, Twitter, YouTube, Instagram* makes the registered Internet users able to express themselves by sending their *feedback*, *even* in a multimedia and visual form: photo and video. The **interexchange of ideas** generates the evolution of thought or the change of the lifestyle launched to the *Environmental protection*.

This last function is considered an element of quality having the double advantage of the feedback on the efficacy of the message and the possibility to respond immediately in case of misunderstanding. This approach is the new strategy of the **Public Administration** concerned to establish a **permanent dialogue** with **citizens**, analysing their needs, their expectations, their degree of satisfaction of the services offered and the information received. A responsible relationship as defined by *Francesco Pira* in which: "*to communicate with the outside world means in primis to be accountable to the social system of reference [...] to communicate with the inside world creates a positive climate and awareness of their role and the sharing of objectives and results*".

Green Tweets
#EnvironmentalCommunication   #QualityofCommunication   #Contents #Environment   #SustainableDevelopment   #Technology #EnvironmentalCommunication   #Credibility   #Detectability   #Ethics #Correctness #Responsibility #Clearness #Userfriendliness #Accessibility #Originality   #Quality   #Information   #Communication #QualifiedCommunicators #Images #Media #Emotions #GoodAuthority #PiattaformadelleConoscenze #Newsletter #BestPractice #Dialogue #Citizens #Web

3.5   Can We Express Environmental Values Through the *Net*?

Communicating the *Environment* through the *Web* is a social duty that needs to focus on a qualified information.

In view of the above, to convey *Environmental messages* through the *new Media* is possible provided that the **quality of information** is constantly checked to prevent the risk of misunderstanding, inaccuracy or *green washing* (see Chap. 5). First of all, we need to clarify the meaning of "value". In this context, it is not intended as an abstract ideal, wished by man (*Max Weber*) but as *everything bringing an important and real meaning for the members of the same social group, smaller or larger as it is* (*William Thomas*; *Florian Znaniecki*). In the case of the *Environmental System* the "public" might coincide with the whole world population.

The social value of the *Environmental Communication* may achieve different goals according to its promoter. The subjects offering services of public utility, Public Administration, aim at satisfying the citizens' needs by a continuous *feedback*. On the contrary, the Private Sector: enterprises, non-profit associations, ONGs, etcetera, focus on the social role of transmitting the *Environmental commitment* to their addressees: consumers, clients, associates, and so on. In both sectors (public and private) the need of transparency and participation have made the stake-holders consider the *Environmental Communication* as a key-issue.

In many cases, *communicating* is meant as a social duty, wanted by the citizens. Here hence the necessity to achieve "*a communication able to become strategy*" (*A. Rovinetti*) respectful of the involved parties: public organizations, enterprises and citizens, with the mutual purpose to offer efficient qualified services. In order to **implement the capacity of cooperating with citizens,** the institutions should:

– devote specific sections to authoritative bibliographies and Websites;
– guarantee a daily updating of the information conveyed quickly by "*errata cor-rige*" (corrigenda) in case of misprint or contents inaccuracy.

To transmit values through the *Environmental Communication* is essential then **to train *communicators*** and **advise the addressees to understand the message** easily, going upstream and trying to avoid any political issues about the "theme" *Environment*. Some editors, nevertheless, are often seeking *sensationalism* more than the truth of facts when dealing with *Environment* news. A good idea is then to select the working *team* on the basis of specific competences through professional and training courses focusing on *Environment* and Sustainable Development issues.

In Italy, on the other hand, the ***Environmental Education*** has entered officially among the compelled subjects in all grades of schools in the school-year 2015–2016. The matter should be treated with an innovative approach involving multidis-ciplinary and discussion on *Environmental themes* and issues to be solved (*problem solving approach*) aiming at stimulating the students to a pro-active discussion allowing them to arise creativity, curiosity, critical sense, open-mindedness, skills to investigate and experiment tangibly the theoretical knowledge. An example similar to Finland school reformation.

A global consolidate practice is now to rely on *primary sources*, which means to contact directly the researchers and professionals of this sector—through *Media* such as: interviews, questionnaires, emails, just to list a few. As underlined before, the com-plexity and technicality of the *Environment* System requires objectivity and credibility of the techno-scientific contents. *Who can, better than a researcher, clarify and deepen the information from his study and interact with colleagues and partners?*

Anyway, a good communicator, also without primary sources, should always rely upon qualified sources of information. The "scientific validity" is found then in the reliability of the author, the institution or the organization. These indirect quali-fied resources are the only ones that can offer further guarantees of objectivity: thank you to prior checks of suitability by experts (*peer review*).

Which role can the Social Network and the New Media play to help the trans-mission of Environmental values in this digital age, dominated by the easiest and inexpensive medium? This question reminds us the concept of ***efficaciousness of communication***. The *Social Media* have got the real capacity to influence, to sur-prise, to move the Internet users, to put them into their shoes, thus creating *empathy*. In the *social platforms*, the subjects can express their perceptions, their interests, their points of view, and *be listened to*. Therefore, if we give the public the possibil-ity to interact with the creators of the *Environmental messages*, this is the first step to become credible by the addressees. In doing that we must be open to accept different points of view, possible emotive reactions, doubts, worries that might arise from the exchange of messages. In a few words: we must accept the debate avoiding the logic synthetized in this axiom *"if it is clear to me therefore it must be clear to the others."*

It is necessary to go further, for instance, investigating the active participation of the addresser of the message as far as the possible "emotions" are concerned (*empathic approach*). A must is "to personalize" the message, adjusting it to the everyday reality of the addressee, to his system of values and the social context. Let us think of a virtual *forum* about *Environmental themes* such as: recycling, energetic efficiency, climate change, rehabilitation, *Environmental management system*, sustainable tourism etcetera. In order to understand what the user already knows on the matter it is useful to listen to actively how much he can learn. To get this goal the addressee must be given the possibility to express doubts and make questions directed to expound some aspects, interact information and clarify the obscure sides of the *Environmental theme*. A *participated approach* to reinforce the awareness of the value *"Environment"*.

Now we can imagine the idea of **Environment** conveyed by the concept of "**Landscape**". The countryside element is a classic example of perception of reality by our senses attributing to its positive and negative values because of a plurality of factors: colours, perfumes, weather conditions, states of mind, souvenirs, and son on, and the *Environmental* ones: biodiversity respect, abusive urban speculation, garbage dumps or industrial plants etcetera. As stated by the *European Landscape Commission*, the mechanisms of perception of the landscape are strictly linked to the *social and cultural values* attributed to it just to demonstrate how man and landscape influence each other. A "*circular relationship*" in which the human being improves both the quality of the landscape and of the *Environment*.

How to facilitate then, in the communicative exchange, the public awareness on the Environment? Trying to make the contents as much as possible usable thanks to some basic communicative techniques. A way to help the communication with the receiver is rephrasing, for instance, the object of the *Environmental Message* by repeating the same "words" of the "user" or their synonyms, without any personal interpretation. The addressee feels listened to and is concentrated on the proposed theme. The *rewording* or *reformulation* points out expressions such as: "you are saying that…", "you want to say that…", "in other words…", "in your opinion…according to you…". Giving explanations and to know the right moment for them could be as efficient. Once again: "from your words I've got the impression/I deduce you have some perplexities/doubts on the matter…". Making the right questions at the right moment is an art not to be underestimated.

In the context of *Social Media*, the **open questions** are the most suitable ones to deepen, to complete and clarify *Environmental Themes*, leaving room to opinions, thoughts, critics and proposals. On the contrary, the **closed questions** to which only specific answers, sometimes pre-packaged, seem to be more limited. They prove to be useful when we are looking for a targeted communication to support other deployed tools. Let us consider the online creation of a *focus group* whose objective is to investigate the importance given by the customers to specific characteristics of energy efficiency of a new electric appliance, launched on the market. A closed question questionnaire might be used as a support of a series of interviews, virtual or in person, aimed at investigating the perception of the targets of communication linked to the ecological characteristics of that product, as in the following real example, used for a market survey:

How should a product be, in order to be considered really ecologic?

A. *it must not endanger the health of the consumers;*
B. *it must have a reduced impact when using it;*
C. *it must have a reduced impact when producing it;*
D. *it must be composed of recycled material or produced with renewable resources;*
E. *it must be totally or partially biodegradable;*
F. *it must have a reduced impact during the whole lifecycle (e.g.; low-emissions of pollutants in the atmosphere, low energy consumptions, possibility to recover materials of the end-of-life products, etc.).*

Green Tweets

#Quality #Information #EnvironmentalValues #Planning #Programming #PeerReview #ImpactFactor #Medium #Media #EffectiveCommunication #SocialMedia #Empathy #Involvement #Landscape #Environment #SocialValues #CulturalValues #OpenQuestions #YesandNoQuestions #FocusGroup

3.6   The Environmental Messages, Are They Always Understood by Their Addressees in the Variegated and Immaterial "Ecosystem" of the *Net*?

To check the degree of comprehension of an *Environmental message* increases the recipient's satisfaction.

At this point of my dissertation, a spontaneous question arises. *The Environmental message conveyed by an indistinct mass of Internet users, constantly seeking the highest connection speed, doesn't it risk to be mistaken?* The main added value of the *Net*, just the one to reach thousands of "users" in a few seconds, might prove to be a "Achilles' heel". The risk to disseminate false or incorrect news can be associated to the innumerable sources of "noise" that may distort the contents of the *Environmental message* when decoding them by each single receiver. Let us think about the communication conveyed through *Facebook* diary or about the cryptic messages of a *tweet* (on Twitter platform) that makes the *Internet* users to invent multimedia *haiku*, a Japanese word, being used since the seventeenth century, which identifies a short text based on a sensorial language to capture a feeling or an image.

The **Social Network**, in fact, **interacts with us** at any time during the day. Therefore, the so called "noises" of communication proliferate. *Semantic problems*: multi meaning words, false friends, etcetera; *psycho-social problems*: personality, educational level, education, culture, etcetera; *additional to the physiological ones*: hot, cold, illnesses, etcetera; *the technical ones*: bad ventilation, road works, network connectivity problems etcetera, and *the professional ones*: professional tasks requiring our full attention without any lack of concentration. All that may often bring about a superficial or partial reading of the message, if not its refusal, in extreme cases. Here hence the potential risk to compromise seriously the original contents, that is the "what", the very object of the *Environmental Communication*.

The spread on the *Net* of a *bad Environmental Communication* can lead to disapproval and protests that may rapidly spread out on a global scale. In the springtime of 2014, for instance, the *World-Wide Fund for Nature* (WWF), the largest world organization for Nature conservation was strongly criticized on the Net for its alleged support and sponsorship to a famous chain of American amusement parks offering acrobatic performances by using marine mammals, bred in captivity (killer whales and dolphins). That engendered a "chain" reaction spreading out rapidly on the *Net* thus activating innumerable criticisms by the other *Environmental organizations* and by numerous *Internet* users on *Tweet*, messages and petitions on *Facebook* and other *Social Media*. WWF had to intervene quickly supplying with personalized answers each single interlocutor and activating a specific section of FAQ (Frequently Asked Question or Q&A Questions and Answers) in which they tried to prevent further criticisms with their own "line of defence". An example of **Multidirectional Communication Plan** with a capacity of prompt answer, more and more essential elements to get out of the *World Wide Web* "jungle".

We must not forget the *feedback*, that is the "*outcome of information*" which is an integral part of the process of the *two-way* communication, from the addresser to the addressee. On the *Net*, it is not by chance that the interaction between subjects through the exchange of *emails*, discussion forums and virtual "showcases", represents an essential step to check the correct decoding of an *Environmental message*, in order to correct any possible "noise" and amend mistakes of communication. Also, the **examination** of the number of "**like**"—a typical expression of approval, in the most popular *Social Network*, is represented by a hand with its thumb up—and the **connected comments** in the *posts*—a textual message to be edited expressing an opinion, a comment that interacts with other *Internet* users inside a shared space on the web, called Facebook "wall"—are efficient to check the degree of approval or comprehension of the *Environmental message*. But they can even disseminate the *Environmental commitment* through projects, information campaigns and action plans. The great flexibility of this instrument can create a link to other *Internet* sites or to documents or images inserting their address or creating new active words (*hot words*) or *hashtag*, a series of words preceded by the symbol (#). In this way, they allow hyper textual links to join a single word to topic and subjects concerning it, expanding and deepening its meaning.

The "***Billion Tree Campaign***", launched in 2006 by the UN *Environmental Programme* to sensitize the public opinion on the themes concerning the global heating, the sustainability, and the risk of the biodiversity loss, represents a good example of interactive *Environmental Communication*. The main goal of the project, originated by the idea of a 9-year-old German child, *Felix Finkbeiner*, is to plant the major number of trees at a global level, to counteract the dangerous effects of carbon dioxide. A logo and a coloured, imaginative and effective graphic show the visitors the representation of the Planet Earth, inspired by the world of animation and comic strips, culturally linked to a moment of fun, game and entertainment to devote to in the free time. An original counter tells, in real time, the precise number of planted trees thanks to the contribution of the Project and how many trees are still to be planted to achieve the *target*, which changes years by years. A very detailed map then allows the visitors to visualize in which *region* of the world the major number of planted trees in concentrated. Furthermore, animations, audio and video contributions, invite the caller to participate himself to the project. In doing so, each "person" can be transformed into a *follower* or better into an *ambassador* of the project, with a further opportunity to donate for that cause, or to attend fee-free laboratories (*Academy*) organized at local level by a *team* of qualified communicators, selected by the Project.

A communicative commitment that has "given its results": so far more than 15 billion of trees have been planted. Thanks to the **Environmental campaign** "**Stop talking, Start Planting**" based on posters showing children putting their hand over the mouth of Heads of state and worldwide decision-makers. A result that has largely exceeded the initial target aimed at planting only one billion of trees within 2007. This fact makes us state that the *Social Media*, can determine the success of an *Environmental Communication* campaign, provided that we are carefully monitored and constantly guided by experts of communication. The risk of the "noise of communication" must not be forgotten, anyway. Luckily on the Net there is less and less jeopardising of news. An important step is then to reflect about the correctness of the comprehension. On the other hand, as *Albert Einstein* reminds us: "*if you can't explain it simply, you don't understand it well enough*".

Green Tweets
#Communication   #Net   #SocialMedia   #Speed   #Risks   #BaselessNews #IncorrectReports   #IncompletePiecesofNews   #Haiku   #Tweet #WorldWideFundforNature #WWF #CommunicationPlan #Multidirectional #WorldWideWeb #WWW #Noise #Like #Comments #Hotwords #Hashtag @BillionTreeCampaign @FelixFinkbeiner @AlbertEinstein

3.7   Communicating on the *Net* Is a Question of Image

The universal communication of images arouses emotions.

Communicating the Environment may be also "*a question of image*". The *Environmental Communication*, in fact, does not express itself exclusively with words (*verbally*). New *Media* help us to communicate efficaciously through visual or multimedia elements. A well-known saying recites: "*a picture is worth a thousand words*", without diminishing the essential role of the words and the alphanumeric key which are the components of the language. An image or a video, however, may have a strong impact onto the public and be the bearers of important *ecological values*.

Latest statistical studies, conducted in the United States, show how each person is interrupted every 8 min, by a message received on the mobile, the Laptop, the *IPad* or any other multimedia device. Considering an amount of 3000 advertising messages per day, we can understand how the combination of the written text and images is widely used to elicit emotion and *empathy*. A binomial which influences us without our awareness. The image effectively goes beyond the linguistic barriers and allows an almost international understanding, though interpreted subjectively according to a personal style and taste, as reminded by *Umberto Eco*, as a semiologist (*The strength of images*).

"*The image has got languages unknown to the reason of the words*", *says Frédéric Lambert*, a semiologist at the *French Press Institute* (*Université de Paris 2— Panthéon Assas*). This statement derives from a long debate opposing psychologists all over the world about the **comprehension of the image** as a **conventional process**. A convention followed by the majority which derives mainly from their lifestyle and way of thinking, that is their social and cultural background.

The **communication** through images is **immediate**, **simple** and **effective** while conveying feelings, concepts, lifestyles and values. It goes with the key role of digitalization when it is requested to do more than one thing quickly and simultaneously to achieve the target before the others, both in the professional and free-time field. Furthermore, **visualizing a concept by images** is simple and easy to be remembered. An element used by the *Net* to catch the Internet users' attention and push them to examine in depth the *Environmental themes*. But this is not always true. The extraordinary number of images on the *Net*, amplified by the *Social Network*, might reduce the "catalyst" function. We often skim thousands of photos edited *on line* without dwelling on the contents or on the context where they are taken. It is just the immediate feeling or impact of our eye on the image to make the difference, then. **Which strategies are supposed to be implemented?**

For example, we could, through *social games*, exploit the *Net* capacity to involve people actively when disseminating the *Environmental message*. Spreading out the image of an eco-system like the *barrier reef* pointing out the negative effects of the pollution and the *climate change* through the visual comparison of "before" and "after", may help to improve awareness in the public of the world.

In April 2016, a piece of news spread out informing that the ***Great Barrier Reef***, in Australia, considered one of the natural wonders in the world, was going to die by a whitening effect which makes the coral more vulnerable to parasites thus leading it to death. The piece of news underlined that the majority of the Great Barrier Reef is in danger of death. Furthermore, more than 90% of the total is suffering from some level of "*whitening phenomenon*". The alarming message can be transmitted visually by this two-sided drawing (Fig. 3.2), herewith below. It gives a sudden idea of the ecological ongoing disaster, strictly linked to the climate change and the abnormal concentrations of carbon dioxide in the atmosphere which is partly absorbed by the oceans (30% of the total), raising the water temperature. A rapidly changing phenomenon that may cause serious problems in the food chain, worldwide. Just think that one quarter of the existing species of fish is visiting the Coral Reefs in their life. The traditional multicolours of the enchanting ecosystem "barrier

reef" (on the right) change into a "desert" of whitish skeletons of coral (on the left), thus interrupting the food chain of all marine species depending on it. The precious submarine ecosystem can be then transformed from a "coloured world" into a "black and white world", if the temperature increases of only 2 °C.

It is therefore not by chance then, if the scientific world and some eminent associations and figures, like H.S.H. *Prince Albert II de Monaco*, with His Foundation are supporting scientific expeditions, raising awareness campaigns and events on the matter through symbolic visual elements. Recently, the Prince Albert II Foundation has promoted the Fourth International Workshop on Oceans Acidification and established, in 2017, an annual event totally devoted to the marine biodiversity preservation, the **Monaco Ocean Week** (www.monacooceanweek.org), whose information campaign is represented by a special ambassador: a seal swimming in the ocean. The **Monaco Explorations Expedition**, a three-year global circumnavigation, started in July 2017, aims at doing research and raising public awareness on the essential role played by the oceans in ensuring the survival of our Planet. In this context, the international scientific **Antarctic Blanc Expedition** (www.antarcticblanc.com), sponsored by the Prince Albert II of Monaco Foundation and the Yacht Club of Monaco, has proved in early 2018 the effectiveness of symbolic communication thanks to PINGI, a penguin-shaped soft toy, appointed as the official mascot. Thanks to this special medium new generations, and any Internet user, were able to share the environmentally friendly experience of each crew member while taking Environmental measurements and making observations of the Antarctic wildlife aimed at verifying the impacts of Climate Change in pristine wilderness. Last but not least, a fine visual storytelling is the basis for **two recent publications** on Climate Change: **Climate Change** and **Le petit livre du changement climatique**, in English and French, edited by Penguin and Dunot Publishing Companies, under the high patronage of H.S.H Prince Albert II of Monaco and H.R.H. Prince of Wales - Authors: Tony Juniper e Emily Shuckburgh (Fig. 3.3).

In 2013, the *European Agency for Environment* carried out an *on line* **photographic contest** open to everybody called *Waste smART competition* in which it was requested to communicate, in a creative and innovative way, about the serious problems of an urban waste disposal through images, multimedia contributions

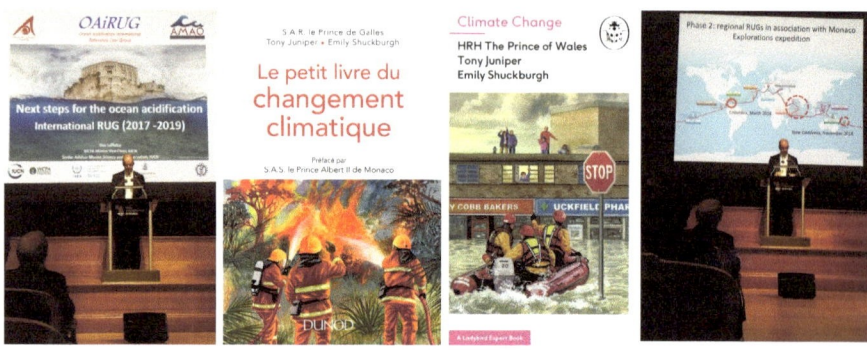

Fig. 3.3 (From Left to Right): the *Forth International Workshop on Oceans Acidification*, Musée Océanographique de Monaco, 17th October 2017, Promoted by the *Prince Albert II of Monaco Foundation* (Twitter @FPA2) www.fpa2.org and the *International Atomic Energy Agency* IAEA www.iaea.org; Speech of Sir *Dan Laffoley*, World Commission on Protected Areas Marine (WCPA) Vice Chair and International Union for Conservation of Nature (IUCN) www.iucn.org Senior Advisor, presenting the *Monaco Explorations* expedition (Twitter @MonacoExplores), started in July 2017—Photo by *Maurizio Abbati* (Photographer) © 2017; two recent publications on climate change, in English and French, edited by Penguin and Dunot Publishing Companies, under the high patronage of H.S.H *Prince Albert II of Monaco* and H.R.H. *Prince of Wales*—Authors: *Tony Juniper* e *Emily Shuckburgh*

Fig. 3.4 "*Seagulls feeding place*" by Stipe Surac (Copyright Holder) © 2013—winner of the *Waste smART competition* www.stipesurac.com—Developer: European Environmental Agency www.eea.europa.eu

and any other form of art. The winner for the photography section was *Stipe Surac*, Croatian nationality, who produced a picture that hit the technical jury for its "impactful" strength. Figure 3.4, as shown below, called "***Seagulls feeding place***" was shot in a garbage dump of *Zadar* in Croatia and shows a bunch of seagulls, the bird symbol of the Mediterranean Sea and protected species on the

Croatian territory, while they are fighting each other greedily for the "junk food" produced by man.

A single image embraces a lot of themes to which much food for thoughts is linked, among them we remember:

- protection of biodiversity and wildlife;
- *Environmental impact* of waste dumps;
- waste management (e.g.: substituting open waste dumps with underground waste dumps or with incinerators, regulated by law, could be a solution);
- toxic effects of plastic in the marine *Environment*;
- increase of aggressiveness of some animal species, attracted by the open damps to the detriment of local species;
- transmission of virus to man or other germs through the contact with flying animals or other animal species visiting regularly the waste dumps, or indirectly through the sting of insect vectors (e.g. mosquitos).

In the springtime 2016, the **International Agreement RAMOGE** (acronym: St. RAphaël, MOnaco and GEnova)[1] launched, through their *Internet Website, a photographic competition* to celebrate its 14th anniversary, as one of the most important driving force of the Barcelona International Convention for the Protection of Marine *Environment* and the Coastal Region of the Mediterranean, originally Convention for Protection of the Mediterranean Sea against Pollution (1976).

Since its very beginning with the high patronage of H.S.H. Prince Rainier III of Monaco (1970), RAMOGE (and RAMOGEPOL, since 1993) has been considered a major scientific, technical and legal tool dealing with the protection and prevention of the marine pollution, in an extremely relevant geographical area, as far as marine biodiversity is concerned. The coastline among South of France (Région PACA), Principality of Monaco and Italian Flowers' Riviera, includes the most important Sanctuary of Cetaceans of the Mediterranean Sea.

Among the shots in final, currently available on the web, inspired to the common theme: "*The Mediterranean in its multiple aspects*", we are reporting herewith (Figs. 3.5 and 3.6) two examples in which it is clear the importance of the image communication, aimed at making the public aware of the biodiversity of the RAMOGE coastal landscapes areas and of the man-made impacts.

Let us try now to *explain in words* the two **finalist photographic works** at the prestigious competition RAMOGE. Regardless to pure aesthetic evaluations: that is if the work is liked or not.

In "**The swimmers**" the photographer Kathrin Hoyos draws the public's attention on the key element characterizing the whole area under the International Agreement: The Mediterranean Sea (Fig. 3.5). She has done this in a very creative and stimulating manner. The total absence of the landscape and the sea in the

---

[1]A precious instrument of administrative, juridical and scientific cooperation among France, Principality of Monaco and Italy for the preservation and protection of the coastal areas between Marseille and La Spezia (Italy)—a sub-regional application of the Barcelona Convention for the Mediterranean Sea Protection of 1976, modified in 1995. RAMOGE has been based in the Principality of Monaco since its creation on the 10th May 1976, when the parties signed the RAMOGE Agreement at the Palais Princier (Prince's Palace). Since 1993, ratifying RAMOGEPOL Plan, the RAMOGE Agreement has been applied to high seas.

Fig. 3.5 "The swimmers", by Kathrin Hoyos (Copyright Holder) © 2016, a Monaco resident photographer, a special mention to the third category: "Coastal landscapes of remarkable beauty", *International Agreement RAMOGE* Photography Contest 2016. www.kathrinhoyos.com, www.ramoge.org

Fig. 3.6 "Colours of the seashore", by Valeria Serra (Copyright Holder) © 2016, winner for the second category: "Waste"—*International Agreement RAMOGE* Photography Contest 2016. www.ramoge.org

foreground, with all the chromatic shades, from turquoise to dark blue, catch the first sight attention, suggesting a question: *is that a painting or a picture inspired by the Impressionists' technique?*

A crystal water, in a shallow and sandy bottom, might influence the observer's fantasy plunging him in a middle-of-nowhere tropical atoll where two swimmers, just about visible by comparison with the vastness of the water element, are venturing. Therefore, thinking of the geographic area of the Agreement we naturally wonder how a tropical landscape can be found at those latitudes. Nevertheless, all that is real. We are soon astonished to learn, from the phrase note that the photo has been shot from her terrace, situated in the heart of Monaco Principality, one of the highest population density in the world [17,973 (1°) inhabitants per square kilometres]. A man-made *Environment*, technically speaking. *But does it point out the good health of the Mediterranean marine eco-system? Or does it proof the so-called phenomenon of "tropicalization", ascribed to the global warming that helped tropical species to grow perfectly in the Mediterranean basin?* Consequently, the autochthone species such as the *Posidonia Oceanica*—a typical water plant of the Mediterranean which many other marine species depend on for their survival—have been jeopardized.

In *"The colours of the seashore"*, the artist Valeria Serra focuses on the foreground that catches the public's attention by the effect of contrast with the very title of the work, suggesting the representation of natural landscapes characterized by crystal water, green hills, blue sky and picturesque chromatic effects of the shore as a consequence of the specific composition of the sand (Fig. 3.6). This is the result of a very long process of erosion involving stones and molluscs, during the centuries. One of those heavens of biodiversity such as: The Pink Seashore at *Budelli* in *Sardinia* or the *Black Seashore* of *Stromboli (Eolie Islands)*, or the Red Seashore at Santorini, in Greece. Nothing of all that. The seashore in question is coloured because of the presence of innumerable man-made objects: bottle caps, toy fragments, plastic containers and thongs. Plastic objects left in the sea and "given back" to the main land by the constant wave energy of the sea and their impossibility to be destroyed by the Nature until after a hundred of years. *Will these objects be colouring the shores and the sea bottom in the future? Which risks can the marine habitat (fish, seagulls etc.) or men encounter by the ingestion of them? What can we do in order to prevent the presence of (macro) plastics in the sea? How can we solve the problems of the sea pollution by (micro) plastics?* A real danger for the marine eco-systems and for the human beings—tiny particles of plastic material generally smaller than a millimetre.

The **debate** on the **efficaciousness of the visual communication**, photos and paintings, started in the United States where it spread out, already in the eighteenth and nineteenth centuries to pass on the well-known shots documenting the melting of the glaciers, immortalized by *Gary Braasch* (www.braaschphotography.com) a renowned environmental photojournalist and writer whose aim was to keep detailed records of natural changes and global warming, worldwide. Thanks to his expertise and exclusive photo reportages (around 40 years), he was able to communicate the audience the importance of preserving the Planet ecosystem and to fight against any threat involving Nature, through aerial and close-up techniques. His major project: *World View of Global Warming* (www.worldviewofglobalwarming.org) is an inspiring form of photojournalism playing a perfect example of visual storytelling willing to increase the eco-awareness and to make a key-contribution to the scientific research.

In recent times, the series of naturalistic documentaries *Planet Earth*, are real homages to biodiversity of Planet Earth offered by **BBC**, *British Broadcasting Corporation* (www.bbc.com/earth/world), the well-known authority for the UK broadcasting.

To sum up, the images can affect our perception of *Environment* and the Sustainable Development basically in two ways: (1) *Influencing the way to consider the Environment* (2) *Building up the keys of our degree of knowledge* about what we consider an *Environmental problem* (*Robert Cox*).

About the first aspect, the images contribute to our perception of the Nature as the *Environment* interacts with the human beings. Of course, the shot is the result of a *choice of the lens angle*, of a specific visualization, determined by the visual proportion on the TV or PC screen (4:3 or 16:9). In some cases, it depends on the *effects of post-production*—the complex operations following a photo shot with the objective to improve the exposition, the contrast, the sharpness, the colour saturation, etcetera. In other words, what we give the public is however a *subjective representation of Nature*, an "*artefact*" by man (*Robert Cox*).

If, for instance, we think of some pictorial masterpieces or photos concerning the uncontaminated landscapes of the *Yellowstone Park or the Rocky Mountains*, deliberately deprived of elements referring to man's activities, we could think of a positive representation of the Nature essence and a celebration of its biodiversity. The search of a natural "*sublime*" that some currents of thoughts (*Gregory Clark, Michael Halloran, Allison Woodford*, 1996) interpreted in a negative sense, underlines how, behind the apparent ecological message, there was a "rhetorical" allusion—an explicit invitation to conquer those territories still governed by the laws of Nature. They forgot that the "Western wilderness", in both approaches, had been already inhabited, thousands of years before, by human native tribes: the American Redskins (*Kevin De Luca, Anne Demo*, 2000). Without taking a position in the debate, what is important to emphasize here is the strength of the image that can guide our perception of the *Environment* and of the effects that man produces on it.

But not only this. **The image can arouse simultaneous sensations**, triggering an exchange of ideas influenced by multiple factors: context, cultural *background*, role in the society: average citizen, politician, doctor, architect, engineer, lawyer, and so on. Besides, a single picture may evoke, in each of us, other innumerable images or documents or sayings or concepts that nowadays the *Net* and the *Social Network*, make available. A "*flux of multiple contradictory discourses*" is then created (*De Luca, Demo*) and it "forges" our interpretation of the image. It is essential to understand "*how that image fits into larger ecosystem of images and texts*" (*Dobrin & Morey*, 2009). We will discover that some subjects have got, thanks to the *images*, such a strength to be considered, themselves, *medium of communication*.

It is the *case of the polar bear*, already crowned, by writers and film-makers, as the symbol of the **North Pole**. This animal is, first of all, associated to the *global overheating*, and therefore linked to the concept of *Climate Change*. Consequence of the visual impact of the images that show the animal fighting against the melting polar pack. We see the bear while is looking for food or compelled to swim for distances much longer than usual: a circumstance that increases the risk of hypothermia, tiredness, and drowning, jeopardising the very survival of the species. A visual element that generates strong emotions such as: worry, anxiety, uncertainty, sadness. This is what is called technically "***visual condensation of symbols***". A single image can provoke multiple sensations arousing the souvenir of fundamental values which are the basis of the human civilization, in those who are observing it (Graber, 1976).

About the second aspect, our **perception of the image** is necessarily **conditioned by a series of elements** that push ourselves to consider a specific *Environmental aspect* as critical and, as such, worrying for the whole ecosystem. A meaningful example for everybody is the theme of the *Climate Change* now in the public domain and associated to "***global warming***" and other phenomena such as the glaciers melting, the drastic reduction of the polar icecap in both Poles (North and South). A decade ago the *Climate Change*, if known, would have been synonym of "challenge" with respect to the extreme natural events. But why this? Basically, for the lack of images to document, at that time, the effects of the *Climate Change* on the Planet Earth, and for the difficulty to disseminate them on a large scale through the *World Wide Web*. The same thing could be said for other *Environmental impacts* shown in photos which rose, for example, the American public awareness on the devastating effects of the massive discharge of hydrocarbon in the waters of the Gulf of Mexico, caused by a dramatic accident on the oil platform *Deepwater Horizon*, in April 2010.

Green Tweets
#Communication #Environment #Pictures #Videos #EnviromentalValues #Multimedia @UmbertoEco @FrédéricLambert #VisualCommunication #Immediacy #Efficacy #GreatBarrierReef #CoralBleaching #PhotoContest #RAMOGE #InternationalAgreement #Efficiency #VisualCommunication #BBC #BritishBroadcastingCorporation #KeytoInterpretation #LevelofKnowledge #PolarBear #NorthPole #ClimateChange #GlobalWarming

3.8 Conclusions and Reflections: The Eco-communication on the *Net* Thinks Increasingly "Bigger"

The "*netiquette*" in the *Environmental Big Data* Age.

From the considerations and examples reported so far, it follows that the written messages and images can, through the *Net* and the *Social Network*, communicate the *Environment* and document and testify the effects of a more or less Sustainable Development. This contributes efficaciously to the comprehension of the *Environmental issues* and to create our own vision of the Natural World.

Without any prejudice to the doctrinal debate that has both criticized and defended the validity of the *mass culture* of which the *Net* is the *medium* and "mouthpiece", already well-argued (*Umberto Eco*), we must acknowledge that the *Environmental Communication* is to coexist and adapt to the new *Media* offered by the *Net*. It will be the "*communication operator task*" to investigate wittingly which resources are at disposal; which level of criticality is to encounter in communicating specific information instead of others and which are the possible solutions to facilitate the process of listening-understanding the *Environmental message* in the huge *arena* of Internet users.

We would like to imagine that the *Net* as a ***theatre group*** in which thousands of actors (stakeholders) are playing simultaneously all over the world and whose script is being written through their interaction. Next to improvisation, it is necessary to rely upon a "*control room*" or better to more stage directors who can offer the "*actors*" suitable tools meant to balance the *traditional acting* to *scenic effects* in order to make the "*audience*" understand the true sense of "the play" on the whole. Means of communication are the main public stage through which the audience becomes aware of the *Environmental problems* and the way they are conveyed, contextualized and solved, as suggested by *Anders Ansen*, vice-director of the Mass Communication Research Centre of the Leicester University.

Out of metaphor, the *Environmental Communication* through the *Net* and the *Social Network* must respect the same rules of ethic, correctness, right information, enclosed in one word: ***netiquette***. A Neologism linking the English word *network* with the French word *etiquette* that is a set of rules of computer etiquette, for a good living together on the *Net*, thus avoiding the so called "***communication pollution***", indistinct messages producing only confusion and not knowledge.

The *Environmental issues* need a particular responsibility in guaranteeing the coherence in the mutual exchange of information from which may depend political and social implications, if not the very peaceful cohabitation among peoples. A spontaneous question arises: in the *Mass Information Society which cultural action must be done, so that the new Media can convey Environmental values?* (*Umberto Eco*).

This question is of great importance and even more meaningful if we consider the increasing number of *Environmental data* in the Net. We can say that the ***Environmental Communication "thinks increasingly bigger"***. The technology evolution makes the creation of ***Big Data-bases*** interconnected each other possible. Against a large demand of information, the technology evolution makes the creation of **bigger *Data Bases*** interconnected each other, to meet the public's requirements.

The ***Big Data*** are then born using unconventional technologies in order to elicit the information quickly in spite of their volume and contents. Mega *containers*, measured in *Zettabytes*, billions of *Terabytes*, uniting elements coming from different sources not necessarily structured in Data Base such as: images, emails, Global Positioning System—GPS and *Social Networks*, gigantic *Data base* serving Public Institutions, research centres, private entrepreneurs and so on. All that has raised, and continues to raise, new debates into the doctrine. There are questions to be answered on how to manage and treat these giants of information in order to guarantee the quality of the *Environmental Communication*.

In this *"Media ecosystem"*, easily accessible thanks to the *Wi-Fi* technology, each **piece of information** is transformed into a correlated "**resource**" to others according to its characteristics. The typical *medium* of connection is always **semantic**: The *Semantic Web Services*—SWS (*Berners-Lee*, 2001). Because of a series of contributory factors as, for example, sensible data or lack of human and economic resources, necessary to manage the *meta-data* and guarantee a qualified public access (*Open Data*), most of the information is covered by intellectual property rights, patents and other means of control. This clashes with the increasing trend of transparency in the communication, the public one included.

Big Data about the *Environmental issues* represent then a big challenge. The multidisciplinary of the matter does not help in extracting data (*Data Mining*) to transform them from a "raw material" into useful information. Generally, we cannot forget that it is very difficult to assemble large amount of data unless they are associated to the source description or their contents, subjected to a continuous updating to ensure a suitable interpretation and reproducibility, above all in the scientific field (Bechhofer, 2010). From this point of view, the *Net* can offer a lot of potential for success, also thanks to the *Cloud* technologies. For this purpose, there are already relevant examples of *Environmental data processing* which constitute a consolidated reality as, for example, **The Environmental Virtual Observatory pilot** (EVOp) project (www.evo-uk.org) created by the *Natural Environment Research Council*, in Great Britain and the **EarthCube** project (www.earthcube.org), created by the *National Science Foundation* in the United States. Nevertheless, to make them be operative they need a system apparatus like *SOAP—Simple Object Access Protocol*, and *REST—REpresentational State Transfer*, able to acquire, analyse and communicate the various models of data in a very efficient way. In doing that a particular attention is focused on the *information traceability*, on their *protection from external attacks* (*hackers*) and on the *operating flexibility*. Very important elements for the *Internet* users that are facilitated in visualizing the available data and interact with them, adapting the system to the professional or amateur requirements. Targets that are all achieved through the Web management.

The complexity of the *Environmental issues* and of Sustainable Development can be truly rationalized, if not simplified, through the creation of *Big Data* functional to single needs provided that *complete meta-data* of definitions and any other useful information are supplied, to make the understanding and the key concepts easier. An essential element in the *Environmental Communication chain*. A commitment that involves more and more governments, private companies and non-governmental organizations, convinced users of the *Big Data*. This attitude is a source of further questions concerning, above all, the legal protection of the huge amount of information that cannot help being subjected to the national and international legal System restrictions *to protect the rights of an individual and the minors*.

It is the essential to "translate" the *Big Data* into efficient *Media* of communication, considering their compatibility with the tool *Internet of Things* (IoT). A factor not taken for granted but always requested as a consequence of the increasingly number of international partnerships dealing with the *Environment* and Sustainable Development at a political, economic and social level. The data must not be considered as "final results", detached from facts, but the "main resource" or the starting point to improve the knowledge of the *Environmental issues*, to deepen the contents and to find out the criticalities in order to look for the most efficacious instruments to overcome them (*Cukier*).

Green Tweets

#Environment #SustainableDevelopment WrittenMessages #Pictures #SocialNetwork #PopCulture #Net #Medium #Media #Netiquette #Database #BigData #Zettabyte #Terabyte #EcoMediaSystem #Information #Resource #SemanticApproach #SWS #SemanticWebServices #DataMining #CloudComputing   @EVOp   #EnvironmentalVirtualObservatoryPilot #PersonalRights #RightsofMinors #IoT #InternetofThings

Bibliography

Anan Dilin, Agatha P., *Internet of Things and Big Data: Predict and Change the Future*: Article (English edition), EFY Enterprises Pvt Ltd, 25th January 2015.

Balzaretti Erik, Gargiulo Benedetta, *La comunicazione ambientale: sistemi, scenari e prospettive* (The environmental communication: systems, scenarios and perspectives), Milano, Franco Angeli Edizioni, 2011, pages: 256.

Braungart Michael, McDonough William, *Cradle to Cradle*, London, Vintage Books, 2009, pages: 192.

Bechhofer S. and others, Why linked data is not enough for scientists, 2010 IEEE Sixth International Conference on e-Science, IEEE (Dec. 2010) pages 300-307.

Clark Gregory, Halloran S. Michael and Woodford Allison, *Thomas Cole's Vision of 'Nature' and the Conquest Theme in American Culture,* in *Green Culture: Environmental Rhetoric in Contemporary America*, eds. Carl G. Herndl and Stuart C. Brown, Madison: The University of Wisconsin Press, 1996, pages 261.

Cox Robert Cox, Pezzullo Phaedra C., *Environmental Communication and the Public Sphere*, Sage Publications Ltd, Los Angeles/London/New Delhi/Singapore/Washington DC/Boston, 4th edition, 2015, pages: 422.

Dobrin Sidney I., Morey Sean, *Ecosee: Image, Rhetoric, Nature*, State University of New York Press, 2009, pages: 340.

DeLuca Kevin Michael, Demo Anne Teresa, *Imaging nature: Watkins, Yosemite, and the birth of environmentalism*, essay, independent *online* publication, 2000.

Eco Umberto, *Apocalittici ed Intergrati—comunicazioni di massa e teorie della cultura di massa* (Apocalyptic and Integrated—mass communication and theories on mass communication), Milano, Bompiani/RCS Libri, 2016, pages: 385.

Umberto Eco, Augé Marc, Didi-Huberman Georges, *La forza delle immagini* (The power of images), Milan, Franco Angeli, 2015, pages: 89.

Gaines Ann, *Tim Berners-Lee and the Development of the World Side Web*, Hallandale (Florida, USA), Mitchell Lane Publishers, 2001.

Graber Linda, *Wilderness as Sacred Space*, Washington, American Association of Geographers, 1976, pages 124.

Pira Francesco, *Il web come strumento strategico per una nuova comunicazione ambientale* (The Web as a strategic medium for a new environmental communication), scientific article, Milan, Franco Angeli, 2011.

Qualman Erick, *Socialnomics: How Social Media Transforms the Way We Live and Do Business*, Toronto (Canada), John Wiley & Sons, Second Edition, 2012, pages: 336.

Rovinetti Alessandro, *Fare Comunicazione Pubblica—Normative, Tecniche, Tecnologie* (Make Public Communication—legislation, techniques, technologies), Rome, Comunicazione Italiana, 2006, pages: 172.

Turckle Sherry, *Alone Together: Why we expect more from technology and less from each other*, New York, Basic Books, 2017, pages: 384.

Turckle Sherry, *La conversazione necessaria. La forza del dialogo nell'era digitale* (The needed conversation. The power of the dialogue in the IT era), Torino, Einaudi Editore, 2016a, pages: 447.

Turckle Sherry, *Reclaiming Conversation: The Power of Talk in a Digital Age*, Penguin USA, 2016b, pages: 436.

Watzlawick, P., *Il linguaggio del cambiamento. Elementi di comunicazione terapeutica* (The speech change. Elements of therapeutic communication), Milan, Feltrinelli, 1977, 2013, pages: 168.

Watzlawick, P., Weakland, J.H., Fisch, R., *Change. La formazione e la soluzione dei problemi* (Change. The creation and solution of a problem), Rome, Astrolabio Ubaldini, 1974, pages: 176.

Watzlawick, P., Beavin, J.H., Jackson D.D., *Pragmatica della comunicazione umana* (The Pragmatic approach of human communication), Rome, Astrolabio Ubaldini, 1971, pages: 288.

Sammartino McPherson Stephanie, *Tim Berners-Lee: Inventor of the World Wide Web*, Minneapolis, Twenty First Century Books, Lerner Publishing Group, 2009, pages: 112.

Isaac Thomas William, Znaniecki Florian, *The Polish peasant in Europe and America; monograph of an immigrant group*, Andesite Press, 2015, pages: 360.

Web Site List

Australian Institute of Marine Science www.aims.gov.au

Billion Tree Campaign www.plant-for-the-planet.org

XL Catlin Seaview Survey www.catlinseaviewsurvey.com

Environmental Virtual Observatory pilot (EVOp) created by the Natural Environment Research Council, Great Britain www.evo-uk.org

Earth Cube, conceived by the *National Science Foundation*, United States www.earthcube.org

Fondation Prince Albert II de Monaco www.fpa2.org

International Atomic Energy Agency www.iaea.org

International Union for Conservation of Nature—IUCN www.iucn.org/

Kenneth Kukier, *Data, data everywhere*, The Economist, 25th February 2010 www.economist.com

Piattaforma delle Conoscenze (Knowledge Platform), by the *Ministero dell'Ambiente e della Tutela del Territorio e del Mare* (Italian Minister of Environment and Protection of Territory and the Sea), *Agenzia per la Coesione Territoriale* (Italian Agency for the Territorial Cohesion) and European Union—European Fund for Rural Development www.pdc.minambiente.it

RAMOGE Agreement www.ramoge.org

World Commission on Protected Areas—WPCA www.iucn.org/theme/protected-areas/wcpa

Communicating the Environment Artfully. Ciak, Action!

4

Abstract

The ability of images to communicate eco-messages linked to the *Planet Earth* comes from the prehistoric times of mankind. The modern high-tech society has accustomed us to conventional and non-conventional means of communication which reinforce any *Environmental* message. Furthermore, the artistic abilities of *Man*, together with the new creative techniques and *ICT technologies*, allow us to play association games among words, images, colours, shapes, lines, spaces and signs. This multidimensional approach addressed to the audience is able to give rise to strong emotions that can really produce a social change towards a more sustainable and participated lifestyle, provided to follow the right eco-interpretation. The visual-artistic language is considered one of the most efficient and quick tool to communicate and raise the awareness on *Environmental* issues, meeting the constant human need to interpret Nature. Communicating the *Environment* means designing architectural structures able to be integrated and interact with the natural dimension, understanding problems and proposing possible solutions that can ensure a peaceful coexistence between *Man* and *Nature*. On the front of design, the research of new sustainable materials and the rationalisation of the packaging enable to communicate the *Environmental* message even through objects.

© Springer Nature Switzerland AG 2019
M. Abbati, *Communicating the Environment to Save the Planet*,
https://doi.org/10.1007/978-3-319-76017-9_4

a

"Every artist dips his brush in his own soul, and paints his own Nature into his pictures."
Henry Ward Beecher
American politician supporter of the Darwin Theory of Evolution
(Litchfield 1813 – New York, 1887)

(**a–c**) *SCART il lato bello e utile del rifiuto* (the bright and useful side of waste), Waste Recycling Project aimed at transforming waste industrial materials into pieces of art: (**a**) *Marylin Monroe* by Antonella Prasse [made of beads, stones and buttons]; *Lucio Dalla* by Federico Niccolai [made of rolled up pieces of tissue]—Location: HERAmbiente stand Ecomondo 2016; (**b**) *The Cobra* by Yllli Kalivaci (**c**) *The Peacock* by Vittoria Lapolla—Location: Ecomondo 2015, Fiera di Rimini (Italy)—photo by Maurizio Abbati (Photographer and Copyright Holder) © 2015–2016. www.ecomondo.com, www.scartline.it

Reading Proposal: In Between Images and Words

Have you ever thought about how many uses could your dustbin waste have? Most of the "waste" in the landfills ready to be disposed of, in fact, contain "noble" materials obtained as a final product of complex and expensive production chains. Recently, the increasing *Environmental* awareness of the public opinion has offered new forms of **down-cycling**, that is the transformation of the raw materials into new materials of lower quality; but also of **upcycling** through the extraction of useful components and their reuse without further changes. The strong symbolic connotation of both the eco-gestures to artistic creativity has, therefore, encouraged the spread of new forms of "regenerated" art and design thus giving a new life to materials considered "dead" according to the overcome economic model *"from cradle-to-grave"*. The effectiveness, from the communicative point of view, is ensured and it occurs at different levels, as we are going to deepen hereafter in *"Communicating the Environment Artfully. Ciak, Action!"*. On the educational and training front, who is realizing design objects and works is led to reflect and to be aware of *Environmental* key issues as the one to prevent pollution and the fight against an excessive production of waste. It is not by chance that more and more frequently specialization courses even at university level, are held in order to make students and professionals acquainted with the art techniques using recycled materials from household and industrial waste. On the aesthetic and symbolic level, the works of art represent "materialized" *Media* equipped with a strong communicative meaning. On one hand, the visual impact, often inspired by the shapes of Nature, and the harmony of shapes and colours of the works create a special empathy with the observer who is pushed to know their eco-compatible story. On the other hand, the usefulness of the *design* objects derived from the reuse and recycling of different elements testify the great potentiality of the raw materials whose "life" is thus multiplied almost indefinitely ensuring new and unthinkable uses. All that is meant to influence the public to change their lifestyle towards more sustainable forms in respect of the three "R": *Reduce waste, Recycle materials, Reuse objects*; as if it were the end of a film entitled *"The revival of the three R"*.

4.1   Eco-strength of the Images

> Communicating through images enhances the environmental message and facilitates the dialogue, the listening and the participation thanks to the strength of empathy.

The images have the great skill to be understood at an international level even with some slight shades of interpretation according to different variables not only cultural but also social and generational (*Umberto Eco*). For example, the phenomenon of the reduction of the ozone layer better known as "ozone hole". Since the early Eighties of the last century it has been causing, in close proximity to the Arctic and Antarctic, a continuous and inexorable thinning of the stratospheric ozone, the natural protective screen of the sun's ultraviolet rays type B, dangerous for the ecosystem Earth. If we had shown some geoengineering graphics of this phenomenon highlighting multicolour spots in the year Seventies, nobody would have understood the real meaning of those stains, more or less extended, which correspond to the ozone layer depletion.

The *communicative skills of images* have always been well known since the prehistoric men who tried to communicate by drawing and carving figures on the rocks (e.g.: *Lascaux and Chauvet* Caves, France; *Cueva de las Manos and Altamira*, Spain; *Balzi Rossi* Caves, Italy; *Kakadu Caves*, Australia). Interpreting and finding the right key joke of the visual narration is not something easy; it is much more complex and it has been the subject of researches since the Years Seventies of the last century, when the **Semiotics** (born 10 years before) spread its investigation on the *visual texts*.

Regardless the traditional subdivision, in the doctrine, between *figurative semiotics*, linked to the representation of the world of images and the *plastic semiotics* dwelling on the matching of colours, lines and spaces, etcetera, the **Environmental and Sustainable issues** communicate through a series of sequences able to convey messages and build a genuine iconographic narrative. The images, in fact, may contain all the useful elements to understand what is being watched, that is time, movement, space, symbols and conventions.

Not surprisingly, that the images and symbols linked to *Environment* reproduce the shapes and colours of Nature. Just to make an example: it is not a coincidence the choice of "*green*", in its all nuances, as a *visual metaphor of the Environment*. It is the direct association of the mind linking that colour to sensitive feelings: quietness, energy and relax. These characteristics make the pigment the most relaxing one, as the *studies of colour psychology* state.

The logical-associative capacity of the human brains united to the iconographical conventions rooted in the culture and traditions of the Western Cultures lead to connect the "green" colour to further concepts: hope, rebirth, springtime, youth and faith. But **green** is above all the natural world. Its nuances for the human eye characterize almost all the vegetable species, some animal species and elements of nature (e.g.: the water source is often green in colour, because of the effect of algae and microorganisms or plays of light). This is why green is the colour and *symbol of Environment and Sustainability* (Fig. 4.1).

To tell the truth, this social convention does not correspond to the reality. As some scholars remind us provokingly, in fact, the green pigment is not "eco" at all.

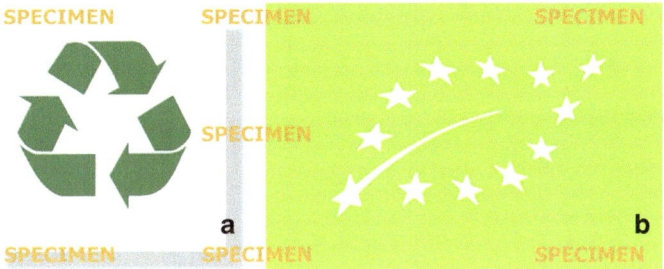

Fig. 4.1 (**a**) International symbol of recycling drawn by Gary Anderson © 1971 [Specimen]; (**b**) European Union logo and labelling for organic products [Specimen]. www.ec.europa.eu

Most of the paper or plastic objects covered in green are often toxic and difficult to dispose (*Michael Braungart—William McDonough*).

The images and symbols therefore are, for their own nature, the result of conventions. This does not prevent them to be an efficient *medium* of *Environmental Communication* as *Ernst Gombrich*, an Austrian art historian, naturalised British, and other scholars underline. This is shown by the artistic representation of the natural elements. A tree, for instance, can be represented by a single line outlining the contours. A *silhouette* is thus created, in green, though being far from reality, becomes its representation. It recalls the concept of "tree" by connoting it to the forest eco-sustainable management or the reuse of recycled materials, to guarantee the preservation of trees and eco-system. Finally, the tree can be considered as the metaphor of the eco-system Earth or the commitment by world government in favour of the *Environment*.

Not to mention actual *figures of speech* communicated through visual texts as much as the written ones. The representation then of one or more leaves, flowers or fruits (elements linked to the tree by a relationship of dependency) can be associated to the concept of "tree" itself or "forest" or "protected area" or "sustainable management" or "ecosystem" or "*Environment*" but also of "freshness" "genuineness", "naturalness" if referred to biological products (Fig. 4.2).

We are facing real *metonymies* (association by mutual dependency) and *visual metaphors* (association by analogy) that can reinforce the *Environmental* communication, drawing the audience's attention on easily *recognizable elements* by the public, based on their own cultural and conventional knowledge. They are introduced in a context with forms, colours and lines completely new and unpredictable but enriching the *Environmental* message with more narrative route.

The element of surprise that catches the attention and changes the point of view of things can be achieved by a *series of images telling a story*. Next to the classical *forms of visual narration* whose subjects and objects are communicated according to a logical structure, the most effective forms of communication are those which offer new narrative schemes. All that is quite clear in the ***cartoonists' graphic*** form with or without very short texts. For some years *Environmental* and Sustainable issues have been represented through sketches and comic strips (Fig. 4.3).

In Fig. 4.3, we can speak [...] literally about an eco-efficient sense of humour thanks to the cartoonist ability to stress the major environmental topic of Climate

Fig. 4.2 Some examples of environmental logos: FSC—Forest Stewardship Council. www.fsc.org (Specimen); Eco Logic, Swerve Design Group for the University of Toronto Book Store, Canada (Specimen); United Nations Forum on Forests www.un.org/esa/forests (Specimen); World Environment Day; ISO 14001 certification www.iso.org (Specimen): European Green Capital www.ec.europa.eu/environment/europeangreencapital (Specimen); Covenant of Mayors www.covenantofmayors.eu (Specimen); EU Ecolabel www.ec.europa.eu/environment/ecolabel (Specimen); Confident in Textiles www.oeko-tex.com (Specimen)

Fig. 4.3 Comic strip: © Green Humour—Author: Rohan Chakravarty (Copyright Holder), cartoonist based in New-Delhi (India) whose aim is to bridge the gap between Nature and Man through his refined sense of humour which gets right to the eco-point—Award: winner of the WWF International President's Award 2017. www.greenhumour.com @GreenHumour (Facebook) @thetoonguy (Twitter) green_humour (Instagram)

Change affecting the biodiversity at the Poles. A clever balance between sketches and dialogues leads the audience to be entertained and to ask questions on environmental issues. How can a penguin meet a polar bear? Why is the sea ice melting? For what reason has a polar bear got a tan, looking like a panda? Therefore, in this particular case, nature humanization helps us to put ourselves in the shoes of animal species.

Fig. 4.4 Comic Strips for Environmental Sea Pollution by *Alecus* (Ricardo Clement) a Mexican-Salvadorian artist and cartoonist. www.alecusdibujos.blogspot.it

In Fig. 4.4, the cartoonist has drawn his attention to the serious problem of waste marine pollution through three sketches in sequence which make the image dynamic, strengthening the major environmental problem, without a further word. According to recent studies, the ocean would have more plastic than fish, in approximately 30 years, if no one is going to do anything. The human impact on Nature could be so strong to destroy the benefits of the ecosystem. Would you enjoy to be submerged by all kinds of human waste while lying on the beach? Where does this human marine waste come from? What can we do to reduce or rather remove it?

It is equally clear that from the visual element to the text something is missing. "*There are things that cannot be represented*" even though their literary *genres* born with the specific purpose **to tell the visual image, *ekphrasis*** (*Umberto Eco*). The present hyper-technological society, above all the youngsters, is accustomed to a type of reading using images and words while the old generations still prefer the use of paper and TV report. The visual communication is therefore necessarily interconnected to other forms of communication. We talk about "*plural writing*", that is using increasingly different approaches based on languages and communication, *semiotics* (*Frédéric Lambert*).

> "*The point of views of the audience and the reader are predominant, they are creators of the work, the open work* (an oxymoron because usually a work is something structured, closed) *to their interpretation,* [...] *a flux of continuous evolution even though not all the interpretation attempts are valid*" (*Umberto Eco*).

This is clearly the position in the **world of art**. In painting, for instance, recurring themes have been passed on (since the Middle Ages) characterized by pictorial diagrams, easily recognizable, thus helping the interpretation of a specific meaning. We think of the Christian tradition and the depiction of the Saints' lives often compared to figures linked to the natural world. But not only this. We can remember the *lion* matched with both St. Jerome and the evangelist Mark. Or even the flower of *iris* which traditionally represents Jesus, the Virgin's purity or the coat of arms of the French Kings or the historical *Medici* Family of Florence.

The interpretation of the depictions, during the sixteenth and seventeenth centuries, created a new discipline still studied today: the **iconography**. Thanks to the detailed description by some "fathers" of biology who lived in the eighteenth century, as the Swedish botanic *Karl von Linné* (*Carolus Linnaeus*) and the French mathematician and naturalist *George Louis Leclerc, Comte de Buffon*, the "ecologic" narration began, still in embryo, to load each single iconographic detail with a lot of meanings and symbolisms. Even though the artistic "message" at that time was extremely anthropocentric, recognizing the man his role of dominator of the natural world, the link between "human" and "natural" was already clear. Great interest was drawn by any event or "prodigy" of nature from the blossoming of a flower to the coming of a storm.

Starting from the reproduction of the **hortus conclusus** (fenced garden), of Medieval tradition, Flora and Fauna details are the symbols of paradisiacal harmony if not of the Divine presence, the source of Life, the knowledge but also its contraposition between good and evil. Concepts generally depicted through the theme of the Tree, particularly in the Christian culture. Also in the cultures of the near East the tree is often associated to Mother Earth and fertility. Some botanic species such as the Palm tree has been, since the Old Times, the myth of the Sun and the Light, of the Glory and Immortality. As well as the depiction of Vine Plants and Grapes were associated to the Greek God Dionysius or Bacchus for the Romans. Later on, they became symbols of Christ, his sacrifice and Faith in the Christian language. Just to make a few examples.

Nowadays the **iconographic artistic narration** is not as much widespread as in the past but it will be influencing the artistic contemporary production. The dissemination of the new *Media* of communication led to the creation of new iconographic models linked to the modern concepts of "*Environment*" and "Sustainable Development" (*Umberto Eco*). "**Foody**", the **EXPO mascot Milan 2015** (www.expo2015.org) created by Disney is a striking example being inspired by the artistic production of the Sixteenth Century Italian painter **Arcimboldo**. The nice "human" face, composed of 11 different kinds of vegetables and fruit coming from all over the world, is a textbook of the positive synergy among the Countries of the World called to face the challenges on food, in order to ensure good and healthy food for everybody, source of energy for life. *Environmental* and sustainable key issues, treated in the Universal Exposition held in Milan, 2015.

Finding points of contact between iconography and semiotics is then inevitable in the narration of the visual and artistic creation of *Environment* and Sustainability, even if the debate on this type of relationship which links the two disciplines is still the product discussion in the doctrine (*Umberto Eco, Omar Calabrese*).

In our dissertation, it is important to underline that the images, both depicted in a piece of art or in advertising messages, comic strips or through a video track, are able, on their own, to communicate endless messages according to their gateway, that is from the point of view chosen by the observers. Above all, when the purpose is to convey *Environmental* and *Sustainable* issues, the debate that arises is how to prevent or foresee the wrong interpretations of what is communicated. We therefore use the technique of the "***anchorage***". The image is accompanied by an explanatory written text of variable length, according to the function to be achieved (e.g.: newspaper article, advert *headline*, caption, etc.). The words act as "anchor" to interpret and lead the addressees to one or more correct meanings and interpretations that is coherent to the objectives chosen by the creator's or communicator's image.

This is evident in the *Environmental Communication Campaign* commissioned by the **Surfrider Foundation Brazil** (Fig. 4.5), based in *Rio de Janeiro* (Brazil), committed in the preservation of the coastline biodiversity. The image of a bomb thrown onto the soil might lead the spectators to develop endless interpretations that could affect negatively the strongly "*Environmental*" soul of the message wanted by its creators. The textual contribution "*the sea level is increasing; we are all in danger—do something against the Global Warming Do it now.*" can better define the context of the message pointing out the analysed *Environmental* issue: The Climate Change. Considering strictly the watery shape, the bomb becomes metaphor of the serious danger that *Man* and *Environment* are running. The global raise of the oceans has particularly increased in the last 100 years, due to the unusual melting of the glaciers and the ice of Poles. The **bomb glassy appearance**, maybe sculptured in the ice, can redesign the structure of the landmass of the Earth by submerging the main coast megalopolis of the Planet, as New York itself.

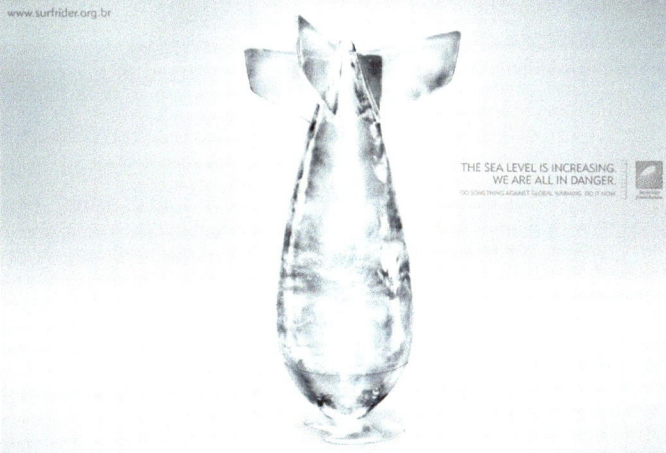

Fig. 4.5 "Bomb" Campaign: Copyright holder: *Surfrider Foundation Brazil—Communication* Agency: *Script* (Copyright Holder)—*Designers: Ricardo Real, Joao Paulo Medeiros, Guilherme Figueira, Marcos Falcao, Luiz Ramos.* www.surfrider.org www.script.net.br

The tendency to overcome the classic idea of the piece of art is shown in the galleries, considered increasingly ephemeral. It is matter of fact that the art works nowadays need a critical text to be fully understood. It would not be a surprise if the new art works of the twenty-first century, were advert spots, films of TV programmes.

The strength of the images is remarkable. Through visual Media we know only the information they convey ignoring the omitted ones. "*The image has a potentiality of alienation and falsehood but also it has great possibilities of evocation, of poetry*" (*Marc Augé*). In order to reduce the risks of "non-comprehension" of images, there is only education, learning and knowledge on how to interpret visual elements.

Most of the Media images are created to show, to surprise and to modify the point of view. "*The apparition of an image, regardless of its 'power' and efficacy, 'invests' us therefore strips us [...] each viewing calls into question and brings all the knowledge into the game*" (*Georges Didi-Huberman*). The function of the image, by its very nature, is to modify our thought, that is to renew our language and our knowledge of the world.

The visual communication about *Environmental* issues can give concrete form to "impacting" images or through scientific representations in the form of diagrams and info-graphics. Some scholars (*O'Neil, Smith, Nicholson-Cole*) remind us that *the visual narration* focussing, for instance, on the problem of the global overheating can generate two opposite reactions rising from the same element of surprise. So here is the *polar bear* looking desperately for food in an ocean not covered by its polar cap anymore, can draw the attention on the *Environmental* issue as well as "communicating" a sense of powerlessness, in face of the *Climate Change*. If not, in extreme cases, it can arouse a sense of nuisance, matching that image to the many others being displayed through the countless *Media* available.

The graphic representation of the *Environmental* data, on the contrary, seems more useful to raise awareness on the *Environmental* issues but, in some cases, they can produce the opposite effect, that is to resize the importance. In fact, the theme of *Climate Change* is so global that it is hardly perceived at a local level unless extreme weather events take place.

Nevertheless, part of our doctrine (*Nicholson-Cole*) is convinced that also the visual "narration, the images of known places, where the addresses live, are more effective. Another part of the doctrine (*O'Neill, Hulme*), underlines that a communication based exclusively on "local" images is less impactful since it conveys mainly messages to the group who lives in that territory.

Furthermore, the decoding of an *Environmental* message through graphic-visual elements is equally influenced by a series of social, cultural and behavioural factors affecting each single addressee. For this reason, new approaches to the "*language of images*" has been developing in the last years thus involving single citizens or groups or associations discussing through the visual element: a photo, a video, a GIF [Graphics Interchange Format], graphics processing, etcetera.

All that brought about the development of ***Photovoice***, as the American researcher *Caroline Wang* called it. A methodology of participated research-action combining photography and the public debate in order to point out the community's needs thanks to the direct witnesses of the members, communicated by their telephoto lens. A "public" exchange of different visual points of view that can trigger the social change and influence lifestyles. Words and images interact in order to communicate and they do it in a very effective way thanks to the digital development and to the *Social Media* which can transform themselves into *lab of psychological-communicative research*, strictly linked to their everyday routine.

Such a kind of methodology, developed in the Nineties, allowed a whole community living in the *Nunavut* territory, the most Northern area in Canada, to check through the camera the effects of the *Climate Change* on their lifestyle. A dialogue-path producing a detailed report called "*Climate Change and Health Community Photovoice Research Project*", conceived and directed by *Qaujigiartiit Arctic Research Centre* (www.qhrc.ca). This is the result of a series of reasoned forum which the community took part to, focuses on increasing the awareness of the singles about how to face the most important *Environmental* issues. They were conveyed through photos made by the citizens and enriched with information, data, observations and possible solutions.

Another useful technique is the ***Photo-elicitation***. A new method of interviewing based on the interaction between interviewer and interviewee through visual elements: photos, videos, paintings, drawings, comic strips, graphs able to convey **sensorial "*output*"**. A target of this method is to evaluate the way men react to the visual communication on the basis of different factors such as: age, social aspects, knowledge acquired or foreground, professional role, etcetera. Scientific studies have proved the efficacy of this technique to face very complex and great *Environmental* issues. The visual representation of concepts, problems and circumstances emotionally difficult to handle, establishes empathy with the addressees which facilitate the comprehension, the participation, the exchange of ideas and the problem solutions.

The ***Environmental Awareness Campaign "Delete?"*** (Fig. 4.6), promoted in June 2016 by *Prince Albert II of Monaco Foundation* and *Opera Gallery* is based on the communication skills carried by visual elements. The major theme of preserving biodiversity as a key element for the survival of the Planet Earth is conveyed by an effective artistic message combined by a short interrogative statement, used as an "eco-slogan". "Delete", in fact, is internationally used to identify the key of a PC keyboard able to permanently eliminate any selected element (text or/and image). The combination between the written text and the visual representation of some of the most relevant threatened species on our Planet. Thanks to the artistic talent by the artist *Thierry Bisch* (www.thierrybisch.com), a faithful reproduction of the selected animals created empathy with the audience, first in the Principality of Monaco and then worldwide: Montreal, Singapore, Milan, Geneva and New York.

Fig. 4.6 Fig. 4a): "Delete?" 2016 Environmental Awareness Campaign on endangered species and biodiversity loss by Prince Albert II of Monaco Foundation www.fpa2.org and Opera Gallery www.operagallery.com; Fig. 4b): Ailuropoda-Melanoleuca, 2016, oil canvas, 200×160 cm © Thierry Bisch; Fig. 4c): Panthera Tigris – 2017, oil canvas, 200×160 cm © Thierry Bisch; Fig. 4d): Rothschild Giraffe, 2016, oil canvas, 200×160cm © Thierry Bisch; Fig. 4e): Gorilla Gorilla, 2016, oil canvas, 200×160 cm © Thierry Bisch

The colours of the animal species fade on a sketch, highlighting the very real risk of losing forever a "key piece" of our ecosystem. What would the world be like without the tiger, the sea turtle, the gorilla, the elephant, the polar bear, the rhinoceros, the panda, the giraffe, the zebra, the lion, the black panther, the leopard, the sea horse, the cheetah, the puma and so on? What shall we do to prevent the extinction of those animals? A highly complex environmental issue conveyed by a skilful and reasoned use of images.

The above-mentioned methodologies are the main development of the "*communication through images*" on the assumption that, as confirmed by the doctrine, we can communicate correctly the *Environmental* message only through an interactive approach.

An effective communicative impact can be ***the assembly of images*** or videos called ***editing***. The combination of images apparently far from one another for their subjects and contents, following a specific logic, or in a casual way, stimulates our point of view and contributes to our *mind openness* (*Aby Warbourg*). As we are

Figs. 4.7 and 4.9 Wild Mediterranean Coast at *Cap Martin* (Côte d'Azur)—*Promenade Le Corbusier*—Maurizio Abbati (Photographer) © 2015; Recreational Harbour under construction in the Western Ligurian Riviera, adopted by a colony of cormorants—Maurizio Abbati (Photographer) © 2017

Figs. 4.8 and 4.10 Tianzi Mountains in China; "Bosco Verticale" (The Vertical Wood), designed by *Stefano Boeri* for *Boeri Studio* (Copyright Holder)—Milan, Italy, photo by *Davide Piras*. www.stefanoboeriarchitetti.net

going to show herewith, the combination of image symbols of the wild (Figs. 4.7 and 4.8) to the man-made intervention (Figs. 4.9 and 4.10) make the recipient feels both the differences and the interrelations between Nature and Man works which often take their inspiration from the Natural World to recreate a "Man-made ecosystem".

To decide which visual elements to communicate and which to delete foreseeing their potential capacity is a choice of communication that affects the point of view, the way of thinking and the awareness of the addressee. The new *Media* can amplify the effects of "*censorship or non-censorship*" also for the *Environmental* and sustainable issues thus drawing the audience's attention only on what is "*visible*".

"*The images convey in many ways. They represent the world by presenting it again, arguing it, analysing it, interpreting it, leading the listeners to messages and conventions of any kind*" (*Gianfranco Marrone*). The author wants to underline here how the visual element is able to modify "the *Environment*" surrounding us. This *communicative journey* of the artistic and visual representation is in continuous evolution contributing to the progress of the "narration" of our *Environmental* messages (*Pinotti, Somaini* 2009).

The society in which we live is defined as the "***civilization of images***" through a linguistic code to communicate for example international trademarks, musical video-clip, journalistic and photo services and professional videos.

Green Tweets
#Pictures #Symbols #CommunicationSkills #VisualForms #Semiotics #Iconography #FigurativeSemiotics #PlasticSemiotics #Time #Movement #Space #Symbols #agreement #Green #VisualMetaphor #Metonymy #Environment #Sustainability #Colours #Psicology #Art #Creativity #SustainbleManagement #Ecosystem #Comics #HortusClausus #Ekphrasis #EXPOMilan2015 @Foody #Mascot #Anchorage #VisualNarration #Photovoice @CarolineWang #VisualCommunication #AssemblyofPictures

4.2   The Environmental Push of the Visual Arts

An effective and creative *Environmental Communication* depends on the correct management of the images and of each visual representation.

The ***image*** in many cases can be the most effective and rapid *medium* to spread out an *Environmental* message *for its immediacy, brevity and capacity to arouse emotions*. This is stated by some scholars among them *McLuhan* who recognizes a kind of communicative DNA to the images, composed of visual signs and codes. All this is amplified by the digital instrument. The second anthropological theory (*Lévi-Strauss*, 1966) focuses on the *semiotic capacity of the images*, that is what they represent. So, the image of a *beach*, marred by the presence of waste produced by human beings, will become the metaphor for the pollution issue, and precisely of the impact of macro and micro plastics on the marine biodiversity. The spectator's attention is then drawn and it produces a potential effect of awareness on the *Environmental* matters. A "*domino effect*" that could be able to mobilise the community to promote clearance activities and marine preventing pollution projects.

In view of the above, it is very important to pay attention to the ***Communication Plan***, in order to the correct management of images and each visual element whose semiotic capacity is to be evaluated *a priori*. We must not forget in fact that the vision of the images can influence in a sensorial way not only the eyes but any other single part of the body. Our capacity to "*Communicate the Environment*" depends on the objectives to be achieved. Coming back to the previous example, we should prevent the showing images or videos portraying subjects throwing waste of the marine birds and tortoise carcasses which suffocated because of eating plastic objects, as if they were modern heroes. Furthermore, the images of coastlines completely covered in waste of any kind, toxic included, make the public reflect on the dangers for Man and Nature.

It is clear enough that a relationship is established between image and addressee, that is, an *"educational relationship"* through a network of perceptions involving all the senses thus leading the addressee of the *Environmental* message to experiment. We are talking about ***aesthetic experience*** identifying it as a specific perceptive condition of man that helps him to interact with his *Environment*. Experience, therefore, as a typical factor of the human behaviour (ethic nature) through which he becomes aware of the *Environmental* issues *(Paolo Granata)*.

This can be seen also when creativity and artistic fantasy communicate the *Environment*. Man has always tried to **interpret the *Environment*** around him by developing his own artistic skill. We can refer to the *Totem*, wooden poles by the American Natives, sculptured and painted and showing natural and supernatural events through the figures of animals and vegetables. But also tapestries, masks, sculptures, graffiti, belonging to traditional cultures spread around the world, where elements of nature, more or less stylised, have become the symbol of biodiversity of our eco-system. Not to mention the *great art masterpieces* having Nature as the protagonist, for example the *"Primavera"* (Springtime) by the Renaissance Italian painter *Sandro Botticelli*; the paintings by the Flemish *Brueghel* Family, developing between the sixteenth and seventeenth century; *"Iris"* or the cycle of *"The waterlilies"* by the impressionist French painter *Claude Monet*; The *"Sunflowers"* by the Dutch painter *Vincent Willem van Gogh*; *"Summer Days"* by the American painter Georgia O'Keeffe, etcetera.

*In the **contemporary art***, the *Environmental* themes are generally conveyed by using materials belonging to the natural world such as: wood, leaves, flowers, shells, musk, etcetera or through a symbology of what the artist wants to represent with his work of art. The IT technology opens new horizons to artistic creativity combined to the digital graphic design that can create info-graphic and visual interpretation about interesting *Environmental* information and data. We would like to point out the numerous ***Apps for smartphones*** that can show immediately the quality of the air telling, in the same time, the polluted concentrations or they supply with information about the *Environmental* good practices. The new boundaries of 3D/tridimensional technology or the panels interact with the users. They may underline the risks of the *Environmental* impact linked to the cycle of life of products and services, to the lifestyles or human activities.

We can also convey an *Environmental* message through the use of *raw materials*, considered as waste, which are transformed in real pieces of art with strong *Environmental* value. This is the target of **the SCART *Art and Communication project*®** by *Maurizio Giani*, its creator who is giving evidence in the section *"the interviews with professionals"*.

Fig. 4.11 (**a**) Artistic outfitting outside the *Oceanographic Museum of Monaco* (Principality of Monaco) for the "Taba Naba", exhibition devoted to the Southern hemisphere art (summer 2016)—Exhibition's Curator: Stéphane Jacob—photo by Maurizio Abbati (Photographer) © 2016; (**b**) *ghostnet* sculptures shown in the *Oceanographic Museum Hall of Monaco* (Principality of Monaco) for the "Taba Naba" exhibition (summer 2016)—Exhibition's Curator: Stéphane Jacob—photo by Maurizio Abbati (Photographer) © 2016. www.oceano.mc, www.artsdaustralie.com

Pieces of Art that can have an unusual visual impact like the "*Jidirah Whale*", an impressive sculpture created by interweaving the fishing net waste patiently recovered in the framework of an Australian Project, ***The Ghost Net Art Project*** (Fig. 4.11a, b), strongly driven by the Aboriginal culture always linked to the natural *Environment* by a *symbiotic link* of mutual respect. This art, called ***ghostnet***, derives from the tradition of the Australian Northern coastlines populations: *Aurukun*, *Darnley Island* and *South Goulburn Island*, already used to a kind of reuse or recycle of the fishing nets sunk onto the seabed or on the beaches. These entrapping devices become real mortal traps of thousands of fish, crustaceous, turtles, and any living form caught in the meshes of the net.

Hence, the highly symbolic value of the art objects is resulting from the interweaving of fibres recovered from the fishing net with pieces of glass, plastic fragments, and shells found on the beach. In the same way, useful objects in the everyday life such as shopping bags, clothes and hats are made. A great global and local (*glocal*) communicative efficacy since the whale appeared in the Aborigines' culture which was not missed by *Stéphane Jacob*, a French expert on Australian contemporary art, graduated at the *Louvre* School, curator of the exhibition ***Taba Naba***, hosted in the ***Musée Océanographique de Monaco*** (Oceanographic Museum of Monaco) in 2016, where the *ghostnet art works* displayed.

The sculptures inspired by marine Flora and Fauna, as described above, hit the audience not only for being the metaphor of biodiversity present in the seas but also to convey, through the materials they are made, a high *Environmental* message. A message that can make people reflect about the importance to preserve the ecosystem and our health by not throwing pollutant substances in the sea or abandoning non-biodegradable waste on the seabed or near the coasts.

The presence in the marine *Environment* of macro (>5 mm) and micro-plastics (<5 mm) is, not for nothing, one of the major *Environmental* issues at world level. A scientific study carried out by *Siena* University (Italy) reminds us that the potential Eco-toxicological impact of the plastic waste concerns us both from the point of view of the chemical pollution poured in the water through their degradation due to the marine salinity (e.g.: phthalates, Bisphenol A, polychlorinated biphenyls PCB, DDT, PBDEs poly-brominated diphenyl ethers, alky-phenol, etc.) or their ability to enter into the food chain of hundreds of animal species, among them: birds, turtles, fish, some mammals and man himself. It is then essential to make the public opinion aware through empowering communicative paths.

The **artistic-visual language** must play a training and educational role, leading to the ecological awareness growth of the human thought, as well as to its evolution and increase the critical sense for analysing and evaluating. Of course, on condition that the *medium* used is able to "*hit the edge*" generating "an experience" never felt before or unexpected (*surprise effect*) in the audience. A goal always pursued in the world of Art in which the artistic creativity increasingly aims to impress, to shock the public thus communicating serious and impactful *Environmental* issues in a very original way.

On the basis of these premises, in 2014, the **Chinese Foundation of Environmental protection** devised an innovative project of *Environmental Communication* that matches the artistic creativity with the strength of images. All this in order to create a surprise effect "*hitting the edge*". The awareness campaign called **"Green Pedestrian Crossing"** (Fig. 4.12) aimed at making citizens of the Chinese most important megalopolis to reflect on the importance to protect the *Environment* through a daily commitment of each individual. An unconventional idea and extremely effective, then was developed by the creative *Jodi Xiong* of the *DDB China Group Communication Agency* started as an experiment in *Shanghai*, and repeated in 15 different urban areas in China (for a total of 132 road crossings).

The traditional *zebra crossings* were covered by mega white sheets on which only the *silhouette* of a leafless tree was depicted. On the pavement hedges, special installations acted as "**markers**". The passers-by, while crossing, were led to pass on "inking strips" (two big stamps) that, **releasing a green eco-paint** on the soles of their shoes, were creating a green canopy, for each passage of pedestrians.

The tree therefore "*dresses up*" with its leaves through the footprints left on the canvas becomes metaphor of the importance to encourage pedestrian areas to use "clean" means of transport, like the bicycle, with no emissions of pollutants into the atmosphere. In the meantime, the pedestrians compare their green prints to their daily commitment to protect the *Environment*, as well as to their own **carbon footprint**, literally carbon footprint of *Climate-Change* gas (e.g.: hydro-fluorocarbon, perfluorocarbon, nitrous oxides, sulphur hexafluoride etc.). It is curious to know that the same calculation is made for an organization or for a product. What is traditionally communicated through a stylised image of a human green footprint.

Fig. 4.12 (a, b) *China Environmental Protection Fund*—DDB China Group Communication Agency (Copyright Holder). www.ddbchina.com, www.cepf.org.cn/en/

A virtuous example of communication that has borne fruit, receiving also the *ADFEST* prize, the most important reward of the advert communication in the Asiatic and Pacific areas. This makes us understand even better the artistic use of the images is an extraordinary *medium* of *Environmental Communication*. Engine of a *cultural ecology* whose target is to bring about the change in the society that can transform itself, at a broader level, into a historical and political evolution. "*Any extension of the human sensorium by technological dilation [but also artistic] has a quite appreciable effect in setting up new ratios or proportions among all the senses*" (*McLuhan*, 1962).

Case Study in Short

When Nature and Art communicate the *Environment* Can human creativity be expressed through a "masterpiece" shaped by Nature? The answer is absolutely positive referring to the *Tree Cathedrals*, a project conceived in 2009 by *Giuliano Mauri*, an Italian artist (born in Lodi, close to Milan) always seeking a direct contact with Nature. He was then considered as the first Italian artist belonging to *Art in Nature*, the International Movement that combines the human creativity to the ecology. *"The Cathedral represents the concept of greatness, giving harmony and sanctity to a specific place; I have always wanted to give substance to that brotherhood linking a specific place to the sacredness of the Earth and to those raising elements called trees"*. This is the core "philosophy" of this unique masterpiece according to his designer. In every art installation, in the North of Italy [*Arte Sella, Borgo Valsugana, Trento*, 2001; *Parco delle Orobie, Bergamo*, 2010 (posthumous work); *Lodi, near Milan*, 2017 (posthumous work)], the basic idea is to conceive living pieces of art, an ever-changing metaphor of the constant relation between both Art and Nature—Man and Nature. Wooden holding patterns in pairs look up into the heavens (12 m high), bending at their ends as to recreate the effect of a *Gothic Church* aisle whose "walls" are made of tree branches that are growing continuously within the area defined by the "pillars". The visual effect is particularly effective and changes over the course of the seasons. The artist, nicknamed "the wood weaver" by *Vittorio Fagone* (Italian art critic) aimed at promoting an exclusive dialogue with the *Environment* without interfering or interpreting it from the human point of view. A *biocentric* approach which has been opposing peacefully to the pure technical and scientific industrial evolution, since the Sixties—Seventies. *Mauri* leaves us, then, a permanent natural inheritance that has forged the landscape, following the natural evolution as fixed by the *Land Art*, an American art movement, developed in the late Sixties, that combines the artistic language to cultural and natural aspects in contrast to the *Pop Art* and *Minimal Art*. Nevertheless, *Mauri* re-creates that creative thinking by strengthening the relation with local, close ideas conceived by the "walking artists", such as *Hamish Fulton* and *Richard Long*. It seems clear from the very selection of locally-sourced native plants such as oak, birch, etcetera; the *Tree Cathedral*, as a final result, invites visitors to join together and meditate both on themselves and on the essential values of the Ecosystem represented by the Tree. Roots, trunks, leaves, branches, flowers, fruits make the tree the universal and ever-present symbol of "Life", as stated in the Holy Bible. A living element that multiplies and evolves in thousands of vegetable species. The tree is always considered an important source of food and a safe shelter for the other living beings. Last but not least, it guarantees the air oxygenation and the CO_2 absorption thanks to the photosynthesis.

Consequently, the *Tree Cathedral*, is even the symbol of well-being, rebirth, biodiversity and respect for the *Environment*, converting itself into a *pure means of communication*, outcome of a social way to express ourselves by sharing a wide range of *Environmental* messages (Fig. 4.13). Link: www.artesella.it/en/ (Web Site).

Fig. 4.13 *Cattedrale Vegetale* by Giuliano Mauri, (2001), Borgo Valsugana (Trentino), Arte Sella (Copyright Holder), Photographer: Aldo Fedele. www.artesella.it/en/

Green Tweets

#Pictures   #Signs   #Semiotics   #VisualCode   #DigitalMedium #CommunicationPlan   #EducationalRelationship   #Perceptions #AestethicExperience #Creativity #Fantasy #Artist #Totem #Tapestry #Mask #Sculpture #Graffiti #Art #Masterpieces @SandroBotticelli @Brueghel @ClaudeMonet @VincentWillemvanGogh @GeorgiaOKeeffe #Infographic #Interpretations #Ghostnet #Tree #EcologicFootPrint #CarbonFootprint #EnvironmentalCommunication   #CattedraleVegetale   #TreeCathedral @GiulianoMauri

4.3   The New Environmental Boundaries of Bio Architecture and Design

Finding the new equilibrium between Nature and Man is a challenge that Architecture and Design have been facing with increasing responsibility and rationality.

Therefore, it is always important to find the most suitable *Media* to communicate the added value in the respect of the *Environment* and *Sustainability*. The *Environment* where we live has been mostly modelled by man and consequently he has changed the Ecosystem according to his needs, both residential, professional or recreational. From an **aesthetic-cultural point of view**, **landscape** encompasses "*asset value, natural configuration and human intervention*" (*F. Magnosi*, 2011). For this reason, it is the subject of legal protection as for example by *The European Ladscape Convention*, adopted by the *Europe Council* in Strasbourg on the 19th July, 2000.

The increasing responsibility of the society against the *Environmental* issues led the architectural sector toward *housing solutions* and *management of the urban spaces* even more eco-compatible. But the "green" nature of a project must be evident since its very beginning. Each architectural element should be in fact at the service of the others and be adapted and coordinated with the ecosystem around in a logic way. "*We must not build the contextualized building, but **build the context***", *Wittfrida Mitterer*, professor at the Faculty of Architecture of the State *University of Innsbruck*, Austria, states, pointing out the role of the *ecological alphabetisation* which the modern architecture should aim at.

A journey that passes through the identification of design priorities meant to improve the *Environment* quality in implementation of an approach "*ecologically correct in respect of the Environmental ecosystem*" as it is defined by the *Italian National Institute of Bio-architecture*. Consequently, we can see an improvement in the lifestyle of the people thanks to the reduction of men's impacts on the ecosystem. In order to achieve this target, it is important to fix clear, simple and rational objectives, promoting forms of cooperation, dialogue and listening among the members of the community who lives in a specific territory.

Hence, the exigency of recreating urban spaces that can make people feel at ease. A square, a pedestrian area, a garden that make people feel as in their own home. But not only because of the most innovative and fashionable design choice, or because a *Jacuzzi* has been installed or because you have a TV 50″ flat screen. The contact of Man and the natural *Environment* increases exponentially the quality of life and its benefit. The **natural sounds** have a **therapeutic effect** on the human system being able to **neutralize stress** and any other tension inside the body. The ethno-musicologist *Marius Schneider* reminds us: "*the greatest therapeutic effect of Nature is given by the pure sound, full of its own vibration energy and in complete harmonic resonance, and not polluted yet by the excessive human and technological presence. It is thus able to clean our ears, and not only this*".

The necessity to design the territory in the full respect of the ecosystem where we live is increasingly urgent. Not just an "eco-friendly" trend or a mere techno-theoretical theory but something real, functional for our everyday life. It is this the mission of some movements in architecture as the *statu nascenti* by *Ugo Sasso* and *Wittfrida Mitterer*. Everything becomes real starting from the choice of "*building, designing, preserving, using good, clean and suitable materials*" to guarantee "*a housing stock whose beauty can be transmittable in its durability; built according to criteria seen by the future generations who are then aware of the recipient's choices made for the time being*" (*Giovanni Pieretti*).

It is well established that the urban architecture can really influence the lifestyles and even on the state of mind of the users. Particularly, a research study directed by the professor of urban planning *Justin Hollander* (*Tufts University, Massachusetts, USA*) pointed out how a well-drawn architectural design improves the quality of life. Along these lines some experiments of **cognitive psychology** are based and they have emphasized the increasing brain activity at the sight of forms of complex and challenging urban architecture such as: a shop, a cafeteria, a restaurant etcetera, according to the studies directed by *Colin Ellard*, a neuro-scientist at *Waterloo University* in *New York*.

A necessary clarification: upstream of each project of **bio-architecture**, there must be a suitable knowledge of the *Environment* and *Sustainable Development*, disseminated on the public opinion by formal and informal methods revolving around the communication chain. A complex operation as much as the one of the sustainability in architecture. *Spiro Kostof*, historian and architect stated: "*the primary task of the architect, then as now, is to communicate what proposed buildings should be and look like*". The preventive phase of "**Edu-information**" and **eco-communication** is then determined in order to understand the added value of the sustainable architecture compared to the traditional one.

What matters is to build a suitable Action Plan, taking into consideration: rational and responsible use of resources; interdisciplinary; empathy with the natural world and study of future scenarios able to foresee and prevent ecological and social disasters considering, for example, the nature of the soil; the trend to hydro-geological instability; the effects of the *Climate Change* and the high level of seismic activity in the region. A path that transforms inevitably the **architect** into **Environmental Communicator** who can contribute to the literacy of both clients and colleagues in an age where the social ethic seems to be neglected. *Mitterer* speaks about "*new Humanism*" of the architecture that from the most individualistic eco-criteria of the *Bauhaus*, interprets space and time making a square, a house and any other space, a place where to socialize. In line with the new systematic vision of the human life developed in the first 10 years of the twentieth century and which sees man in constant relationship with the ecosystem Earth with the view not to dominate it but to preserve it.

The ancient opposition between **mechanist** and **holistic approach**, between material and form, currents of thoughts that alternate from the ancient Greece until nowadays. In the last 20 years, they enrich themselves of an added value.

Thinking not as individuals anymore but as a part of a "net" both **economical** (*new economy*), **social** (*society globalization*), **cultural** (*cultural mainstream*) and **ecological** (*Environment and Sustainable Development*). With reference to the latter, we need above all a solid, basic awareness of the public opinion on the fundamental concepts of living in harmony with the ecosystem. The fact that the sun, for instance, is the main source of energy on the Planet Earth and what is considered by the humans as waste can feed other living species or be reused or create new products.

On the basis of these assumptions, the ***bio-architecture*** and the ***eco-design*** can be really effective. The advent of renewable sources in respect to traditional fossil fuels; the development of hydrogen technology; the coming back to sustainable agriculture and the use of solar energy have given a further boost to the development of ***eco-buildings*** and whole ***green quarters***. Today is not a utopia anymore. We think of the "***Bosco Verticale***"—Vertical Wood (Fig. 4.14), a residential building in the urban area of Milan (near Isola district), designed by *Stefano Boeri* for *Boeri Studio*, in order to regenerate the *Environment* through an unusual shape of vertical urban reforestation. The two towers (110 m and 76 m height) host in the terraced spaces 900 high trunk trees (from 3 to 9 m height) plus more than 20,000 plants and bushes, following a ratio linked to the hours of ***sunshine on the facades*** during the

Fig. 4.14 "Bosco Verticale", designed by *Stefano Boeri* for © Boeri Studio (Copyright Holder)— Photographer: *Davide Piras*. www.stefanoboeriarchitetti.net

day. According to the beauty canons, visual result of architecture, a biological one follows. The wooden area provides a microclimate able to maintain a correct degree of humidity, to absorb CO_2 and to filter fine pollutant particles, besides representing a new habitat for birds and insects. A project that answers "*the human need of contacting nature*". This is the motivation of the *International Highrise Award* given to *Bosco Verticale* by the *Architecture Museum of Frankfurt*, in 2014.

Examples that show that if there is a strong will and responsible competences, "communicating" the ecosystem is possible. The conditions are: (1) be very fond of the life of Nature; (2) go beyond any pre-packaged scheme; (3) "dare" in order to protect the eco-system on the basis of the local bioclimatic conditions, different in any part of the world. "*No ecological communication will be successful if we are not awaking with the crickets singing and think 'we are well because the moon is round'*" (*Giannozzo Pucci*). An urgent necessity if we consider that the time to act is already time out, according to the well-known *Stern Report*. The target we fixed is to reduce the 80% of CO_2 emissions within 2050, compared to the registered concentrations in 2000.

We can reflect on the total glass cover of the building facades, metaphor of lightness and transparency, if it is really efficient from the energetic point of view. Or if, on the contrary, it contributes to the CO_2 production or greenhouse effect, as some architects have stated. Part of the Doctrine shares the second reflection but the *Crystal House* in Amsterdam (Holland), designed by the *Studio MVRDV*, is a perfect example of innovative structure in crystal that enabled to minimize the consumption of material, thanks also the capacity to remake the imperfect *crystal bricks*, thus guaranteeing the energetic efficiency inside the building, powered by a geothermic central below the street level. The question of energetic efficiency is in fact one of the most important subjects of study and a debate in architecture. Both for a new building or for the requalification of the existing real estate, very important in countries like Italy where a little more than the 58% of buildings were built before 1977, year of the first Legislation on energetic building efficiency.

The theme of *energy efficiency* of the buildings has been the subject to specific national and international legislations as well as **Environmental Certificates** like **CasaClima** (KlimaHaus in German, ClimateHouse in English), a system created in 2002 by *Norbert Lantschner*, former manager of the Office "Air Noise" of the Italian *Agenzia Regionale per la Protezione dell'Ambiente—ARPA* (Regional Agency for Protection of the Environment), autonomous province of Bolzano (*Trentino-Alto-Adige*, Italy), in application of the European Directive 2002/91/EC. In 2005, the **ClimateHouse Agency**, a public body able to issue an appropriate **Energetic Certificate** (and an ID plate), subdivided into classes: from *CasaClima Gold* to *CasaClima C*, applicable to any kind of building, on the basis of evaluation criteria easy to understand. Among them: including the efficiency of the shell, the total efficiency, the *Environmental sustainability* of the whole liveable structure as far as the people's benefit. A regulatory model involving designers, craftsmen, and customers that has revolutionized the *eco-real estate business* in Italy, Austria and Germany. Other *Environmental* Certification applied to architecture, at local and international level, are: *Building Research Establishment Environmental Assessment Method BREEAM; International Initiative for a Sustainable Built Environment— iiSBE, Protocol ITACA; The Leadership in Energy and Environmental Design— LEED, U.S. Green Building Council—USGBC.*

Any kind of eco-intervention must be accompanied by a fundamental change of the housing concept rooted in the Western countries in the last 10 years. Starting from the so called *"economic boom"* (Years Fifties and Sixties) until the end of the Seventies, when we became aware of the *Environmental issues.* The majority of buildings had been conceived without considering the energetic impact. Looking for fossil fuel far away, for example, was a common practice, even with higher transport costs, an increase in CO_2 in the air, and often waste of energy because of lack of insulation of fixtures, windows and doors, in increasingly larger flats.

Effectively, the costs to make the existing housing eco-efficient are still very high though the technology market prices are gradually lowering and many Countries foresee incentives and fiscal detractions. This is a fact which we have to compare with in the near future. The new energetic policy of the European Commission has fixed, not by chance, within the 2020, a goal as much ambitious as essential for all the housing of new buildings that is a standard of emission equal to zero [European Energy Efficiency Directive, 2012/27/EU—2015/2232(INI)]. A general increase of energetic redevelopment is to be planned, for example:

- the thermal insulation able to heat or to cool the rooms with less energy waste;
- the production of energetic need through alternative systems as the photovoltaic panels (for the production of electric energy) or solar (to heat the water and the heating plant);
- the heating and cooling plants under the floor;
- the installation of passive plants meant to exploit the thermic changes already existent in nature among sky, water, soil and outside air;
- the exploitation of the energy unleashed by the sea waves to produce, in very good conditions, the so called *talassothermia* (or the Ocean Thermal Energy Conversion system—OTEC);
- stabilization of the internal temperature through the creation of green extensive roofs (suitable for small plants) or intensive roof (suitable for high trunk trees and bushes).

The idea of home comfort has been changing in the last 10 years, as a consequence of much more awareness of the *Environmental Issues.* Nowadays, also the spaces are reduced, rationalized in terms of decor. A practical need (less rooms to clean, less rooms to light and to heat) which fits in well with the above-mentioned interventions of bio-architecture.

A path increasingly "green" involving the parallel sector of the industrial and art *design.* The majority of household electric appliances on the market such as: fridges, ovens, microwave ovens, washing machines, dishing machines, TV sets etcetera, have achieved high energy consumption performances. The *energetic classification* regulated by the European Union (EU) for refrigerators and freezers, for instance, has achieved the very high standard "A+++" equivalent to an annual consumption less than 188 kWh. But there is more and at the end of this paragraph there will be further information about design.

How to communicate efficaciously the eco-revolution which is affecting architecture and design? How much the new Information and Communications Technologies (ICT) are influencing this "green design revolution"? The communicative capacity of architecture is not something new. It was a subject of study in the Middle Ages (only in embryo) and it blew up during the Italian Renaissance and afterwards. At that time, the greatest architects were competing to present the most communicative design to show the cultural, military or religious power of their customers as the big noble families and merchants or the Pope and the Clergy. In modern time the architecture is always at the centre of a complex communication chain, inside and outside, and it includes designers, clients, design managers, builders, suppliers, public authorities and the users of the works carried out. These are just some of the stakeholders. Many other forms of communication potentially *Environmental* are telephone calls, emails, *Internet* and *Social* network. But also: the written documentation such as the hand drawing, the photography and video. With the evolution of the new technologies ICT: The Computer Aided Design (**CAD**); the Building Information Modelling (**BIM**), a software to optimize the planning, the construction and managing of buildings, the virtual and augmented reality and the 3D press.

From the *Environmental* point of view, the most fundamental thing to do is planning in details the contents of a message, the form of communication and the most suitable medium to enhance the *Environmental* characteristics of a building or any structure. Without a suitable Communication Plan, in fact, the average customer might be induced to invest on traditional "goods" as a super equipped kitchen than in interventions of bio architecture or energetic requalification, considering also the family budget limits.

A possible approach is that of the *Media*, already experimented. *EnerGia-Da* design, an experience founded and managed by the architect *Gianni Terenzi* and the engineer *Daniela Melandri* has already been a reality for some years. Between 2005 and 2012, a television format, **TV SOStenibilità** (SOSustainability), produced by *RAI Educational* and the Italian Ministry for Environment, started with the idea to introduce the big audience to international best practices in the field of bio-architecture. A skilled new and original voice, managed by professionals who achieved the goal of catching the audience's attention thus allowing a second television project: *La Mia Bio Casa*—My Bio Home, a series of reportages to see first-hand the technical and legislative aspects for a bio—requalification of the housing spaces. A trend followed also at European level with *ad hoc* planning as Universities and Students for Energy Efficiency (USE EFFICIENCY—Intelligent Energy Europe, 1/06/2009–30/01/2012) whose purpose was to exchange experiences and knowledge among professionals, university professors and students in order to increase the energetic efficiency of the respective university housing.

"*It is useless to invent 'nature in the city', it is enough to make natural the shape of the city: Nature will feel at home soon and will act naturally*" (*Lucien Kroll*). The planning of common green spaces, both private and public areas, equipped so that to help the biodiversity such as: nests for birds and other animals, insect houses useful for maintaining the flora lush, the planting of indigenous trees and bushes or compatible with the climate conditions of the area where the urban space is going to rise. It creates easily, without much waste of human resources and financial, the conditions for the biodiversity to be integrated with the urban dimension, if planned by high qualified professional: biologists, ethologists and geologists and adequately monitored.

These principles have definitely inspired the design of a new urban area extending into the sea in the Principality of Monaco. The *project Anse du Portier* is coming into being in 2025 and it will be created as an eco-quarter, 6-ha diameter, designed by the *Bouygues Travaux Publics* in order to produce the least *Environmental impact* in a marine area particularly sensitive, between the *Larvotto* marine reserve, to the East, and the protected area of the coral seabed of *Spélugues*, to South-East. A **totally pedestrian area** including:

- an underground parking and housing and touristic flats, commercial areas, touristic harbour, all powered 40% by **solar and marine energy**;
- **enhancement of green**, both public and private, in respect of the Mediterranean flora and fauna;
- full **respect of the marine biodiversity** starting from the preliminary phase of the construction with a detailed study of the *Environmental impact* aimed at preserving all the marine species with particular attention to those at risk;
- an ambitious project for **transplanting**, with high-tech equipment, **the *Posidonia Oceanica***, an indigenous aquatic plant present in the whole Mediterranean Sea;
- the creating of a *water soundproofed shield* to reduce the sound emissions particularly dangerous for the marine cetaceans;
- the application of **eco techniques** designed to guarantee the site ecological balance with wall devices easily colonized by marine Flora and Fauna.

A virtuous example of architectural adaptation to the natural *Environment*.

The role of the architecture and architects must be active and participant in the ecological cause avoiding thus an aseptic mechanism, detached from the ecosystem. On this basis, a new humanism of the architectural shapes is growing, based on the *principle of nearness*, a neologism coined by the architect *Lucien Kroll*. A positive vision of a future in which spaces are shared by Man and Nature in a dimension of increasing *eco literacy of the human beings*, respecting biodiversity. Contrary to most part of science-fiction literature by *George Orwell, Isaac Asimov, Philip K. Dick*, and so on who describe the post-modernity age as a dimension compressed by worrying loss of human and natural identity in a town spreading out rapidly and untidily. The so called *urban sprawl*, which unfortunately is already a reality in some quarters of the big towns and the world megalopolis such as: Atlanta, Georgia, in USA; Barcelona, in Spain; Hong Kong, in China; Los Angeles. California, in USA; Milan, in Italy; New York, in USA; Paris, in France; Washington, in USA.

To counter the phenomenon, the most efficient message of many architects seems to be: to know how to *communicate "the beauty"*, not only the aesthetic one created by Man or Nature, but the beauty that involves emotions, memory, traditions, culture and the same natural resources like water and energy. The architecture, in fact, does not deal with what is built but also with the surrounding *Environment* modelled by man. In other words, the architects must have a systemic vision, well known in our ecosystem, that influences on the wellness, health, safety, economy, social aspects, *Environmental* and sustainable quality.

The design of long line of trees of various size in the flowerbeds and urban gardens, for example, in the inheritance of a European old tradition, originally demonstrated for practical reasons: the roots used to mark the roads. The tree canopies could provide shade during summertime, also natural umbrellas in case of rain and produce fruit to eat for many species. But there is something more: **the avenues are carrying an ecological and useful message**. The plants are able to create a microclimate that opposes the rise of the temperatures in the urban areas, especially in summer. We can record even ten degrees less in the trees areas. This is a metaphor of a better quality of life.

Nowadays, the new frontiers of the green technology together with bio-architecture techniques allow the realization of vertical garden, green roofs and green walls indoor and outdoor. As the design of the outside space of the ***Trussardi Caffè*** in ***Milan***, created by the architectural firm: *Carlo Ratti* and Associates and *Patrick Blanc*, an artist, a botanist and a professor internationally renowned. A great visual impact both indoor and outdoor. A crystal transparent showcase surmounted by a vertical garden of 102 m² working as a vegetable roof and giving the impression of being really hang in the air. A patent created by *Blanc* that, exploiting the technique of hydroponic cultivation without using soil, helps the growth of many indigene and exotic vegetable fed by an innovative plant of irrigation. All this can guarantee an inside increased oxygenation of the air and a protection from the external contaminants.

So, this "green" extension of the historical architectural spaces of "*La Scala*" eighteenth century palace in Milan becomes ***a medium of Environmental Communication not mainly as the "channel" for the eco-message but as an eco-message itself***. The outside space of a trendy Cafeteria in the heart of Milan, visual metaphor of a place where to spend a pleasant and relaxing moment is then connoted by the ecological element with a direct contact with the natural world, a micro eco-system generating wellness both physical and psychological in terms of quality and acoustic insulation. Hence the ***metaphor Environment-Health***. A potential *input* to trigger a more sophisticated communication process not only aimed at the marketing but also at the treatment typical of that ecosystem, the most important ecological issues in the form of cultural meetings, vernissage, labs and moments of collective reflections on the main *Environmental* issues.

The **vertical garden**, a new medium that makes the public opinion aware public opinion aware": as shown in Figs. 4.15 and 4.16. The Vertical wall with recycled plastic bottles at G7 for the Environment, held in Bologna from the 11th to 12th June 2017, which witnessed the approval of Bologna Charter, a programmatic document about the policies of Sustainability in line with UN Objectives – Agenda 2030, countersigned by the representatives of the main Italian Metropolitan Areas – Campaign. Or just a decorative element to create a surprising urban environment made of different species of plants and flowers (Fig. 4.17, vertical garden by Gaetano De Bellis). This is the challenge launched by *Blanc* who has "colonized" housing all

Figs. 4.15 and 4.16 Vertical wall with recycled plastic bottles at G7 for the Environment held in Bologna, from the 11th to 12th June 2017, which witnessed the approval of Bologna Charter, a programmatic document about the policies of Sustainability in line with UN Objectives—Agenda 2030, countersigned by the representatives of the main Italian Metropolitan Areas—Campaign #ALL4THEGREEN—Maurizio Abbati (Photographer) © 2017

over the world. Some examples: *Genoa Aquarium* in *Genoa* (Italy), *Athenaeum Hotel* of *London* (*Great Britain*), *Avlabari Station* of *Tiblisi* (*Georgia*), *Brasserie Moritz* of *Barcelona* (*Spain*), *Caixa Forum* of *Madrid* (*Spain*), *Drew School* of *San Francisco* (*California, USA*), *FAAP* of *Sao Paulo* (*Brazil*), *Galeries Lafayette* of *Berlin* (*Germany*), *La Bastide* restaurant of *Los Angeles* (*California, USA*), *Halles Marché* of *Avignon* (*France*), *Parliament of European Union* of *Bruxelles* (*Belgium*), *Sofitel Palm Jumeirah* of *Dubai* (*United Arab Emirates*), *Quai Branly* Museum of *Paris* (*France*), *Via Verde* of *Rio de Janeiro* (*Brasil*), *The Tower 53W53* of *New York* (*USA*) and the *Trio Building* of *Sydney* (*Australia*).

Fig. 4.17 Innovative stage
area of a flowery vertical
garden created by *Gaetano
De Bellis* for the firm
*Botanica Rent, Galleria
Cavour*, Bologna,
springtime 2017—
Maurizio Abbati
(Photographer) © 2017

The **virtual reality** and the **3D models**, on the other hand, encourage our own
"green" ideas in a very easy way and we can see their virtual realization, in a short
time. As the futuristic ***architectural project Manta Ray***, currently being implemented, meant to redesign, in a sustainable way, the urban pattern of Seoul, in South
Chorea. Designed by the Belgian architecture firm *Vincent Callebaut Architectures*
(www.vincent.callebaut.org), the new installations will be built in the heart of the
existing city park, *Yeouido Park*, that already claims a system of water ecological
purification, changing it into a sustainable urban area totally self-sufficient from the
energetic point of view. Starting from the shapes which *Vincent Callebaut*'s team
was inspired by: ***The Boat Terminal on Han river***. A ***wooden structure on lami-
nated timber*** coming from Chorea eco-managed forests, inspired by a ***gigantic***

manta (diamond-shaped cartilaginous fish of the *Myliobatidae* family), able to adapt to the river water level. The huge structure will be *supported by large pillars framed by a network structure in hexagonal mails* reminding another natural element: the cells constructed by the bees for their hives. On the terrace above the structure, there will be *an eco-park placing side by side trees and bushes with glass structures* equipped with polycrystalline photovoltaic cells surmounted by multi wind turbines branched as the tree canopy. Finally, *the work will be completed by a series of fluctuant aquatic barriers*, inspired, in their shape, *by the dolphins' and whales' tails*.

We will be in front of a *real visual mental map*, at the moment only virtual, represented by **architectural structures, metaphors of the biodiversity of our ecosystem** to protect and manage with each of our single gesture. Structures able to "dialogue" with any "user" included those who occasionally go and visit them, or just enjoy them. A multiplicity of emotional reactions is thus provoked that *facilitate empathy ties with the Environment* able to make people understand better the complexity, as well as the challenges to be overcome and the key role they play in ensuring the conditions of habitability of the Planet Earth. **So therefore, it will be the architectural masterpiece to "communicate" the** *Environment*.

Following the "green" wave, also industrial and craftsmanship *design*, has changed completely the perspective in comparison with the original one. During the *Industrial Revolution* (in between the eighteenth and nineteenth century) the Western manufacturing world was caught by an excessive business enthusiasm aimed at producing any artefacts in series, from the car to a little spoon, in order to create a hypothetical fairer society whose access to goods and services was guaranteed to lower classes. A principle not wrong in itself and that could produce effectively better social conditions and an increase in life expectancy. But with a main limitation: the non-prediction of the consequences of such a kind of "revolution" on the ecosystem. Nobody was aware of the impacts on the *Environment* produced by industrialization and the overexploitation of the natural and mineral resources. On the contrary, everybody seemed to be convinced that the Mother Earth could regenerate itself absorbing all kinds of man-made pollution.

Things are going to change especially from the cultural and social point of view in the Years Seventies, when there is a growing awareness of the ecosystem vulnerability. From the industrial point of view, even nowadays the mission is always the same: to produce as much as possible a wide distribution of goods at low prices, without thinking of other "contraindications".

With the increasing demand of eco-compatible goods and services, nevertheless, more and more businesses of small, medium and large size have encouraged the study of new **types of "eco-friendly" productions**. Starting from the *objects and packaging design* that in some extreme cases, have been removed upstream. *Supermarket chains selling bulk goods or drinks on tap* ensuring ecological aspects, savings in price because of the lack of the container are spreading out.

A "greenest" design research in every detail, with *experimentation of new materials* included, is increasingly growing. What prompted, for instance, the chemical industry to convert itself into "**green chemistry**". The excellence of the Italian chemical industry, the *Novamont* S.p.a., has patented and marked *Mater-Bi*®, a kind of bio-plastic, totally bio-degradable, composed of some bio-elements, among them the corn-starch. From this, the creation of some design objects such as *"Aperitivo Bio"* (Bio-Aperitif) by *Pandora* Design, composed of: small glasses, small forks, small dishes, tooth sticks, and the innovative multifunction piece of cutlery *"Moscardino"*, a combination of fork and spoon created by the same *Pandora* designer, winner of the Premio Compasso d'Oro (Golden Compass Prize) in 2001. Or even the *"RGB"* art installation realized in April 2016 for *Eatitaly* by *ZidaLab* innovative reuse Laboratory of Turin that used 800 Mater-Bi® dishes, unsold stocks, to attract interest on waste and on reused materials.

To boost the continuous cycle promoted by the *Circular Economy*, the research of new materials compatible with the *Environment*, shaped by designers does not stop anymore. The students of the main international *Design Academies*, increasingly driven by the desire to create something useful to Man and *Environment*, keep on experimenting new materials that revitalize even organic waste, like pine needles, from which fabric can be produced as the *"Forest Wool"* project by *Tamara Orjola* proves or using new leathers generally considered as industrial waste. The creative and technical skills by the Italian designers *Simone Farresin* and *Andrea Trimarchi*, founders of the Studio *Formafantasma*, based in Belgium, have made a collection of objects and furniture for the well-known fashion designer *Fendi* of superior craftsmanship using vegetable tanned leather made from fish skin such as: salmon, trout, perch, conch shells and bladders of animals.

On the *packaging* front, new more and more compact shapes, 100% biodegradable, recycled, and functional are being created. To overcome the well-known principle *"from Cradle to Grave"*, at the basis of the traditional production chain, the new industrial designers increasingly tend to give value to packaging ensuring its reuse and recycling in order to reduce the *Environmental* waste. In this sense, for example, *eco-friendly* containers in recyclable cardboard of any shape and size have been produced already. This is the multiuse packaging (Fig. 4.18a–d) for milk bottles and other agri-foodstuffs created by designers *Adrián Froufe* and *Chucho Nieto* for *Cocina de Ideas*, brand: *100 × 100*, devoted to bio-food meant to put local farmers through urban consumers. Another unusual packaging *Clever Little Bag*® (Fig. 4.19), eco-designed by *Yves Behar* for a famous brand of the sports fashion industry. The result is the utility of a "cool" sack made of recycled fibre reusable many times. The peculiarity is the inner packaging essentiality, in cardboard recyclable, able to preserve the product until its purchase, occupying the minimum volume of space and lowering the transport charges with a reduction of 10,000 tons of CO_2, and the paper saving of 8500 tons of paper, thus eliminating the traditional bulky boxes used for the shoes marketing.

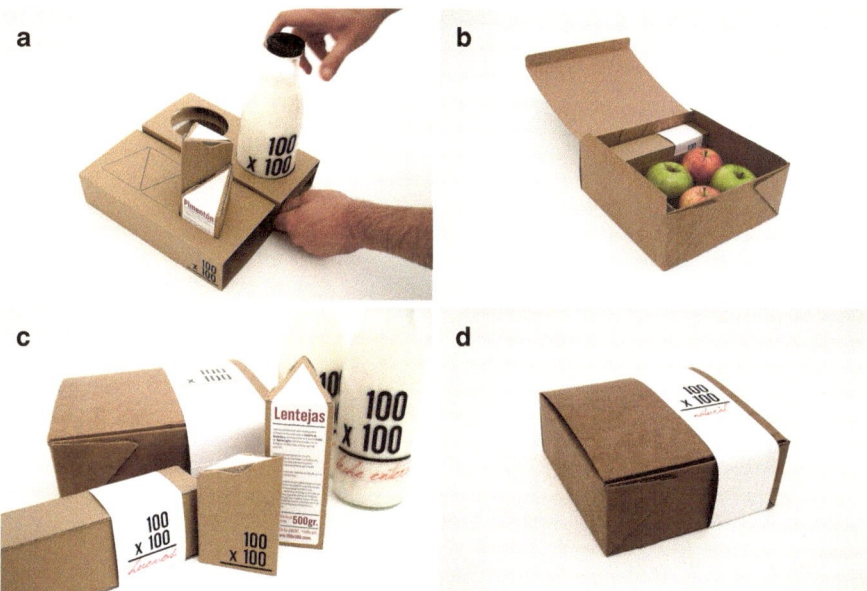

Fig. 4.18 (**a–d**) Multiuse Packaging for organic food, made of corrugated cardboard—Developer: *Cocina de Ideas* © 2011 www.cocinadeideas.net/en/work/—Brand: 100 × 100—Creative and Art Directors, Designers: *Adrián Froufe* and Chucho Nieto (Copyright Holders) www.cocinadeideas. net

Fig. 4.19 Clever Little Bag® eco-designed by Yves Behar for PUMA (Copyright Holder) @yvesbehar (Twitter)

Finally, an example of functional furniture *design* as the portable chair in decorated cardboard called *Dutch Design Chair*™ (Fig. 4.20) created by the Dutch designers *Tim Várdy* and *Suzan Bergman* in 2011 able to catch the attention of many international headlines of magazines and reviews specialized in design and tendencies of fashion and style: Casa Vogue, Glans, India Today Home, and so on. *Dutch Design Chair*™ highlights the creative use of cardboard, a **flexible**, **strong** and **recycleable** material which becomes a ***medium of Environmental Communication***, itself.

Fig. 4.20 Dutch Design Chair™ created by the Dutch designers © Tim Várdy and Suzan Bergman (Copyright Holders): open (on the left); folded (on the right). www.dutchdesignbrand.com

Green Tweets
#Bioarchitecture #Design #Landscape #Ecosystem #Environment #SustainableDevelopment   #SoundofNature   #SressTherapy #StatuNascenti   #Ecofriendly   #Psicology   #EduInformation #EcoComunication #EnvironmentalCommunication #HolisticApproach #Economy #SocialAspects #Culture #Ecology #GreenBuildings #Ecodistrict   @BoscoVerticale   #Nature   #EnergyEfficiency #InformationandCommunicationTechnologies   #ICT   #Biodiversity #Nearness #EcoLiteracy #UrbanSprawl #Trees #TreeLinedStreet #MicroClimate   @CafféTrussardi   #VirtualReality   #EcoMessage @MantaRay #MindMap #Metaphor #Biodiversity #Empathy #Planning #Design #Packaging #Testing #GreenChemistry #NewMaterials #EcoSustainability

4.4   Cinema, Music and the Environmental Message

Can the allegoric strength of the plot, images and musical notes increase the audience's awareness on the main environmental issues to protect our ecosystem?

Cinema production of *Hollywood* contributes to emphasize the sensibility toward the *Environmental* issues through new forms of communication seeking the emotional involvement associated to *Environment* and Nature. The great *Environmental* themes, for their own nature, have fascinated, from the very beginning, the film makers for their potential immediate hold on the public: floods, glaciations, earthquakes, volcanic eruptions and tornadoes. The stage fiction of cinema, together with the creativity of expert communicators, are able to go further the scientific theories, simplifying complex themes and emphasising the man's impacts on the *Environment* through a *mix* of forms of communication: images, signs, words, writing, music, sounds and special effects. All these can be reproduced outside the movie premises, anywhere there is a device that can be connected to the *Net* thanks to the digital revolution of the last 10 years.

The **biocentrism** is by now a reality. A cultural current that marked a real revolution in the world of cinema, opposing to an *anthropocentric vision* (e.g.: "*The Planet of Apes*, directed by *Tim Burton*, 2001) and the *eco-centric one* (e.g.: **Avatar**, directed by James Cameron, 2009). The first meant to represent Nature only through the human point of view. The second was based on deeper values of the *Environmental ethics* that govern all the living beings on this Planet, "*placing itself in their shoes*". The *stimuli* that, in the twenty-first century, brings us closer to the *Environmental* cause are countless. They have been documenting the constant change of the ecosystem Earth because of Man, through visual contributions from widespread Network. **Google Earth**, a georeferenced service, is an example. Originally created by the *Central Intelligence Agency* (CIA) of U.S.A. allows you to view the evolution of natural *Environments* and the urbanized-industrialized areas, with precise details thanks to satellite and aerial photos.

For example, the increasing attention by the public opinion to the preservation of the ecosystem on the Planet Earth gave birth to a specific project able to "mapping" some of the most important biodiversity hotspots. Among them, the *Great Barrier Coral Reef* which covers 2300 km of sea-land, miles off the North-Australian coastline. Since 2012, the Internet users have been able to plunge themselves, virtually, in it to the discovery of that extraordinary and colourful submarine wildlife, thanks to the partnership between Google and a qualified team of biologists, film makers and underwater photographers as part of the *Catlin Seaview Survey*, a company specialised in the visual monitoring of the ocean sea beds for scientific research purposes. More than 50,000 wide-screen and geo-localised high definition pictures aimed at catching the attention of both the general public and the scientists are spreading world-wide, to make them aware of the importance of preserving such a natural heritage. An extraordinary *means of communication* which is shared among thousands of people acting as a bio-indicators about the state of the seas.

An **eco-centrism** that caused, then, a "green revolution" in the way of dealing with these matters through the screenplay, the way of acting, the scenery and the soundtrack. Even during the complex film production, in fact, less *Environmental* impacts can be produced and on what the business world defines technically as the **Life Cycle Assessment** (LCA). This is the objective of the *Environmental* Certification developed in this sector. **Edison Green Movie** is the first European protocol of *Environmental* sustainability for cinema, based on the initiative **Azzero CO_2** (www.azzeroco2.it), an *Environmental* energetic consulting company by

Legambiente, the electric Edison company and the production company *Tempesta*. Reduced emissions of CO_2; use of energy from renewable sources; mandatory recycling on the set; eco compatible practices with respect for the territory where they are shooting scenes and filming; "reusable equipment" such as: glasses, cutlery, dishes, cups etcetera used by actors, operators and technicians during their meals have being exploited so far. These are the priorities already tested by relevant film production companies as in the film "*Il capitale umano*" (*Human resources*) directed by the Italian film maker *Paolo Virzì* in 2014. The film went down in history not only for its box office hit but also for being the first eco-sustainable film in Europe. *Edison Green Movie* was introduced 2 years before at Cannes Film Festival, a protocol completely translated into concrete actions that, if carried out properly, may lead to reducing emissions equal to 1120 tons of CO_2 (Fig. 4.21).

Cinema as a "*green revolution*" is not to be underestimated if we consider that the cinematographic *Media* are the first sources of information and ecological "training". In many cases, the world of Nature, that is the *Environmental* issues, are idealized by the Cinema according to the rules of the wellbeing or ecological *feeling-good*. All this draws the attention to the topic. At the cinema, a very effective *medium* is represented by the strong emotions of satisfaction, enthusiasm, pleasure, interest, but also anguish, fear and anger. Even in the *eco-film* the atavistic struggle between good and evil, between value and disvalue, wrong and right are always present. Already found in other expressive forms as literature and philosophy from which quite often the cinema productions take a cue.

C'È UN NUOVO CINEMA ALL'ORIZZONTE.

Il protocollo Edison Green Movie porta al cinema l'attenzione per un utilizzo corretto e consapevole delle risorse minimizzando l'impatto ambientale dei film. L'obiettivo è quello di produrre film sostenibili, usando soluzioni energetiche eco efficienti e pratiche green in tutte le fasi della produzione.

Fig. 4.21 "*C'è un nuovo cinema all'orizzonte*" [There is a new cinema on the horizon]. Advertising Campaign of the project *Edison Green Movie* by Edison S.p.A., leading Italian Company on the energy market (Copyright Holder). Its main objective is to minimise the environmental impact in film production, boosting energy-efficient and *eco* best practices. @edisongreenmovie (Facebook)

Of course, a single film cannot determine a radical and immediate change of the way of thinking of the world community. That would be a *utopia* to be worthy a science-fiction film. On the other hand, there are factors that may hinder the correct comprehension or the "decoding" of the *Environmental* message by the eco-cinema communicator. Each cinematographic project of TV series follows a specific *Communication Plan* that foresees many audience's reactions. A landscape, a tiger, an ocean and any other element of Nature may arouse or assume meanings, by association, thousands of connotations *"into the larger ecosystem of images and texts"* (*Dobrin & Morey*, 2009). A great deal depends on the ***socio-cultural context*** the audience live in, besides their family *background*. All that generates various **responses**: haring what communicated (**dominant position**), total disagreement (**opposition**) or partial sharing (**negotiation**). In any case, the goal to "speaking about *Environment*" is achieved. Another characteristic typical of the cinema and television *Environmental Communication* is the capacity **to create empathy** between audience and protagonists of the film plot or TV series as stated by *Roger Ebert*, a cinematographic critic of Chicago Sun-Times. Everything thanks to a skilled use of: images, sounds, music and symbols.

The enhancement of the relationship between Man and Natural Ecosystem, in the cinematographic narration, is the result of a particular evolution that leads a lot of film makers to stage and dramatize *Environmental-ethic values*, surprising the audience with special effects, visual aesthetic and characterization. A ***visual rhetoric*** was thus created aimed at persuading of the existence of *Environmental issues* while giving a subjective representation of the *Environment*. In film like ***Grand Canyon*** (directed by *Lawrence Kasdan*, 1991), ***Jurassic Park*** (directed by Steven Spielberg, 1993) or ***Avatar*** (directed by *James Cameron*, 2009). Nature is generally associated to a positive image defended by "positive heroes" being on the side of "Good".

Next to this *"filmic choice"*, **Nature is seen as a *"stepmother"***, a monster, tending to punish Man who dares challenge it or not respecting it. The theme of the challenge between Man and Natura Forces, on the other hand, deepens its roots in the Ancient Greece. The main characters of the Greek tragedy, in fact, are often at the mercy of the natural forces because of their *"hubris"*, i.e. excess, arrogance, pride and abuse of power. They are guilty, therefore, of committing the sin of *Hybris* (ὕβρις), a *topos* (recurring theme) in the Greek tragedy.

The *Environmental Communication* on the big screen (***cinema***) or on the small one (***television***) tends to exaggerate, both in a positive or negative way, the relationship between Man-Nature. A *Media* hyperbole responding to market principles with the target to achieve the best result ever in terms of tickets sales. Nevertheless, amplifying the *Environmental Communication* addressed to the great audience triggers a *virtuous domino effect* making people be aware and making them reflect on the *Environmental issues*. It is clear that the cinematographic narration influences, consciously or unconsciously, the way of thinking and the lifestyles in the whole Planet Earth. The widespread dissemination of ***mainstream films***, doubled and subtitled in various languages, produces effects of *"Word of mouth"* similar to *Internet*. It is no coincidence that some authors like *Pat Brereton* talk about *"creative imagination"* and *"ecological fantasies"* referring to the big Hollywood productions.

On the other hand, the analysis of the **ecological ethics**, takes us back to the well-known American *Western* films in the years Sixties. Despite they are the emblem of an anthropocentric vision where the white Man conquers and dominates the Nature, on the contrary, in these films there is a "green soul". The sharing, more or less peaceful of boundless lands suitable for cattle and sheep rearing, pushed *cowboys* to ride for thousands of kilometres looking for the few water resources available. A familiar theme to the Classical economic theories by *Thomas Maltus* and *David Ricardo*: the lack of raw materials of natural origin. A current topic that makes us reflect on the necessity to manage the water element in a sustainable way in order to defeat social or armed conflicts among the geographic areas with abundant water and the dry and desert areas.

The creativity and the research of new forms of *Communicating the Environment* brought the film directors to use a lot of approaches to the matter and created the curious trend of the **lifeboat ethics**. Famous films like **Wall-E** (directed by *Andrew Stanton*, 2008), **All is Lost** (directed by *J.C. Chandor*, 2013), **Captain Phillips** (directed by *Paul Greengrass*, 2013) and the same **Gravity** (directed by *Alfonso Cuarón*, 2013), Academy Awards in 2013, emphasize the fragility of the ecosystem Earth, seen as a spaceship threatened by pollution and by any other form of man-made degradation (Figs. 4.23, 4.25 and 4.26). The protection of our natural resources according to their quantity and time of reproduction is firmly reiterated until Nature and Religion become one thing like in the recent film **Life of PI** (directed by Ang Lee, 2012) [Fig. 4.33].

The **idealization of Nature** goes also through the **sublime catastrophic vision** of a science fiction film as it is the case of **The Day after tomorrow**, directed by *Roland Emmerich* in 2004. Just one of the few films that makes the audience reflect and debate on the topic by using a perspective of the *Climate Change* emphasized in the stage illusion. The protagonists are the powerless victims of a spectacular glaciation without having the possibility to prevent or foresee it. But this seems to be in the foreground. This k-message is prevailing: men are subdued to natural forces (*Salvador Norton*, 2011). Its objective was a world success and we can say that it *hit the edge*. A film that came out, in one hundred and ten nations simultaneously, and soon it was placed at the top in the major parts of the Countries with a total box office income of 544 million dollars, in spite of some critics (Fig. 4.22).

Nevertheless, some film productions deal with "*natural phenomena*" without using an *Environmental-ethic* message. Nature is sometimes presented as a "stepmother" which tries to destroy man independently by his actions. This happens for example in **Twister** (directed by *Jan de Bont*, 1996), **Volcano** (directed by *Mick Jackson*, 1997) or **Dante's Peak** (directed by *Roger Donaldson*, 1997). Some critics like *Campbell, Wheatley, Ivakhiv*, underline that in those films any reference to the *Environmental protection* is missing but, on the contrary, the communicative will to drag the audience into the pure "*adrenaline pleasure*" prevails to make people experiment apocalyptic scenes. A perspective that irresistibly attracts a part of the audience.

In other cases, the **primary Nature** or **Wild Nature** hosts the protagonists; heroes and heroines, that make the natural world as their sanctuary, both at a cinematographic fiction or documentary film level. "*An inconvenient truth*", directed

Fig. 4.22 Poster of *The Day After Tomorrow*, Twentieth Century Fox © 2004

Fig. 4.23 Poster of *Captain Philips*, Columbia Pictures © 2013

by *Davis Guggenheim*, whose main character is the once Vice-President of America *Al Gore* points out the evocative emotion of Nature linked to a nostalgia of an uncontaminated *Environment* experienced by *Al Gore* in his childhood, now destined to a constant and inexorable implosion. The narrative communication, reinforced by an innovative graphic and new filming techniques like the *time-lapse*, is a stimulus for a large audience allowed to travel quickly in time and space. It is not by chance that the documentary film about *Gore* is considered the milestone that drew up a new age of **"eco-ethics" Cinema**.

Another example is ***Wall-E***, an animated film shot in 2008, by *Pixar Animation Studios*, in co-production with *Walt Disney Pictures*, and directed by *Andrew Stanton* (Fig. 4.25). The adventures of the main character *Wall-E*, a robot whose charge is to clean a Planet Earth abandoned by men because it has become unliveable by the unrestrained consumerism and tendency to waste. It derives then a narrative metaphor of an ecological message. The need to educate old and new generations to take care of our Planet, trying to reduce waste production and food waste.

The theme about the respect of Nature and its strengths is the basis of a series of films taken from literature like Hemingway's novel ***The Old Man and the Sea***, where we cannot help remembering the famous interpretation by *Spencer Tracy*, protagonist of the same film directed by *John Sturges* in 1958, very effective *Environmental Media*. ***All this lost***, written and directed by *J.C. Chandor* (2013), whose main character, the famous American actor *Robert Redford*, must face a stormy ocean in a *crescendo* of tension communicated through a soundtrack keeping the real sounds of the wildness. Just to point out the link between Man and Ecosystem. Though it is risky, the safety and the survival of Man takes place when he finds a right balance with the Natural *Environment* and he respects it. As ***Cast Away***, film directed by *Robert Zemeckis* in 2000, where *Tom Hanks* plays a new *Robinson Crusoe*, main character of *Daniel Defoe*'s novel.

In the above-mentioned examples, the filmic narration is shown from the nature point of view and not *vice versa*. Mother Nature requires the respect of its rules pushing Man to behave humbly on the basis of the highest values of sustainability. A concept underlined in other two recent film productions: ***Captain Philips*** (2013) directed by *Paul Greengrass* and played by *Tom Hanks*, and the award-winning ***Gravity*** (2013), directed by *Alfonso Cuarón*, whose main protagonists are *Sandra Bullock* and *George Clooney* (Figs. 4.23 and 4.26). It is the view of the Earth from the space dominating the scenes of this science-fiction film to communicate the beauty and the fragility of our Blue Planet. Triggering deeper reflections of ethic nature on the importance to respect and safeguard any form of life. It is therefore crucial to live in harmony with the ecosystem without dominating Earth and Universe. A journey of inner growth and of "rebirth" of an ecological conscience more mature that relives in the ***pathos* of the emergency landing** by doctor *Ryan Stone* (*Sandra Bullock*), a biomedical engineer in the space mission, that she must face in order to avoid getting lost in the empty space.

What happens if the animals themselves are communicating the values of sustainability and ecologic ethics? In the movie scene, more or less recent, there are many examples of this kind. We cannot miss out to remember the animation feature film ***Bambi*** (1942), a classic of *Disney* production where all the main characters are animals and they embody their own character and form of communication: the human language. Such a kind of experiments have divided the critics. Since the films are addressed to a young audience, somebody thinks it is a good *medium* to make them aware on ecological issues thanks to the empathy created

Fig. 4.24 Poster of *March of the Penguins*, Warner Independent Pictures, National Geographic, Canal + © 2005

Fig. 4.25 Poster of *Wall-E*, Disney Pixar © 2008

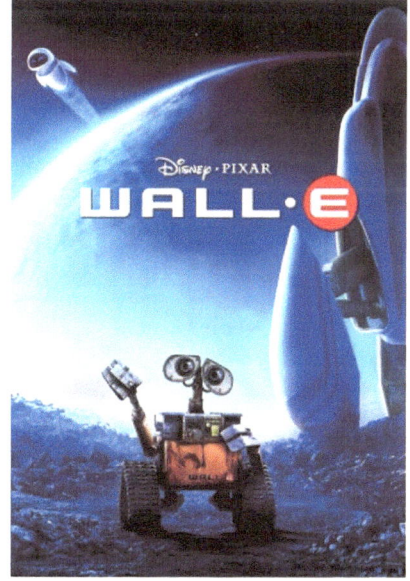

by the humanized animals. Others criticize sharply the ethic and ecological function considering it an attempt to minimise the importance of the natural world reduced to a mere *fiction*. A sort of "*branch*" of the human world where the animals are felt as *soft toys* copying communicative forms and ways typical of the human beings.

In any case, when the characters communicate with their own language, the messages are more effective for the big audience. ***The Penguins march*** (2005), a documentary film created, directed and produced by the French filmmaker *Luc Jacquet*, has shown how the reproductive practices of the Emperor penguin in the Antarctic ice could catch the audience's attention. A very patient work that lasted more than 13 months, finally rewarded (Fig. 4.24). Despite the criticism by a part of the public's opinion about the inadequacy of the comparison between man and penguin, with some attempts to humanise natural behaviours dictated by instinct, the ecological message was greatly effective from the communicative point of view. The empathy by a part of the audience, especially the young people, toward the penguins of the film narration aroused emotions while focusing the attention on the importance to protect the uncontaminated places as the ones described by the documentary.

It is the case of the documentaries, made both by experts of biology or ethology or simple amateurs. The difference is made by the communicative point of view and the knowledge of conveying enthusiastically and with simplicity the natural world. All this is the result of research, studies and observation on the sites together with a right deal of creativity in the mounting and filming. But in some cases, a key element of effective communication could be the *narrating voice* that, for **David Attenborough**, the famous documentarist of the **British Broadcasting Corporation** (BBC), is often associated to his image, becoming soon a reassuring familiar face: "*the Nature's friend*".

The series **Life on Earth**, become a *bestseller* at the end of the years Seventies, is a turning point in the *Environmental Communication* of documentaries. The direct personalized dialogue of the anchor-man towards the audience is of great importance in order to make the complex concepts of the Natural Sciences accessible to all. These concepts are made "digestible" thanks to the narrative creativity in many **BBC** series and in other production companies such as the **National Geographic** and **Disney**. A new kind of programme is thus created that is a smash hit on TV above all, called **edutainment** [*education + entertainment*], neologism used for the first time by *Robert Heyman* in 1973. A carefully-judged mix of communicative skills meant to educate, inform, experiment and arouse emotions, largely used by the above-mentioned production companies, specialized in this sector.

A trend experimented successfully for the "***blue chip***", film and television production aimed at "educating" the audience to the respect of the natural world with spectacular effects based on the surprise and some *Media* device to reinforce the effectiveness of communication such as: melodramatic temporal elements, moments, climaxes, innovative shootings in high definition or 3D. All this in total absence of political stance and scientific controversies and even Man. The illusion of being in an uncontaminated Nature is conveyed like new *Robinson Crusoe* landed on an unknown and lost island. A *win win choice*, that led to series like **Planet Earth** (2006) or **Blue Planet** (2001), produced by the BBC, to become *best seller* worldwide.

A new communicative approach that inspired also a series of **historical-natural** films developed first in India (e.g.: *The Last Migration*, 1994; Shores of Silence, 1999; *Vanishing Giants*, 2000, etc.) for both vegetal and animal species threatened by a non-peaceful coexistence with human communities. Elephants and sharks are an example.

Following this line of interpretation, on the other hand, cinematographic productions, independent, have started a rich vein of documentaries and films with a strong eco-sustainable message underlined by amazing shooting techniques or by specific choices of screenplay and editing. This is the case of the documentary film "**El abrazo de la serpiente**" (2015) directed by *Ciro Guerra* (Fig. 4.27). This Colombian, Venezuelan and Argentinian co-production is able to achieve its ambitious target: making the audience aware of the precious biodiversity of the Amazon forest by deleting a communicative element particularly effective for the Equatorial ecosystem: the colour. The Amazon fauna and the flora, in fact, use mainly chromatic nuances to send messages of different nature from danger signals to soothing and sexual symbols. In this natural kaleidoscope, the exploratory trip of an American ethno-botanist looking for a very rare medical plant, the *yakruna*, is told in an evocative black and white colour. The film-director's choice, besides pointing out the

flashback of a past story he produces and unexpected effect of awareness on the ecologic theme of the dramatic loss of a forest heritage unique in the world due to the Man's indiscriminate exploitation that has been subtracting Amazon areas to the Nature. We must reflect on the statistical data: between 2012 and 2015 the deforestation has increased of 75%.

To make people aware of the *"Environmental cause"* may be effective through "more conventional forms" but as much as effective as for example the reportage shaped on the journalist model, that turns into a documentary film. This the mission of "**Before the flood**" (2016) a cinematographic as well as a television project edited by the National Geographic by which the famous Hollywood actor, prize awarded, *Leonardo Di Caprio* committed in *Environmental issues* wanted to offer the big public a precious testimony to clarify on *Climate Change* (Fig. 4.28). *Di Caprio*, acting as an international special Correspondent, visiting the most meaningful sites from the natural, scientific and decisional purposes a series of video-interviews to scientists, activists, politicians and witnesses with the purpose to represent an instant "snapshot" on the situation of the *Climate Change* and on its causes and possible solutions to mitigate the effects.

In the last years the deep interview conducted by a qualified moderator, is one of the most used instruments of research not only in marketing and advert world but also in *Environmental* and *Sustainable Development* issues. This method that aims

Fig. 4.28   Poster of *Before the Floods*, National Geographic © 2016

at the quality and objectivity of contents, allows to reach the "heart" of the questions by the "live voice" of the experts in that sector giving the public different "reading keys" of the *Environmental* message according to the interviewees' professional research field. It is not without a reason that we have shaped on the focus group, our special section called "*The interviews with professionals*".

Also, the TV series can be transformed into *Environmental Communication* instruments. This is the case of "***Collapse of the Oceans***" (2016), an episode of the series Years of Living Dangerously, produced by the film director *James Cameron* and the actor *Arnold Schwarzenegger* (Fig. 4.30). The fragility of the eco-system and the vital role of the oceans to guarantee the survival of the human species itself is the core of a cinematographic "journey" towards the Australian Big Coral Reef, guided by an exceptional mouthpiece: the actor *Joshua Jackson*.

A contemplative vision of the marine biodiversity relives in the documentary film "***Océans***" (2009), directed by *Jacques Perrin* and *Jacques Cluzaud*, a result of a 4-year film shooting (Fig. 4.29). A production of a great effect made by a professional team. Researchers of the scientific world, operators of underwater shooting, sound engineers, assembly editors and composers have contributed to the realization of this film that enables the audience to plunge themselves in the ocean depths of the Planet Earth. The faithful reproduction of sounds and colours, with a few human narrative interferences, and the high-resolution images realized with new

shooting techniques arouse strong emotions and make people aware of the importance to protect the eco-system from the concrete threats of pollution (Figs. 4.27,
4.28, 4.29, and 4.30).

But voicing Nature not always produces clear and effective communicative
results. Think of **Grizzly Man**, a documentary film directed by the German film-
maker *Werner Herzog* in 2005. The movie deepens, in an unusual way, the relationship between Man and the Wilderness represented by the *Grizzly* bear (Fig.
4.32). The strong desire by the human protagonist, *Timothy Treadwell*, an environmentalist explorer (who really existed) who spends the summertime in the
Katmai National Park (Alaska) to live "peacefully" together the biggest bears on
the Planet Earth, conveys one of the highest ecological value: the full respect of
Nature without domineering it. A message that is, in reality, full of contradictions.
A strict analysis, in fact, shows that the main character is critic against the hard-
ness of the natural world and quite often he is tempted to dominate it, trying to
communicate and become "friend" of the bears through a language exclusively
human. A choice that will be fatal to him after the bear mortal attack involving
also his fiancée. A coming back to the concept of Nature as a "stepmother" which
could arouse hostile attitudes against the natural world in a part of the audience,
if it is not led by its own *critical sense*.

A very effective ecological message is, on the contrary, the one by **Avatar**, the
award-winning film-event directed and produced by *James Cameron* released all
over the world in between 2009 and 2010 (Fig. 4.31). The extraordinary special
effects and the science-fiction reconstruction of a new Planet, **Pandora**, where an
uncontaminated natural ecosystem communicates effectively a strong ecological
message, praised unexpectedly by a historical non-profit organization protecting

Fig. 4.29 Poster of *Océans* Disney Nature
© 2009

animals: *People for the Ethical Treatment of Animals* (PETA). A high impactful visual narration (later emphasized by the 3D version) that opposes Nature to Technology. The natives (the Na'vi), respectful of the natural world in which they live, in opposition to an army of conquerors pushed by economical and imperialistic interests of exploitation and destruction of that eco-sustainable symbiosis. Their idea is to exploit the energy properties of *unobtainium*, a ferrous crystal that could solve serious energy problems of the Earth, the real "Pandora's vase", compared to the creatures of the alien Planet.

Fig. 4.30 Poster of *Years of Living Dangerously* National Geographic © 2016

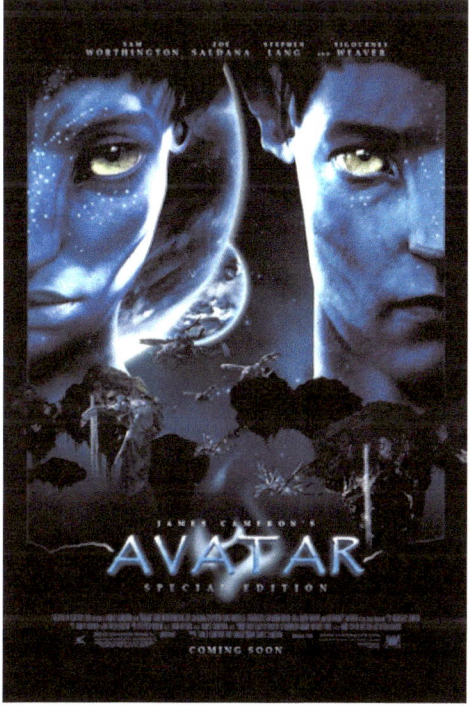

Fig. 4.31 Poster of *Avatar*, 20th Century Fox © 2009

Nevertheless, the "*heroes of the good*" are the humans that identify themselves in the body and soul to the local population and they are at the forefront of the fight to defend each single creature living in the "Amazon rainforest" covering *Pandora*. Among them, and first of all, the former handicapped marine *Jake Sully*, a sincere and fearless heart, called to substitute his brother, who died prematurely, in a project allowing the reproduction of a double (*avatar*) of oneself with the same physical features of the autochthone population, exemplary icon of the *Environmental ethics*, embodied by the princess *Neytiri*, whom the protagonist falls in love with.

A Nature that from being hostile changes itself in a great Mother allied with those who are fighting and defeating the cynical and materialistic army considering Pandora a deposit to be exploited at the expense of its inhabitants. In this respect, the final alliance between the Na'vi and the other exotic creatures, "technologically" perfect, living on the Planet are emblematic. Their only purpose is to preserve their Planet for the future, metaphor of the man's commitment to protect the ecosystem Earth with real actions and not only words. A message that has been well conveyed and that has made this film one of the most successful *eco*-**film** of the whole movie history.

Our reflections, then, make us be aware of the main problems in communicating the *Environment* through Cinema: the fact that we are not always ready to "translate" what the filmmaker wants to communicate. We often try to improvise as new self-taught film critics. This leads not to notice the *Environmental* messages present in the small and big screen. "Reading the image" courses should be considered starting from the school benches while learning to read and write. From a naturalistic, ecologist, humanist, moralist, scientific, aesthetic, utilitarian and negativist approach.

The relationship between Man and Nature and Man and Animal is often in contrast on the basis of behavioural rules to which we are subjected being a part of a specific society. Culture, religion, convention, belief or superstition influence inevitably our way of feeling the ecosystem even though we are living in an increasingly multicultural dimension. Everything depends on how is the sequence of images perceived by the **editing manager** that sometimes plays with their capacity to arouse emotions to give an image of Nature from a certain point of view more than another one (*James Elkins*).

The point of view can really determine a different communicative result. The same that brought the vision of *zoo* like prisons for the animals as the consequence of unscrupulous colonialism and the extravagant *mania* of exploiting Nature and its scientific cataloguing. Nowadays, on the contrary, they are generally considered the image of symbolic places for their natural biodiversity, last paradises of endangered species and eco-sustainable values of a responsible management of our Planet. In this direction, the communication campaigns are based, promoting bio-parks, safari or theme parks as **Disney Animal Kingdom** or **Sea World.**

The scenes and images showing the Ecosystem Earth are they leading the audience to a greater eco-awareness? Susan Sontag, a critic, suggested, in 1979, before the digital age, to pay attention to the use of images in order to avoid their becoming abstract entities of a virtual world, without a suitable explanation.

Beyond the commercial and marketing interests influencing often the filmography, but also the modern photography, only the pictures able to involve emotionally the audience make the difference. They are easy to be memorized and influence those who look at them to be more active and less passive. Some authors, among them *James Elkins*, underlined that most of the public feels the irrepressible need to take a picture that can be memorized and remembered with the passage of time. In the interconnected and globalized world where we live communicating visually the protection of the *Environment* is now extremely easy and so exchanging ideas through visual or verbal messages. The problem is that the destruction of the ecosystem or the extinction of species of animals and vegetables seem to be often "less appealing" by the big public if they are not correctly communicated. The "*man of the digital age*" hides himself behind a virtual dimension made "authentic" by the new technologies of communication, called **augmented reality** that can change, for example, the car dashboard into an interactive digital screen.

The images that run in a film are abstract entities unless they are followed by tangible actions. A set of colours, shapes, lines, thoughts, hopes of an Ecosystem Earth where Man and Nature can live in the reciprocal respect. It is up to us acting on this sense and make it possible to last for a long time and not only as an immaterial *avatar* of the Planet Earth to project as a *documentary film* in a 100 years. New *eco-films*, able to shake the audience's consciences and make them more eco-responsible in their everyday life, are welcome.

No wonder that the film companies are looking for new expressive forms that, through metaphor and visual hyperbole, stimulate provokingly the audience' senses in order to encourage the reflection and awareness on the big *Environment* and *Climate Change*. Without any romantic claim to *revolutionise the world* with **eco-friendly cinematographic projects** released both on the small and big screen and aiming at making people know about innovative aspects of Nature. From the fight for survival, involving Man and Nature, to the contraposition between the unrestrained consumerism and the preservation of the natural resources of the territory strictly linked to sustainable management. But it becomes also the metaphor of the protection of culture and traditions of the local populations who live in the respect of the ecosystem where they stay.

An approach in continuous evolution that does not undervalue even *the communicative capacity of the musical notes and the singing*. A lot of composers, in different ages and still nowadays, took their inspiration from Nature and its biodiversity as we have deeply investigated in the interview of the award-winning Maestro of piano, *Niccolò Ronchi*, see dedicated section to "Interviews with Professionals". *At present, can the different musical genre like rock, pop, hip-hop and country convey real environmental messages? May a concert be eco-managed and communicate good environmental practices?*

In many cases, it is quite difficult to find common elements among musical bands belonging to the *mainstream* dealing with the *Environmental issues*. There are singers or bands who have become *testimonials of ecological campaigns* such as: *Pearl Jam, RadioHead, Bon Jovi, Laura Pausini, Gianna Nannini, Jovanotti* and so on. But this is not a novelty anymore. We cannot forget the *Live Aid* rock concert by the Irish singer *Bob Geldof* held in 1985 simultaneously in two Countries: at Wembley Stadium in London and at the JFK Stadium of Philadelphia, USA.

Can music, se per se, be a medium of Environmental Communication? Surely these topics have been dealt with for a long time and a special word has been coined "**eco-musicology**" to identify this discipline, investigating and examining on the relationships among music, culture and Nature on complexity (*Grove Dictionary of American Music*). A dimension which takes care of both the song texts and the performances pointing out the communicative, creative and persuasive aspects of music.

The music we usually listen to **through the Net**, we watch on the small and big screen as well as on the stage of *live* concerts is greatly created kilometres and kilometres distance from us. Then the following question:

- *Can a music composed of in a territory and in a social context completely different from ours, effectively "communicate" the Environmental issues of the place where we live?*
- *How can we communicate at best the Environment through the "making music"?*
- *Are more efficient the well-known international singers or the local bands?*
- *What can influence more the lifestyles of the public opinion in order to make them more "green"?*

Questions not to be deepened in this dissertation but we have liked to share with the readers for further reflections and cues.

It is certain that music can narrow those who listen to it to the world of Nature. This is confirmed by the numerous composers, authors and sing-song writers, and singers of yesterday and today. From *William Gardiner*, a British composer of the nineteenth century to *Ferde Grofé*, an American musician and composer, author of the *musical suite "**Grand Canyon Suite**"*, in 1932, soundtrack of the homonym Disney short film, Academy Award in 1958; from *Ani DiFranco*, an American guitarist sing-song writer, Author of "***Your Next Bold Move***", a courageous reflection on the evil of our modern society. To *Pete Seeger*, an American composer and sing-song writer "father" of the American folk music, author of *"This Land is your Land"* and also protagonist of a spectacular concert at the Lincoln Memorial of Washington (USA) singing in duet with *Bruce Springsteen* to celebrate his 90th birthday. To *Chris Cornell*, a member of the American rock band *Soundgarden*, author of the *Environmental "Hands All Over"* in 1989. From *Gianna Nannini*, author and singer of *"Maremma"* from her album *"Per forza e per amore"* (Whether you like it or not), in 1993, inspired by the sounds of Nature of the Tuscany countryside. To *Giorgia*, singer of *"Mal di Terra"* (*Earth sickness*) included in the Album *"Stonata"* (Off-key), 2007 pointing out the human devastating impacts on the *Environment*. To *Laura Pausini*, singer of *"Sorella Terra"* (Sister Earth), included in the album *"Primavera in anticipo* (Early Springtime) in 2008, celebrating the beauty of our Planet. And finally, to *Peter Gabriel* composer of *"Down to Earth"* in 2008, soundtrack of the animation film WALL-E, mentioned before.

Green Tweets
#Filmproduction   #Environment   #SustainableDevelopment   #Cinema #Television#Film#Filmography#AugmentedReality#CommunicationExpert #Image #Signs #Words #Writing #Music #Sounds #SpecialEffects #Biocentrism #LifeCycle #CycleAssessment #GreenRevolution #FeelGood #Ecology #Ecofilm #TelevisionSeries #SocioCulturalContext #Empathy #VisualRethoric   #StepmotherNature   #GreatMother   #Mainstream #EcoogicalEthics   #Western   #LifeboatEthics   #ScienceFiction #FilmingTechniques #TimeLapse #Media #EcoMessage #DocumentaryFilm #Edutainment #Bluechips #EcoMusicology

4.5   Conclusions and Reflections: When Superman Dresses Up in "Green"

The strength of the Visual *Environmental Communication*: the sensitive superpower.

The *image capacities* and any **artistic visual creation** to communicate the *Environment* is a reality, as we have pointed out in this chapter. On the other hand, communication is *that symbolic interactive instrument we use to build the Environmental issues and negotiate the possible solutions offered by the society* (*Robert Cox*). This emphasizes the fact that the visual narration in many cases has an immediate impact on the public that triggers reflections, debates and idea exchanges focussing on the *Environment* and *Sustainable actions*.

In fact, that the *visual representation of the eco-system* is considered by the communication experts particularly effective to influence the lifestyles into more eco-sustainable ones. Particularly impacting was the awareness campaign *"Destroying Nature is Destroying Life"* promoted by the *German Environmental Advocacy Group, Robin Wood* (www.robinwood.de) that, since its birth, in 1982, has been engaged in protecting the forest eco-systems fighting against any form of pollution and attempt to destroy them by Man.

The graphic designers and illustrators of Grabarz & Partner (www.grabarzund-partner.de) have focussed exclusively on a visual communication by creating a multi-level narration particularly meaningful. **The larger image of some animal symbols of biodiversity**: a baboon, a deer, a polar bear is going to fade, intersecting with the representation of a *zoomorphic eco-system* strongly affected by the destroying strength of Man.

- The **baboon**, metaphor of the natural world, is breaking down together with the tropical forest that covers his body in flames voluntary generated to facilitate the deforestation.
- The **deer** "ambassador" of the natural world seems to disappear together with the coniferous forest that covers its body, devastated by Man's unconditional exploitation.
- The **polar bear**, symbol of the natural world, seems to vanish together the artic eco-system invaded by bulldozers and drills dominated by an oil platform in flames, source of pollution and main threat for the biodiversity.

Very dramatic situations that make people to reflect on the necessity to protect the *Environment* and manage it in a very sustainable way, sanctioning those who do not respect the international regulations established for that purpose.

Many readers may think: what can I do to prevent all this? We need a super hero's intervention! May be a super hero like *Superman* that besides being one of the first fancy super heroes (1938) had the merit of conveying the first ecological messages at world level. *Karl-El*, coming from *Krypton Planet*, as known as *Clark Kent*, a simple journalist in an American metropolis, gets strength mainly from the sunrays, thus underlining the big energetic potentialities of the main Star of our Solar System. A "solar strength" "well expressed by the colours of his "uniform": red and yellow but also blue able to absorb the yellow energy of the Sun. All his super powers are therefore "Zero emissions" and his important task is to protect unconditionally the Planet that hosted him and to preserve its eco-system: The Planet Earth.

Apart from this film digression, each of us can transform himself into a super hero, in an eco-Superman, even without the super powers of the comic strip character created in 1933 by Jerry Siegel and Joe Shuster. He can do it by adapting his simple everyday gestures to the *Environmental cause*. The rhetoric influence of images can be exploited to make us become authentic eco-influencers able to influence the others and convince them of the utility of following *Environmental good practices* and fight against any threat contrary to a sustainable management of our Planet. In a few words, let us put into practice the ecological effectiveness through our choices, ideas and creativity opening our minds to sustainable innovation.

Probably it means to come back to being more "humans" and to rediscover ourselves as the suitable inhabitants of the Planet Earth, evolutionarily formed to belong to our ecosystem together with the other living being. *This does not mean to live in a pre-technological reality (Michael Braungart, Wiliam McDonough)*. The evolution of the **eco-compatible High-Tech** applied to Architecture and design, shows how it is possible to design buildings, quarters, touristic structures and even industrial plants in perfect symbiosis with the surrounding *Environment*, for a **reciprocal benefit exchange**. In this logic of the *Environment protection*, the **natural parks** are left the task to manage some areas of our Planet to the benefit of our life quality.

In the same way, it would be desirable to create, at an international level, a "**green list**" of **bio-compatible materials** for the architecture and design to be preferred instead of toxic substances. From this point of view, the objective is not to be "less dangerous" but to find instruments to be "*in perfect harmony*" with our Planet (*Michael Braungart, Wiliam McDonough*).

- *Therefore, why cannot we put on the market only Environmental low impact products?*
- *Why shouldn't we allow an easy identification using "clever" labels able to give us complete and detailed information about the materials used and their recyclability of reuse capacity?*
- *Why don't we think of the traditional structured forms of the buildings of object and transportation vehicle designs to improve the Environmental performances so that they become useful to the vital cycles of other living beings?*

What is important is not to stop communicating on these issues and to inform continuously on the status and quality of our ecosystem to push the public opinion to act responsibly in its regards as shown in the new film produced by the once American vice-president *Al Gore*, **An Inconvenient Sequel: Truth to Power**, directed by *Bonni Cohen* and *John Shenk*, presented in *world premiere* at *Cannes Film Festival* in 2017.

Green Tweets
#images #ArtisticCreation #EnvironmentalCommunication #VisualStorytelling #GraphicDesigners   #Illustrators   #Biodiversity   #AwarenessCampaign #EnvironmentalOrganisation   #EnvironmentalProtection   #Architecture #Design #HighTechnology #EnvironmentFriendlyMaterials

Bibliography

Attenborough Sir David, Blewitt John, Nimmo Harriet, *Media, Ecology and Conservation: Using the media to protect the world's wildlife and ecosystems* (Converging World) (Converging World Series), Green Books, 2010, pages 160.

Augé Marc, Eco Umberto, Didi-Huberman Georges – postscript by Gianfranco Marrone, *La forza delle immagini* (The power of images), Milano, Franco Angeli – Comunicazione e Società, 2015, pages 89.

Badaloni Federico (foreword by Resmiti Andrea e Rosati Luca), *Architettura della Comunicazione* (Architecture of Communication), Indipendent Publication, 2016, pages 173.

Baldacci Cristina, *Artigiani del Paesaggio* (Artisans of the Landscape), ArteDossier, Giunti, January 2012, pages 30-35.

Berenton Pat, *Environmental Ethics and Film*, Oxford, Routledge – Taylor & Francis Group, 2016, pages 245.

Bergman David, *Sustainable Design – A critical guide,* New York, Princeton Architectural Press, 1st edition, 2010 1956, pages 144.

Braungart Michael, McDonough William, *Cradle to Cradle*, Vintage Books, London, 2009, pages 192.

Campbell Vincent, *Framing environmental risks and natural disasters in factual entertainment television*, Environmental Communication 8 (1), 2013, pages 58-74.

Cox Robert Cox, Pezzullo C. Phaedra, *Environmental Communication and the Public Sphere*, Sage Publications Ltd, Los Angeles – London – New Delhi - Singapore – Washington DC – Boston, 4th edition, 2015, pages 422.

De Fusco Renato, *Dentro e fuori l'architettura. Scritti Brevi (1960-1990)* [In and Out architecture. Short Essays] Editoriale Jaca Book S.p.a, Di fronte e attr. Saggi di architettura, 1992, pages 272.

Dobrin I. Sidney I, Morey Sean, *Ecosee: Image, Rhetoric, Nature*, State University of New York Press, 2009, pages 340.

Eco Umberto, *Apocalittici e integrati* (Apocalyptic and Integrated – mass communication and theories on mass communication), Milano, Bompiani, rist.2016a, pages 385

Eco Umberto, *I limiti dell'interpretazione* (The limits of interpretation), Milano, La Nave di Teseo, 2016b, pages 240.

Eco Umberto, Opera Aperta (The Open Work), Milano, Tascabili Bompiani, 2013, pages 329.

Gombrich Leonie, *The Uses of Images: Studies in the Social Function of Art and Visual Communication*, Londo – New York – Paris – Berlin, Phaidon Press, 2000, pages 304.

Gombrich Ernst H., *Breve storia del mondo* (Brief history of the world), Salani Editore, 2012, pages 338.

Granata Paolo, *Ecologia dei media* (Media Ecology), Franco Angeli, March 2015 – Cultura della comunicazione, pages 160.

Hollander B. Justin, Sussman Ann, *Cognitive Architecture: Designing for how we respond to the build environment*, Routledge, 2014 pages 183.

Hulme M., O'Neil Saffron, *An iconic approach for representing climate change*, Glob. Environ. Change 19 (2009), pages 402-410.

Impelluso Lucia, *Nature and its symbols*, J Paul Getty Museum Pubns, 2006, pages 383.

Ivakhiv Adrian, *Ecologies of the Image: Cinema, Affect, Nature*, Waterloo, Canada: Wilfred Laurier University Press, 2013, pages 435.

Ivakhiv Adrian, *Green Film Criticism and its Futures, Interdisciplinary Studies in Literature and Environment (ISLE)*, 2008, pages 1-28.

Kastner Jeffrey, Wallis Brian, *Land and environmental art*, London – Barcelona – Milan – Tokyo – New York, Phaidon, 1998 pages 204.

Kroll Lucien, *Buildings and Projects*, Rizzoli, 1988.

Lévi-Strauss Claude, *Antropologia strutturale* (Structural Anthropology), Il Saggiatore – Collana La cultura, 2015, pages 392.

Lévi-Strauss Claude, *Il pensiero selvaggio* (The wild thought), Il Saggiatore – Collana La cultura, 1964, pages 330.

Magnosi Francesco, *Il diritto al paesaggio – Tutela, valorizzazione, vincolo, autorizzazione* (The right of the landscape – Protection, development, restriction, authorization), Exeo Edizioni, 2011, pages 352.

Manella Gabriele, Mitterer Wittfrida, *Costruire sostenibilità: crisi ambientale e bioarchitettura* (Building Sustainability: environmental crisis and bio-architecture), Milano, Franco Angeli – Collana Sociologia Urbana e Rurale, 2013, pages 146.

Marrone Gianfranco, *Corpi Sociali. Processi comunicativi e semiotica del testo* (Social bodies. Communicative processes and text semiotics), Piccola Biblioteca Einaudi, 2001, pages 400.

Marshall McLuhan Herbert, *La Galassia Gutenberg – Nascita dell'uomo tipografico* (The Gutenberg Galaxy – Birth of Typographical Man), Armando, Roma, 1976.

Mastrilli Pamela, Nicosia Roberta, Santiello Massimo, *Photovoice – Dallo scatto fotografico all'azione sociale* (Photovoice – from photo shoots to social action), Franco Angeli, 2016, pages 144.

McLuhan Marshall, *Gli strumenti del comunicare* (The Communication tools), Il Saggiatore, Garzanti, 2015.

Nicholson-Cole Sophie, O'Neill Saffron, *"Fear Won't Do It" – Promoting Positive Engagement with Climate Change through Visual and Iconic Representations*, Tyndall Centre for Climate Change Research, University of East Anglia, Norwich, UK, Science Communication, Volume 30 Number 3 - March 2009, pages 355-379.

Nicholson-Cole Sophie, *Representing climate change futures: a critique on the use of images for visual communication*, School of Environmental Sciences, University of East Anglia, Norwick NR4 7TJ, UK, Elsevier, 2004, pages 255-273.

Norton T., Salvador M., *The Flood Myth in the Age of Global Climate Change*, Environmental Communication: a journal of Nature and Culture, 2011, pages 45-61.

Pedelty Mark, *Ecomusicology: Rock, Folk and the Environment*, Temple University Press, U.S. 2012, pages 229.

Pinotti A., Somaini A., *Teorie dell'immagine. Il dibattito contemporaneo* (Theories on the image. The contemporary debate), Cortina Raffaello, 2009, pages 280.

Polidoro Pietro, *Che cos'è la semiotica visiva?* (What is the visual semiotics?), Carocci Editore – Bussole, (ristampa) 2016, pages 126.

Wang Caroline, *Using Photovoice as a Participatory Assessment and Issue Selection Tool* – A Case Study with the Homeless in Ann Arbor, Retrieved Maggio 2017, pages 179-196.

Wang C., Burris, M. A. *Photovoice: Concept, methodology, and use for participatory needs assessment. Health Education and Behavior*, 24(3), 1997, pages 369-387.

Wheatley H., *Beautiful Images in Spectacular Clarity: Spectacular Television, Landscape Programming and the Question of (Tele)visual Pleasure*, Screen 52(2), pages 233-248.

Wiseman Carter, *Writing Architecture – A practical guide to Clear Communication about the Build Environment*, Trinity University Press, 2014, pages 240.

V.A., *Paesaggio Urbano* (Urban Design), Maggioli Editore, Rivista bimestrale, Anno XXV - July No. 4, August 2016, pages 128.

V.A., *The grove Dictionary of American Music* (Book Series), New Grove Dictionary of American Music, Oxford University Press, 2014.

Web Site List

Ballocchi Andrea, *Comunicare l'edilizia green, un'idea che piace* (Communicating the green housing, a brilliant idea), TEKNECO, www.tekneco.it

Berchmans M. Britto, *Focus group*, included in Franco Lever - Pier Cesare Rivoltella - Adriano Zanacchi, *La comunicazione. Dizionario di scienze e tecniche* (The Communication. Dictionary of sciences and techniques), www.lacomunicazione.it

Casa Clima – Klima Haus www.agenziacasaclima.it/it

Christopher Reed, *Animals speak color – a new exhibition reveals how they acquire the language and use it*, www.harvardmagazine.com

Douin Jean-Luc, *"Océans": deux ans de préparation et quatre ans de tournage* ("Oceans": two years of preparation and four years of filming), Le Monde, 28th February 2011, www.lemonde.fr

Giliberto Jacopo, *Clima, Di Caprio e il "Punto di non ritorno" per salvare il pianeta* (The Climate, Di Caprio and the "Point of no Return" to save the Planet), Il Sole 24 Ore, 25 th October 2016, www.ilsole24ore.com

GhostNets Australia: www.ghostnets.com.au

International Highrise Award http://www.international-highrise-award.com/en/

Istituto Nazionale di Bio-Architettura (Italian National Institute of Bio-Architecture): http://www.bioarchitettura.it

Giuliano Mauri – official website: www.giulianomauri.com

Patrick Blanc Vertical Garden: www.verticalgardenpatrickblanc.com

Qaujimajatuqangit Health Research Centre: http://www.qhrc.ca

V.A., Sulla riva dell'Adda la "Cattedrale Vegetale" di Giuliano Mauri per Lodi (On the river of Adda, the "Tree Cathedral" by Giuliano Mauri for the city of Lodi), Culture, ADN Kronos, www.adnkronos.com

Communicating the Environment in the *Green and Circular Economy*

5

Abstract

Can products and services truly communicate the Environmental message? Is the traditional economic model able to encompass both concepts of Environment and Sustainable Development, enhancing natural resources? These are some of the key-questions at the base of the *Green Economy* and the most recent *Circular Economy,* the "green revolution" of the economy as a result of strong socio-cultural change that pushed the corporates to implement eco-sustainable actions as well as to find new communication tools to communicate their *Environmental commitment* in a very efficient, eco-responsible, transparent and shared way. In the modern economy, still linked to the business-oriented approach, the *Environmental compartment* is an integral part of the economic system on the tail of the *Triple Bottom Line* model based on the trinomial: Population, Planet and Profit. *But is it sufficient to prevent the risk of "sprinkled with green" products and services that really have nothing to do with an eco-sustainable management aka (as known as) "green washing"? The process and product certifications are they able to guarantee the communication of Environmental quality to the consumer-user? Can the marketing tools be efficient Media to communicate Environmental and sustainable values? And if so, which forms and within which limits? The promotion conventional and non-conventional tools (guerrilla marketing) have raised the "green" flag, but are they really effective?* These are some of the questions we will try to give an answer in the *"Chapter 5—Communicating the Environment in the Green and Circular Economy".*

© Springer Nature Switzerland AG 2019
M. Abbati, *Communicating the Environment to Save the Planet*,
https://doi.org/10.1007/978-3-319-76017-9_5

"A whole new way of thinking is needed in order to solve the problems we have created with the old way of thinking"
Albert Einstein
German Physicist and Philospher, naturalized Swiss and American
(Ulma, 14th March 1879 – Princeton, 18th April 1955)

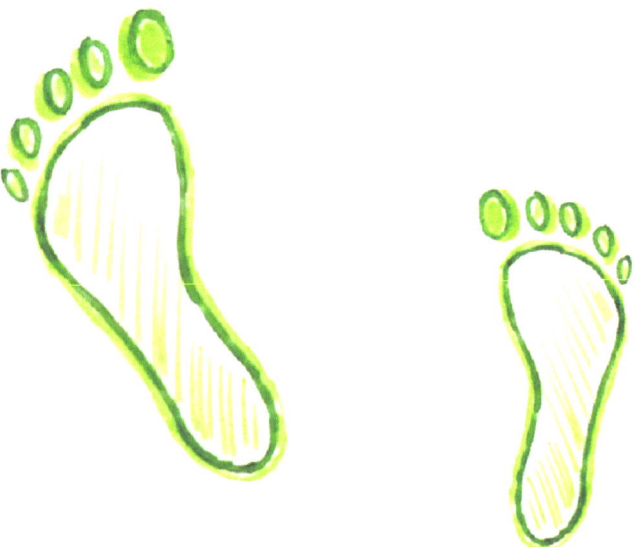

"Green Footprint", sketches by Maurizio Abbati (Copyright Holder) © 2017

Reading Proposal: In Between Words and Images

Is it possible to "measure" the availability of the Environment at our disposal? This is the key question at the base of the information and understanding process on environmental issues. It is essential to have a vision of the real risks we are going to face in order to communicate correctly and achieve a consensus among the public opinion toward the necessary actions to contrast them. After a long course of studies and researches, more than 20 years ago the **Ecological Footprint** was set up. It is a tool of environmental accountancy, scientifically proved, able to calculate the demand and offer in terms of environmental services based on data and indicators. These tools, if reinforced, can have a strong communicative value as we are going to see in *"Chapter 5—Communicating the Environment in the 'Green' and Circular Economy"*. It is not by chance if the "footprint" becomes a human footprint, visual metaphor of any action originated from Man that affects the ecosystem around represented by the "green" colour, at its turn, metonymy of the Environment System with all its nuances. It is also the symbol of a methodology that let us take a social and economic value to the main ecological assets. Goods with economical value necessary to a specific territory-community to produce natural resources and to neutralize the produced waste, especially the carbon dioxide (*Carbon Footprint*). Food, natural fibres, seafood product, wood-and-wood derived products, urban green areas, these are some of the environmental assets considered on the demand side. On the offer side other elements are considered such as: the bio-capacity of a territory, its productive eco-capacity in terms of agricultural areas, grazing lands, forests, fishing areas and built-up areas. The *"green footprint"* is increasingly associated to a numeric number linked to the territory, stated in global hectares (standardized unit) that shows the health status of the ecosystem and the real risk to overcome the ecological deficit. A datum that encourages the exchange of ideas, the awareness and the sharing among all the parties interested in building a sustainable management plan whose future is directly linked to the community choices and their political and institutional representatives. The **Ecological Footprint** thus has become a tool of information at our service, largely used in the scientific, business, governmental and institutional fields as stated by *Mathis Wackernagel* and *William Rees*, researchers at the University of British Columbia (USA), who created it in 1990.

5.1   *Green* and *Circular Economy* and the Environmental Message

The economic "green revolution" communicates values in order to protect the eco-system.

The *Environmental products* and services are functional elements in respect of the *Environment*, both as resources or when they are transformed in waste, at the end of their life. This is one of the key-concept of the *"Blueprint for a Green Economy"* report, issued for the British Government in 1989 by some of the most skilled *Environmental economists* of the time: *David William Pearce, Anil Markandya* and *Edward Barbier*. In this "pioneer" document, successively updated in 1991 and in 1994, the neologism "***Green Economy***" was coined for the first time. An economic model, strictly linked to the concept of *Sustainable Development* that promotes *the improvement of "human well-being and the social equity while significantly reducing Environmental risks and the ecological scarcities"*, as defined by the *Environmental Program* of the United Nations (UNEP): *Towards a Green Economy—Pathways to Sustainable Development and Poverty Eradication*, in 2011.

We should not forget, in fact, that the Sustainable Development implies the contextual examination of various problems linked to the social inequality, of any kind, and to poverty. They are the main obstacles to the growth of a society and they prevent from guaranteeing the ecosystem protection in which the community lives. This has been strongly stated, at international level, by fixing the **Sustainable Development Goals** by the *United Nations* and the principles by the **Paris Agreement on Climate Change** (COP 21, November 2015). Two key documents that must be considered as a single framework for the humankind and the Planet. *"Education holds the keys to sustainability"* as **Irina Bokova**, Executive Manager of **UNESCO** (**U**nited **N**ations for **E**ducation, **S**cience and **C**ulture **O**rganization) reminded us during the Climate Conference held in Marrakech (Morocco) in November 2016 (COP22).

Education is an important sector increasingly connected to communication so that some scholars from Latin America have coined the neologism ***edu-communication*** [*International Congress "Communication and Education"* (San Paolo, 1998) and the *International Seminar "Communication and Education"* (Bogotá, 1999)]. This neologism perfectly reflects, in our opinion, all the criteria requested today to those who communicate the values of *Environment* and Sustainability. Therefore, the *edu-communicator* of the twenty-first century must be able to convey the *Environmental message*, supplying with more knowledge on the matter those who receive it (***educational function***) and, in the same time, be able to teach them *Media* skills. A real tutor of the *Environmental message* interpretation and eco-educator" (*Geneviève Jacquinot*, a researcher and professor at the Sorbonne University 8 in Paris).

A fundamental approach to the *"Green Economy"*, developing through an action plan concerning economic, legislative, technological and educational issues and aimed at creating investment opportunities, growth and employment by introducing "green" sectors of production of goods and services. All of them responsibly "focused" to achieve eco-sustainable goals without affecting levels of economic and social welfare. Reduction of energy consumption and natural resources; cutting *Climate-Change* emissions; containing the waste production; promotion of sustainable consumption patterns and sustainable *management*. Just to make some examples. The involvement of the governments, the representatives of the economic world and the main legal-social groups, therefore, seem to be inevitable.

The "*green*" revolution of the economic thought has its origin from the radical change of the point of view. ***Can we give an "economic value" to the Environment and Natural Resources?*** Of course, yes we can. The evolution of the former economic theories of the eighteenth century (classic theory: *Smith, Ricardo, Malthus*) pointed out as the natural resources are available only within a specific *consumption capacity* beyond which Man cannot go without provoking an imbalance in their capability to regenerate, both renewable and non-renewable resources.

The main issues linked to the *Environment* such as the atmospheric pollution (smog) or the waste as a consequence of human acts, though not economically quantified can produce direct or indirect damages to people, things and business. Therefore, the compelling necessity to consider them and calculate them with economic theories. Hence, the concept of **negative externality** that is the cost of the *Environmental damage* caused by Man to the Ecosystem.

An economic "*green*" system anything but static. Still in evolution but "linear" and with meaningful *Environmental impact*: function-based as follows: *raw materials > planning > production > distribution > consumption > waste*. But something has been changing in the last years. The European Commission, in fact, has launched a new *circular economic approach* called **Circular Economy** beginning just from an act involving actions of communication. A publication whose title is: * ***Closing the Loop—an EU Action Plan for the Circular Economy Communication*** (acknowledged on the 2nd December 2015).

The main objective of this document is to convey a new environmental message able to influence the lifestyle of European citizens. The need to make people

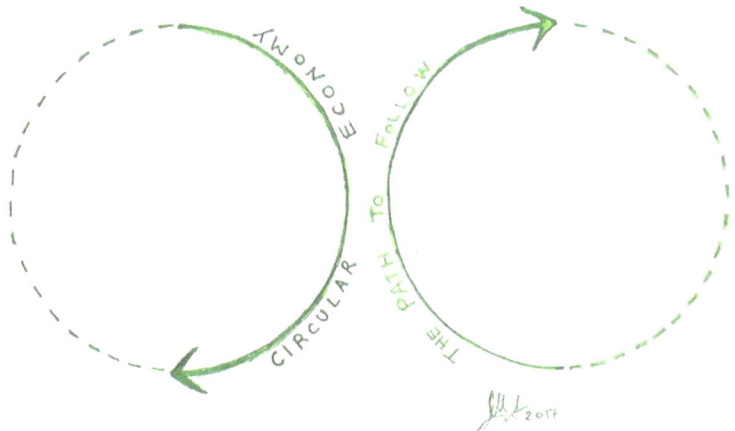

Fig. 5.1 Circular Economy: the Never Ending Cycle of Life—Maurizio Abbati drawing (Author and Copyright Holder) © 2017

understand the value of goods also in their end life to extend as far as possible their function of utility through the rescue, recycling and reuse as if they were raw materials lowering drastically the quantity of waste. *"In Nature the waste does not exist. The only species on the eco-system Earth to produce it is Man"*. This is one of the key-messages by the video realized by the official communication campaign to support the ***European   Action   Plan*** (www.ec.europa.eu/environment/circular-economy).

Analysing it from the mere communicative point of view we can link the visual narration of the message to a stream of k-words as follows:

📶*Circular   Economy*: ⊙Resources ⊙ Eco-design ⊙ Prevention to Waste Production ⊙ Reuse and Recycle of Products ⊙ Resilient Society ⊙Resource Efficiency ⊙ Job Opportunity ⊙ Business Profitability ⊙ Reduction of CO_2 Emissions ⊙ European Union Future.

The ***visual language*** chosen by the European Commission is particularly incisive and underlines that the waste drew on the sketches can be easily recognized by the audience as a resource. Hence, a bitten apple, a fishbone, a piece of pizza, a *laptop* to be scrapped, an electric bulb, a bottle, redesigned in a comic strip shaping together the symbol of Euro, metaphor of the economic opportunities which may derive from this new economic approach on the territory of the European Union. Its identity is then reinforced by the *visual simile* of the waste/resources that, by forming a circle, transform themselves into the stars of the United Europe flag.

The choice of the symbols is very impactful, as they went public, have become well-known and understandable as the one which identifies the recycled or recyclable material and the vegetable metaphors: the green leaf and the apple-shaped World to identify the eco-design and the prevention to waste production as an instrument to save the Earth eco-system.

In this social-economic change process the sectors of production and consumption are directly involved. From natural resources, for example the forests, in fact, raw materials come out as: paper, pellet, cellulose, and so on, used largely by the productive system to create goods and services. The firms have therefore the need to communicate efficaciously their eco-sustainable actions to the public. Their communication has two focuses to fulfil: (1) improve the reputation to fight the competitors; (2) to convey their own *Environmental and Sustainable commitment* and good practices in the various phases of the goods production and services to the consumers and stakeholders such as: suppliers, transporters, distributors, outsourcers and so on. As already mentioned, in the models of *Green* and *Circular Economy* each single actor "works the system" with the others to increase the awareness about the *Environment* and the *Sustainable Development*, strictly linked to the success of any single plan or communication campaign.

Green Tweets
#EnvironmentalProducts #EnvironmentalServices #Environmental Economists #GreenEconomy #Wellbeing #SocialEquality #Environmental Dangers #Scarcity #SustainableDevelopment #Ecosystem #Environmental Message #Educationalfunction #Ecoeducator #GreenEconomy #CircularEconomy #SustainableConsumptions #SustainableManagement #GreenRevolution #EconomicValue #AirPollution #NegativeExternalities #EnvironmentalDamage #RawMaterials #LifeStyle #Nature #Keywords #VisualLanguage #Comics #Graphics #Metaphor #Environment #CommunicationPlan #AwarenessCampaign

5.2 Communicating the Environment Through Products and Services

Giving a value to the environmental commitment through a coherent and honest *eco*-message.

"The Environment and the Economy are really both sides of the same coin. You cannot sustain the Economy if you don't take care of the Environment. If we cannot sustain the Environment, we cannot sustain ourselves", stated the Noble Prize *Wangari Muta Maathai*, the Kenyan environmentalist and biologist who entered the history for being the first African lady to get the Nobel Prize for Peace (2004). In this statement, we can identify two fundamental items: in Economy and in the *Environment*, we need to be honest and coherent.

The **Environmental Sustainability** is a process in continuous evolution integrating to the social and economic aspects by involving each of us, citizens and consumers. From our choices of consumption, it may derive a more sustainable production chain. Provided that we are correctly informed from the corporate body with a clear, transparent and dialogue-open communication despite its *business oriented* nature, oriented to the Company's results. All that to avoid the threat of **green washing**, that is misleading advertising. The engagement of the consumers to prefer eco sustainable compatible products and services and packages of sustainable tourism is increasing largely. The final goal is to improve the quality of life in respect to the *eco-system*. A concept not to be mistaken with *green consumerism*, the uncontrolled purchase of everything that recalls *Environment* without thinking whether it is ecological or not.

We must always remember that the present economic model is not completely balanced from the *eco*-sustainable point of view because the *"business oriented"* attitude tends to prevail over any other aspect, the "green" ones of *Environment* and

Sustainability included. Solving *Environmental issues* may affect the firms with heavy costs even though they will be counterbalanced by the creation of new business opportunities and new job places as a consequence of a hint towards a **World Green Circular Economy**.

In any case in the corporate sector we tend to think about **Corporate Sustainability**. And this is a great success anyway. It is in fact a turning point in the cultural and economic thought as stated by the **United Nations Conference on Sustainable Development** (Rio +20), held in *Rio de Janeiro* in June 2012. In order to ensure the achievements of green economies we need to put in place measures and incentives at governmental level able to operate not only in the short and long period but, above all, in the medium and long one. A new way of thinking the scenarios where a **qualified Environmental Communication**, that is based on data and indicators internationally understandable and traceable, can play an essential role to inform on the state of progress of the Sustainable Development and on the effective *Environmental policies*.

We are facing now new models including the *Environmental*, social and economic dimension as it was introduced in the Nineties with the **Triple Bottom Line** (TBL) or **People, Planet, Profit** (3P), having the target or carrying on *business*, considering more the annual turnover than the income. The "green" branch is not an independent sphere but it works closely with the other two spheres: economic and social. The new tendency is what we call *corporate sustainability* or *sustainability business* which means simultaneous management of the economic, *Environmental* and social *trade off* (balanced or unbalanced situation) thus developing more sustainable models of business.

A new approach which reinforces the traditional well-known "*magic triangle*" to achieve a successful company management: quality, integrity and honesty. Keeping a high quality of "*eco*-sustainable" products and services against reasonable prices (even highest than average) will become winners if the consumers/clients are satisfied with their performances and will fully understand the added value represented by the "*Environmental issue*". A cause that must be pursued in any stage of the chain production also on condition of giving up offers only apparently advantageous but contrasting with their own values (*Shel Horowitz*).

The American firm *Stonyfield Farm* (www.stonyfield.com) represents an **example of sustainable chain production** based on the above mentioned values. The strong commitment of *Gary Hirshberg*, founder and CEO of this business reality, one of the leaders in the *bio*-yogurt sector in which all the firms of the "chain production" are eco-sustainable. All this without the support of any advertising and marketing campaign. On the contrary, the same very products are the *medium* of *Environmental Communication* through a winning idea that made the turnover increase of 26%.

The use of "communicative" yogurt pot lids in the *packaging* of ordinary yogurt pots, is able to convey messages oriented to the consumers' awareness on themes of global warming, clean energies and sustainable management of the territory.

Last but not least, *the "firm policy", shared by all the stakeholders*, is being communicated in a very clear and transparent way. An original interactive map where the *Internet* users can be informed on the sustainable choices by the suppliers, working both in the United States and in other geographical areas (e.g.: Canada, South America, Europe and Madagascar). They all use organic raw materials, improving local businesses.

The consumers' growing ecological awareness, in fact, tends to make them prefer the products with less *Environmental impacts* both for transportation and use of pollutants but contributing to give more job opportunities and *business* in the territory they live. A socio-environmental view that can make "green" also the "global" is not necessarily contrary to the values of *Environment* and *Sustainability*.

The well-known *globalization*, term first coined in the Nineties to point out a varied and large set of *phenomena* linked to the growth of the economic, social and cultural integration among the various areas in the world, can be communicated in a way as to create positive situations of *"green" bench marking*. Namely all the *Environmental* and sustainable practices of a specific chain production or a whole productive sector or market that can influence other companies, productive sectors or markets and make them more *eco*-sustainable. That is what happened on the American market where more and more *chains of production* have been introducing regulations to protect the consumers' safety and the *Environment*. A *spill over* which originated from the European Union regulations on the matter, in some cases stricter than the ones of the United States.

The good practices spreading out to protect the territory and the ethical and sustainable working conditions is encouraged by the *global market*. The *marketing*, focusing on communication campaigns to show the above stated principles in the *management*, has become increasingly successful on condition that the "*fair and equitable*" products are certified that is they can prove their sustainable management of the chain production, monitored by a third party verifiers.

The ever-increasing attention of the public opinion for the local and global "*green*" allows new forms of eco-communication for commercial goals able to make the world values of the *Environment Protection* a *medium* able to increase the credibility of trade marks, promote trust and reliability among the consumers/users and increase the turnover of sales of eco-sustainable products and services.

Green Tweets

#Environment   #Economy   #Sustainability   #SustainableSustainable #Corporate #Communication #Impresa #BusinessOriented #GreenWashing #GreenConsumerism #CorporateSustainability #TripleBottomLine #People #Planet   #Profit   #   #CorporateSustainability   #SustainabilityBusiness #MagicTriangle   #ProductionChain   #Globalization   #Marketing #EnvironmentalCartification

5.3   *Pure Green* or *Green Washing*? Making the Message Credible: This Is the Question

"Marketing is part of evolution. It helps the world go round and is part of mass communications, so it greases the skids of progress inside and outside the mind" (Jay Conrad Levinson).

The business strategy, based on the values of ethics and cooperation for the companies has proved to be successful. As a matter of fact, they are busy to secure the customers' loyalty as much as possible or to create durable and lasting *partnerships* and *joint ventures*.

In this way, the **Language of Communication** is a precious instrument to amplify the sustainable "soul" of the firms. But to be credible it **must inspire confidence**. The *Environmental commitment*, in other words, must be conveyed continuously and must emphasize the ecological performances of products and services in a very realistic and certified way. In this regard:

– the **voluntary certifications of product** such as *Eco-label, Energy Star, Forest Stewardship Council* FSC, *Der Blaue Engel*, biological agriculture;
– the **process certifications** such as ISO 14001, *Eco Management and Audit Scheme*—EMAS.

They can help the credibility of an entrepreneurial organization. **Further instruments of communication** are able to analyse the *Environmental impacts* along the **Life Cycle** of goods and services such as: *Life Cycle Assessment*—LCA; *Carbon Footprint*; *Water Footprint*, and so on and in the procedure of purchase of "green" products by the firms and public administrations (*Green Purchases, Green Public Procurement* GPP). Finally, strategic instruments are able to report back the effects on the *Environment* by the public policies as the **Environmental Accounting** or the **Balanced Scorecard**.

All these potential tools of *Environmental Communication* are not enough to guarantee the "green" performances, though they are aimed at satisfying the strategy of **Integrated Product Policy** (IPP), adopted by the **European Commission** in 2001 and since then it has been the main eco-instrument of intervention both in the private and public sector. Nevertheless, we must protect the consumers from the *green washing risk*, when we are in the presence of deliberate exaggeration on the *Environmental performances* which are not really "green".

To avoid this kind of risk we need an adequate **Plan of Communication** considering any detail. The use of words and verbal tenses is of vital importance. We must be very careful because the credibility of the company and its trade mark are at stake. Furthermore, target of the consumers' trust is very difficult to achieve and it takes a long time and cannot be missed in a blink.

It is then fundamental to evaluate *a priori* the **Communication Plan efficacy**. The "evaluation" in fact is an activity that allows us to understand the consequences of a series of planned activities. Of course, we must take into account that the activity of communication is complex and to evaluate it means making more than one judgment according to the different points of view of the actors involved and interacting in the communication process. The comparison, the exchange of ideas, and the critical reflections on the efficacy of the adopted measures triggers a virtuous mechanism that improves the communication itself. Going through an experience helps the awareness of our own actions and makes us learn from our mistakes. This is the basic attitude which a Communication Plan should be based on.

Let us imagine we run a *chocolate factory* where we have introduced ethic-environmental values so that the 45% of raw cocoa is made by eco-sustainable companies as far as the working conditions are concerned with equitable wages, a decent lifestyle, education services for the new generations and the total ban of any form of exploitation, childhood included. A result achieved by the direct control of the firm on the suppliers who are asked to follow a specific *Memorandum of Understanding*.

For instance, if we state that:

- Our cocoa is fair and equitable.
- The choice of our cocoa is eco-sustainable.
- Our eco-sustainable cocoa is a synonym of environmentally friendly.
- We certify the organic status of our products, internally.

Somebody could complain because the fair and equitable factories are less than 50% of our suppliers. The eco-sustainable nature of the *cocoa tablets* then will be called into question if adequate and scientific data, eco trademarks or certifications are not provided and this is also for the third statement. Therefore, to use the term "equitable and sustainable", without being able to prove their consistency, is not enough. A proof of the respect of the workers' rights and their equitable wages at any step of the chain production must be shown. The descriptions of the best practices of the sustainable cultivation of the *cocoa plants* enhance the conversion to organic local farming meant to improve the *Environmental quality* of the territory.

In the light of the above, a more credible *Environmental message* could be as follows:

- The 45% of our suppliers can guarantee equitable and sustainable working conditions.
- We are committed to make our products be ecological and sustainable within 2020. Your support is precious to us.
- The choice of our products can give a real contribution to achieve the targets of sustainability and socio-environmental protection in the *cocoa production*.
- The organic status of our products is certified by a qualified independent auditing body.

The new statements could be followed by *documented sources* such as: data, tables, info-graphics, *Environmental planned pilots*, or *Environmental declarations* in order to increase the consumers' awareness on *Environmental issues* pointing out, in the same time, the *Environmental performances* of the products and services to be promoted.

In the "chaotic" world of *Environmental Communication*, amplified by the Net, any tiny detail makes the difference. The eco-messages must be more and more precise and emphasize "surprisingly" the *Environmental advantages* (*Levinson, Horowitz*). In the *green marketing*, the principle by which "who makes the first move wins" is worth, on condition to do it responsibly both during the wholesale and trademark promotion or revival after complaint.

The professional ethic and ecological values help successfully as a consequence of the consumers' willingness to change their habits and prefer that brand or that shop showing a serious commitment in the *Environmental*, sustainable and social field: **the *cause-related marketing***. *Who neglects the global environmental issues as the overheating, the atmosphere pollution, Climate Change, alternative energy sources, but also food safety, protection of workers' rights and other issues?*

Therefore, a factory involved in a serious *Environmental Communication* **improves** also its **image** among its customers, suppliers and employees who are much more motivated. The *management* costs are more "sustainable" because the firm works according to *standard* of ethic-environmental quality and it will be not charged by legal disputes or administrative penalties. Furthermore, the form will be able to consolidate a long lasting relationship with its customers, caring the eco-system. A *driving force to expand the clients'* portfolio and the possibilities of *partnership*.

The quality of contents based on real information and data adapted to the cultural and linguistic background of the addressees is more efficacious than marketing over telephone or user base. An *Environmental effort* with economic advantages as a "natural" consequence and not *vice versa*. In fact, *Merril Lynch* reminds us that *the corporate social responsibilities (CSR) reduce the risks of investments*.

- **Truthfulness of the message:** to maintain the honesty of the statements is a *win-win attitude*. This means to describe realistically the performances of the product. **Examples:**
  - *Congratulations! By using our product or service you contributed to save the ...% of CO_2 [data processed by the Research Institute XY]*
  - *The new eco-friendly packaging allows the reduction of ...% waste [Statistics data processed by the University of XX]*
  - *The energy used by our firm is generated by 100% from renewable sources [Certification released by the Issuing Body XY dated....]*
- **Clear and feasible objectives:** pointing out through the character, the size, the highlighting, or the chromatic effect the targets/objectives that can be achieved by using that product or service. **Examples:**
  - *Our firm commitment contributes to safeguard our Planet and make it happier.*

- *Water resources: we want to improve of ...% the efficiency of the freshwater use and reintegrate the 100% within January 2020.*
- *Energy and climate: we want to reduce of ...% the emissions of CO_2 in all the phases of the production chain within March 2019.*

- **Tangible solution proposal:**
  Examples:
  - *Recycle with three simple steps! You will be able to recover more than 2.8 billion of drink cartons equivalent to a distance of 16 times around the World. You will be able to save more than 1.2 million trees equal to the surface of 4200 football fields. [Lucart Campaign 2013–2016 with TetraPak].*

- **Positive thought**
  Examples:
  - *Recycling rewards you! Put the bottles and cans of our production in the approved biological waste containers smart. Give a gift to the Environment and yourself. Up for grabs discount coupons, complimentary tickets for cinema and theatre or on a fantastic cruise to discover the marine biodiversity.*

- **Accomplished results** or under achievement which follows an improvement in the *Environment quality* and consequently it contributes to improve health and wellness of citizens/communities living in a specific territory.
  Examples:
  - *From 2014 to 2016 we saved tons XXX of CO_2 thanks to the conversion of our traditional transport vehicles into hybrid/electric.*

- **Maintaining the clients' contacts:** dialogue/advising/support/qualified chat blog/customers support [we assume that nowadays the customer is sceptic because bombed by unreliable advertising messages some of them "green"] Therefore to maintain a contact as much "human" as possible that is emotional, popular, able to influence the customers by suggesting and not by imposing good practices or virtuous behaviours.

- **Being able to communicate values** and not only mere economic advantages.

- **Allowing to experiment the *Environmental performances*** of the product at no cost/a service on a *trial basis* in order to create a positive image of the company forward looking to new perspective customers.

In any case, the efficacy of a corporate communication or any communication must be considered along the way. Only by analysing each single communication action through adequate instrument of dialogue with the costumer/user we will be able to identify wrong investments and correct them in progress. This means to assess, case by case, the subjects, the context, the object, the conditions, possible limits, human and financial means and resources available for our "communicating" the *Environment*.

Some authors like *Bezzi* remind us that the communication evaluation is not easy for some reasons. First of all, it is an intangible concept, immaterial and not always "customizable" *a priori*. We refer very often to a ***Communication Agency*** more than an individual with name and surname. Communicating the *Environment* for both parties (producer or service provider and consumer/user) needs a certain kind of flexibility in respect to the objectives set, based on what the receiver expects from the communicator and *vice versa*.

In other words, to evaluate the efficacy, a *good negotiation* is always needed. It is useless, for instance, to promote the energetic saving by *wind turbines* in a poorly ventilated region or whose utility is severely limited by *Environmental* and aesthetical restrictions. Technology moves quickly and new plants are currently under way to produce no wind energy or by exploiting the advantages of an aerodynamic structure able to produce vibrant small vortices, the *vorticity*, instead of traditional three blades, with highly visual impact. This is the main target of *Vortex Bladeless*, a Spanish start-up.

Hence, the importance of conveying right messages at the right moment. Without imposing, of course, an intrusive communication meant to "inculcate" in the consumer's/user's mind the *Environmental product* or *service* performance when the latter has no interest in it: the so-called **marketing pull** or attraction. On the contrary it proves successful to encourage the consumer to get the best and useful information for his real needs through a personal and focused research with search engines such as: Google, websites, directories, advertisements, publications, reviews, word of mouth, *focus group* or discussion groups that is the **marketing push**. The "*green marketing*" authors and scholars pointed out that the marketing push is the only one that can influence the market toward ecological products and services since it is the result of the consumers' personal experience communicated by them and for them (*Roy H. Williams*).

"*Our coffee machine will be able to reduce 30% of the monthly consumptions of electric energy thanks to a rechargeable battery 100% recyclable! You must absolutely try it!*". This commercial message if repeated many time as a TV spot, radio advert, or in paper form, will make the consumer/user feel annoyed because too "aggressive", not balanced to the real interest of the same. Think, for instance, that we are not coffee consumers as a consequence of a personal choice or for health reasons. We will never buy any coffee machine, even ecological.

On the contrary, if we publish, on a specialized coffee-oriented magazine, an article introducing the ecological innovations of our coffee machine, based on authoritative sources: scientific researches, qualified surveys, real verifiable cases, we will be able to promote our ecological product to the right public, lover of the Italian authentic *espresso* or an expert of the various *blend*. A user very interested in getting the best results also at home. *And what is the best way to do it without a high-quality tool able to offer a product at a low Environmental impact and consumption?* It is a good beginning to catch the attention of those who are really interested in the "coffee ecosystem". The possibilities of choice of our products will increase.

Green Tweets

#Environment #Economy #Sustainability #EnvironmentalSustainability #Communication #CorporateCommunication #BusinessOriented #GreenWashing #GreenConsumerism #CorporateSustainability #TripleBottomLine #People #Planet #Profit #Sustainability&Business #MagicTrangle #ProductionChain #Globalisation #Marketing #EnvironmentalCertification #Product #Process

5.4   When Does the Advertising Message Inform on the Environment and Sustainability?

Increasing the customers' *Environmental awareness* means to follow their real needs.

The relationship between corporates and clients is more and more concentrated on sharing common values which are mostly identified, nowadays, with those *Environmental* and *Sustainable*. On the other hand, the same concept of **Sustainability**, new entry of the global dictionary, encompasses simultaneously the principles of *social, juridical and economic ethics* that cannot be ignored even being promoting eco-sustainable goods and services.

The latter, even at higher prices, may offer advantages from the productive and consumption point of view. Thanks to the new *Media* on the Net, the consumers' awareness on *Environmental issues* is at a global scale. It derives then a more easily marketing of everything meant to improve the *Environment conditions* and quality. Products and services called "green" make reduced impacts as far as pollution, generation of waste and gas emissions are concerned. If not, they will increase the problem of global warming and in this respect we talk about "*carbon footprint*".

Reduced consumptions, energetic efficiency, reuse and recycle, all *Environmental values* to be communicated to consumers. *But how can we do all that efficiently?* The corporate bodies in fact have always the necessity to introduce, in a credible manner, their "eco" commitment in their activities of *marketing* in order to carve out considerable portions of the market and maintain them over time. This begs the question. *How much space does the ethic-Environmental aspect take in comparison to the mere business oriented? In other words, how much "green" is the communication with marketing goals?*

Undoubtedly, the increasing interest on *Environmental issues* by the public opinion drives the companies on the following "bio" concepts: ecologic, natural, biodegradable, recyclable, *eco-friendly*, for which *jingle* and *slogan* are created. The inflation of these terms is likely to produce a negative effect if the consumers are not provided with guaranties of the information quality and its sources. It is not a surprise, then, if recently cases of well-known trademarks have made headlines because they had communicated a misleading positive image about the *Environmental impacts* of their products and services. A real phenomenon which everybody can be exposed to, identified by the 1990 English neologism: **greenwashing**, considering that this colour is the metaphor of *Environment* and *Nature*.

Effectively, as we reaffirmed many times in this essay, so far there is no instrument to guarantee 100% the consumer or user about the eco-compatible nature claimed in the corporate market. Nevertheless, there are some "green" actions that the companies can put into practice and, if adequately communicated, can give reliability of the company's *Environment* and *Sustainability* commitments:

- **Let us do the detail work:** use energy from renewable resources, use low-consumption lighting installations; install the excess flow and limiting device, water saver, flow adapters and any other device able to guarantee a sustainable water management; prefer the use of materials producing minor transport impacts at local level; use unpacked products or derived from recycling or reuse to limit the waste; the production of digital documents, in the internal and external documentation, enables duplex printing to minimize the paper consumption; use low *Environmental impact* vehicles for product transportation.

- **Be straightforward and coherent:** the promotional message to be communicated must be the clearest and most detailed one in describing both the corporate commitment of social and *Environmental responsibility* and the eco characteristics of the product or service. Relying on scientific data, authoritative sources, (e.g. universities, or research institutes, etc.), eco trademark, process certifications (e.g.: *Environmental management systems: ISO standard, EMAS, etc.*) and product certifications (e.g.: *eco-label, energy label*, energy-star, sustainable product certification etc.) is therefore an important support to increase the credibility of their accountability. Besides, putting into practice the good practices inside the corporate body and the relationships with the stakeholders (e.g.: suppliers, consultants, distributors, transporters etc.) using recycled materials, recycled paper, low energy consumption machineries, inks with low pollution etc. An eco-management that cannot be betrayed in the *marketing* which must prove it is coherent both with its contents with direct invitations to implement *sustainable actions* and *its forms*, in the case of *paper marketing:* videos, images, reference for graphs, hypertext links to reinforce the eco-sustainable message in the *marketing online*.

- **Share the "green" commitment:** also for the *Environmental Communication* of commercial transaction the principle "*unity is number*" is still made. The more the eco-sustainable commitment is shared by the corporate body the more the commercial potentialities of your products and services increase.

- **Improving the image:** to promote a positive *Environmental* image of your products or services is equivalent to ensuring the effective public relations by analysing the language and the stylistic choice. Any *Environmental* marketing message must be expressed in a clear and unequivocal way. All this is transacted into a visual homogeneity both for the choice of characters, of the text and in compliance with the grammar rules and spelling and for the use of images, logos, trademarks or symbols easily recognizable by the consumers or due to *Environmental* characteristics to be underlined.

- **Transforming consumers into new communicators:** involving the user/consumer by listening to his opinion on the quality and ecological *performance* of the products/services is a communicative *medium* particularly efficacious to initiate a constructive debate. Give the consumers, for instance the possibility to choose among various *designs* or types of packaging or evaluate the satisfaction degree of a specific slogan or jingle by counting the "likes" on the factory website. It triggers a process of communication able to bring out emotions while using the products and services. Making the consumers/users satisfied by meeting their requirements means to make them become "friends" and thus generating a new virtuous circle of word of mouth (a real *network*) about the *Environmental performances* of the product/service (*John Kremer*).

The above mentioned elements might be reinforced, at communicative level, by pictures of pristine mountains and forests; crystal clear waters of rivers, lakes, seas; clear skies; Fauna and Flora symbols of biodiversity. Or even by graphic representations, animations and any other types of special effect linked to the natural world as a result of imagination of creative advertising. Or again by the use of some k-words such as: biological, non-toxic, GM free, biodegradable, natural, eco-sustainable, recycled, nature-friendly, ozone-friendly etcetera.

Another communicative *medium* is the direct reference to the **ecological trademarks** such as: energy star, *FSC—Forest Stewardship Council,* and any other *Environmental logo* certifying the product and the process which, being of voluntary character, need further check case by case on the basis of the reference standards e.g. ISO standards.

But the so called *green* marketing must not go through expensive advertising campaigns. There are **alternative innovative tools** which hits the mark increasing the turnover on account of real and traceable eco-sustainable characteristics. Studies on *packaging* are making great progress by enabling the very same products to communicate the environmental message directly off the shelves in the shops and supermarkets.

The majority of products nowadays are packaged so as to communicate eco-messages informing about the eco-sustainable production chain, the good practices adopted the product being packed or about its proper disposal at the end of its life or even how to recycle it.

Who has not encountered the problem where to put the waste in front of the separate collection containers? It is increasingly frequent to find, on the package, a "guide" with visual elements: graphic and chromatic reproduction of the components to help for the correct destination of paper, plastic or mixed waste recycling and disposal (Fig. 5.1). It may happen to find the factory *Environmental commitment* onto the info-graphic packages (Fig. 5.2) or to suggest the consumer of a product an eco-gesture like the one to turn off the tap while soaping hands (Fig. 5.3).

Sometimes, it is the object itself that becomes the means of communication and cornerstone of the corporate strategy of the product/service promotion. That's the case of the *eco-gadget* increasingly popular at Trade Fair or professional Meetings. An inexpensive opportunity not to be underestimated and that should be coherent to the corporate *Environmental* and *sustainable targets* trying to reinforce the eco-values the corporate body relies upon without being a mere self-celebration of the *brand*.

A *gadget* therefore that is able to experience an action or create an emotion will be the winner. An example is the biodegradable glass shown in Fig. 5.4 launched by the Italian factory *Novamont*, international leader of the "green" chemistry and founder of *Mater-bi*® (www.materbi.com): a bio-plastic family completely biodegradable thanks to its components of vegetable origin united with polyester.

The small glass in *Mater-Bi*® draws the attention for its shape (assimilated to a small coffee glass), for the chromatic choice and the original cover lever that makes us discover its content. The graphic, the "impacting" image of the vegetables (tomatoes) together to the message *"my wished garden"* with *"instruction for*

Fig. 5.2 Particular of food container "Separate collection helps Nature: outer package > plastic; corrugated cardboard > paper; labelled package > mixed waste"; photo

Fig. 5.3 Advert campaign: "Stop the water while using me" T.D.G. Vertriebs GmbH & Co. KG (Copyright Holder) www.stop-the-water-while-using-me.com/intl/

the use". Some seeds and a small disk of compressed turf make the public feel very surprised and, through their emotion and experimentation, they understand the eco-value of the patented product and its total capacity of biodegrade and

Fig. 5.4   Eco-gadget, a small coffee glass in *Mater-Bi*®, photo by Maurizio Abbati (Photographer and Copyright Holder) © 2017

changing itself into organic fertilizer once in the soil. And a final "prize" for the consumer: being able to grow bio-tomatoes on his own garden or home terrace or balcony.

The *branding* based on *Environmental* and *sustainable values* may also **enhance** a specific place from the **touristic and commercial point of view** contributing to improve its ecologic image. Thus triggering a positive *"boomerang"* effect, as it has happened for the City of *Copenhagen*, capital of Denmark. Relying upon a tradition which rooted at the beginning of the twentieth century, the well-known Danish city has been associated for a long-time to the *"two-wheel transportation"* so much to get the first prize as the *most bicycle friendly city*. A virtuous example of urban commitment to help the mobility at zero *"Climate-Change"* emissions.

But the capital of the Danish kingdom went further its own *"green"* image by launching an innovative product able to make the two-wheel transportation even easier to use while respecting the *Environment*. The ***Copenhagen wheel*** has become the new symbol of the city in the world with resulting touristic impact associated to the Danish city (Fig. 5.5). An invention deliberately created for Copenhagen by the *SENSEable City Lab* at the *Massachusetts Institute of Technology* and introduced in world *premiere* in 2009 at the *United Nations Conference* about the *Climate Change*, held this year in the same city. A metal red disk able to change any kind of bicycle into an electric hybrid with pedal assisted, linked to the *Net* thanks to a dedicated *App*. And all that in a few simple gestures.

Then its easy use and the advantage of not releasing hazardous emissions in the *Environment* is based on the focus of the publicity campaign, with a video as much fluent in the narration as in the mounting. A public garden, a green or a protected area are the ideal background to test this technologic application considering that *"the bicycles are one of the best means of transport"*. A message extremely "green" identifying a universal *Environmental value* applied to the

THE
COPENHAGEN
WHEEL

*"Transform your ordinary bicycle into a **hybrid E-BIKE** that also provides feedback on pollution, traffic congestion and road conditions in real-time!"*

Fig. 5.5 Advert campaign *The Copenhagen Green Wheel*—MIT SENSEable City Lab (Copyright Holder) for the Kobenhavns Kommune © 2009. @copenhagenwheel

commercial promotion. A Publicity reinforced by the interactive capacity of the device being able to dialogue with the user/cyclist. In fact, it can convey *Environmental information*, useful for improving the physical performances and steering according to the pedal rpm (*revolution per minute*) and the slope of the bike path. Data that can be exchanged among more users triggering potential processes of communication. An excellent example how to communicate *Environmental values* through an advertising campaign promoting a whole urban area, spreading its "*green*" all over the world.

To sum up, communicating, informing, educating and promoting a product or a service is the great challenge of the new "green" *marketing*.

Case Study in Short

Products communicate their ecologic footprint: In order to ensure the maximum transparency to calculate the *Environmental performances* of products and services, the European Commission published in 2013 the Recommendation No. 179/2013 referring to the use of methodologies to measure and communicate the *Environmental performances* during the *production Cycle* of Life. All this to the benefit of a greater awareness of the *Environmental Plan* efficacy, as well as of the actions useful to implement it and the best Media to provide the documentary evidence to the *stakeholder*s of the responsible commitment in favour of a Sustainable Business Management. Hence, the birth of *Organisation Environmental Footprint*—OEF and the Product *Environmental Footprint*—PEF, both based on the *Life Cycle Assessment* (LCA).

On this basis, the ***Progetto Life PREFER*** (October 2013–December 2016), coordinated by the *Scuola Superiore Sant'Anna* (High University

School of Saint Anne in Pisa) together to six Italian partners among which: *Regione Lombardia* and *ERVET* (*Emilia-Romagna Valorizzazione Economica del Territorio S.p.a*—Emilia Romagna Enhancing Territory, Ltd.), represents a good example of application of the PEF methodology, a *cluster* including eight productive areas of small and medium-sized companies, situated in five of the most significant Italian regions for the *Made in Italy*: *Campania, Emilia Romagna, Lombardia, Piemonte and Toscana.* A particular effort has been made on the communicative aspects, both in progress and in phase of business reporting of the results shared with other three organizational realities in Spain and Romania. After a first investigation phase, involving more than 90 Italian companies, five technical tools have been drawn up, to help the methodology and meant to make an inventory of the use of resources, the production of emission and the degree of the business management efficiency. From the point of view of communication, the visual contribution helps the reading of the final report. "*La tua impronta*—Your footprint" through the use of colour and the tailored graphic to identify the various industrial sectors: paper, textile, footwear, fashion, wood, pasta, tomatoes and wine (see Fig. 5.6). The *Info-graphic, photos, layout, use of bold letters, different sizes of characters* can facilitate the reading of the documents and contribute to draw the readers' attention on the key words and concepts: *Environmental Impact Results*, Activities, Experimentation, *Clusters* and *Best Practices*. For each sector, furthermore, tailored videos have been shot with a strong visual impact where the products are the real protagonists of the visual narration, they tell about themselves helping the audience to get the nutritional or manufacturing information useful for understanding the *Environmental added value* of the product and the whole eco-sustainable chain of production also thanks to its visual and graphic representation (see Fig. 5.7). Link: www.lifeprefer.it (Web Site)

Fig. 5.6 PREFER Project—Report: "La mia impronta" [My footprint] © 2016 by Kairós Studio in cooperation with Videopress, Parma, Italy (Copyright Holder) © 2016.www.kairostudio.it/en/

Fig. 5.7 PREFER Project—video on the North-Italy "Distretto del Pomodoro" [Tomato Distict], by Kairós Studio in cooperation with Videopress, Parma (Italy) © 2016. www.kairostudio.it/en/

Green Tweets
#Sustainability   #SocialEthics   #LegalEthics   #GreenEconomy
#GreenhouseEffect   #Ecofriendly   #GreenWashing
#EnvironmentalCertifications   #EnvironmentalRegistrations   #Process
#Product #Ecocommitment #Communicators #Images #Trademark #Brand
#Ecolabel #RecyclingArea #WasteSeparationArea #Infographic #Gadget
#Mater-Bi®   #CopenhagenWheel   #EnvironmentalFootPrint
#CycleAssessment #EnvironmentalImpact Bestpractice

5.5   Conclusions and Reflections: The *Guerrilla Marketing* Raises the "Green" Flag!

An aggressive approach does not make the marketing "green". The emotion and the rationality are the new "pacific weapons" of the global market.

We live in a global World. This concept is very often seen on newspapers, websites and any other *medium* of communication. The ***globalization*** *of markets* has brought to the progressive opening to the external national markets thus overcoming the logic strictly linked to the political boundaries of each Country. Not only the purely economic aspects, but also other aspects such as those: social, techno-scientific, cultural, *Environmental* are directly involved. Today it is possible, for instance, to send products easily and deliver services in many Countries simultaneously; to move from a Country to another for professional reasons or studies; to send information in a few seconds thanks to the *Internet* Network in more Countries; to get a *feedback* in real time; to initiate debates, exchanges of ideas, etcetera. And all this despite the political boundaries still active and national identities well defined.

We can therefore talk about *Globalization of Information and of Communication*. Besides our personal opinions about the "global" phenomenon, everybody can sense it every day. Thanks to modern technologies of transport and telecommunications, computer technology, our lifestyle has changed and consequently also our relationship with the Ecosystem Earth. It follows that negative consequences from a radical exploitation of natural resources because of the pressing demands of the globalized market made some social groups react violently against, the so called "*no global*" who appeared for the first time in Seattle, USA, at the summit WTO (*World Trade Organization*), in 1999.

From a commercial point of view, a process of **Media trench** has been triggering for a long time and it involves quite often the *Environmental Communication*. With this in mind the *green marketing* is still used by some factories as a "weapon" to fight the "enemy": *market competition*. Forgetting the ethics values, honesty, coherence of the *Environmental policies*, making them an empty container powdered of "*green*", as it is the case of the **green washing**.

Without deepening, in this context, the political theme of the globalization and its consequences, it is interesting to make a few final thoughts about how to remove the "*War Armor*" from the global market. It is necessary to allow the *Environmental* and *Sustainable values* to overcome it and spread them out internationally, regardless of any "*armed conflict*". It is not by chance that the US publicist *Jay Conrad Levinson*, in 1984, coined the neologism **guerrilla marketing** to identify any kind of commercial promotion, low cost but at high impact, using unconventional tools of communication based on the addressees' psychological reactions meant to make them "getting crushed". In other words, their attention is drawn on the product or service or, generally speaking, on the message itself when the consumer/user does not expect it. Just when their **advertising consciousness** is in *standby* and they do not realise they are the target.

Authoritative sources have stated that an aggressive commercial approach has no more reason to exist. The existence and repetition of the messages, any medium you may use, is going to produce the opposite of the desired effect: that is, it pushes the perspective customers away. And this also in the case your products or services have authentic and certified environmental characteristics (*Levinson, Horowitz, Cox*).

The **winning attitude** (**Win/Win**) aims at the **ecological reputation** which is strictly connected to each single detail of the *Environmental Communication*. Bringing the public opinion to believe the message conveyed by your corporate organization is the result of a constant dialogue which is not based only on the traditional "formula" of 5 Wh: *who, what, where, when, why/how* of the modern journalism.

The new *"weapons"* of the *guerrilla marketing* raise the green flag, creating a new equilibrium between emotion and rationality. And if they are well-balanced and respectful of the legislative and regulatory limits may be effective communicative tools in the *Environmental* and *Sustainable* field alongside those traditional. *Who could imagine, for example, to see some statues in Rome wearing anti-pollution masks and road-signs against CO_2 emissions produced by the vehicles circulating in the Capital?* It happened in June 2008 in a campaign of awareness to energetic efficiency promoted by "Earth!", an *Environmental association* newly formed.

Understand the public of consumers/users and stay in constant contact with them; respect each point of view, culture, language and tradition; stimulate the imagination and intellectual ability of the receiver of our message be credible, honest and practical: these are only some of the strengths of the *green marketing*, both traditional or *"warrior"*.

So all this is translated into a lot of *Media* of communication such as: discussion groups, social networks, emails, websites, newsletters, TV, radio, newspapers, magazines, conferences and gadgets, and more than this.

In December 2016, the citizens of Milan are the witnesses of a surrealistic scene. *The facade of a historic building in the heart of Milan (La Foppa small square, by the Moscova metro station) was covered with a deep ice sheet* (Fig. 5.8). Windows, sills, walls of the building looked frozen from the top floors to the basement. The Firemen were soon called; the tenants are poured into the street, shivering in the early hours of the morning; a lady is blocked in her flat because of the iced lock and she is rescued with a car-scale of the Civil Protection. The fact is reported by many news Media. The piece of news says that one of the tenant went on a business trip on the previous month of August forgetting to switch off the air conditioning plant.

The passers-by were staring at it very curious. Many readers and *Internet* users stopped on that surrealistic piece of news. What happened is soon circulating on *Social Network*; very successful *hashtag #PalazzoGhiacciato* (*#FrozenPalace*) is created and becomes a hit *online. Is a cinema set for the sequel "A day after tomorrow"? Is a piece of contemporary art? Is that the effect of the Climate Change?* In each case people are making questions about the consequences of a "lack of attention" which could be done by everybody and which is of serious impact on our economic *budget* with a high energy bill. Furthermore, this phenomenon affected also the town *Environment* or, in a more general perspective, the ecosystem Earth. The refrigerant gases, odourless, colourless and non-toxic for human beings, contribute to reduce the ozone layer if they are released in the atmosphere. Some of the chlorofluorocarbons (CFCs) and hydro chlorofluorocarbons (HCFC) have been banned for many years. It is important then to reflect on the ecological consequences before acting.

So, the target is reached! The *E.ON Italia*, a company primarily acting in the energetic sector, is the maker of this **unconventional advert campaign** against the energy wastage. It has involved several extras and fake firemen, being documented by an official video which has collected more than 5000 hits on YouTube, so far.

A good example of **green guerrilla marketing** which prompted us to reflect on the universal *Environmental value*, as the energetic efficiency, through the experimentation of an event arousing astonishment, concern and anxiety. In the same time, the energetic company has improved its commercial image spreading its own

Fig. 5.8 Iced Palace in Milano—Commisioner: E.ON Italia—Communication Agency: M&C Saatchi S.p.a. Milan, Italy (Copyright Holder) © 2016 www.mcsaatchi-milano.com

commitment in the *Environmental field* to a vast public. A strategy that is the winner in a highly polluted town but sensible to *Environmental aspects* like Milan.

Target reached! Ergo: always remember to use the air conditioning when strictly necessary!

Green Tweets
#GuerrlillaMarketing #NonConventionalCampaign #globalization #Trade #Commerce #feedback #global #glocal @WTO #WorldTradeOrganisation #AdvrtisingConsciousness   #WHquestions   #Media   #Communication #Environment #SustainableDevelopment

Bibliography

Jay C. Levinson, Shel Horowitz, *Guerrilla Marketing goes Green*, Hoboken (New Jersey, USA), John Wiley & Sons Inc, 2010, pages: 236.

Michael Braungart, William McDonough, *Cradle to Cradle*, London, Vintage Books, 2009, pages: 192.

Wangari Maathai, *Replenishing the Earth: spiritual values for healing ourselves and the world*, New York/London/Toronto/Sydney/Auckland, Doubleday, 2010.

David William Pearce, Anil Markandya, Edward Barbier, *Blueprint for a Green Economy*, Oxford (UK) Earthscan/Routledge, 1989, pages: 192.

Frank Prévot, *Wangari Maathai: The Woman who planted millions of trees*, Watertown (Massachussetts, USA), Charlesbridge, 2015, pages: 48.

Paolo Taticchi, Paolo Carbone, Vito Albino, *Corporate Sustainability*, Berlin (Germany), Springer-Verlag, 2014, pages: 350.

Roy H. Williams, *Secret Formulas of the Wizard of Ads*, Austin (Texas, USA), Wizard Academy Press, Austin, 2013, pages: 224.

V.A., *Relazione sullo Stato della Green Economy—l'Italia in Europa e nel mondo* [Report on the State of Green Economy—Italy in Europe and in the World], Stati Generali della Green Economy [States General of Green Economy], Fondazione per lo Sviluppo Sostenibile [Foundation for the Sustainable Development] and Dual Citizen, Rimini (Italy), Fiera Ecomondo [*Economodo* Fair], 2016a, pages: 74.

V.A., *Proposte di Policy del Consiglio Nazionale della Green Economy* [*Policy Proposals by the Italian* Consiglio Nazionale della Green Economy—*National Council on Green Economy*], in cooperation with the in Ministero dell'Ambiente e della tutela del Territorio e del Mare [Italian Ministry of Environment, Land and Sea Protection], Ministero dello Sviluppo Economico [Italian Ministry of Economic Development], 2016b, pages: 14.

V.A., *Towards a Green Economy—Pathways to Sustainable Development and Poverty Eradication*, United Nations Environment Programme, 2011, pages 630.

V.A., *Corso Breve di Economia dell'Ambiente* [Short Course on Environmental Economy], Centro Studi Villa Montesca (Perugia), 2nd Level Master for *European Manager on Social and Environmental Sustainability* (EMSES), Umbria Region and European Social Fund, 2006, pages: 33.

Web Site List

United Nations Sustainable Development Goals. https://sustainabledevelopment.un.org

Martino De Mori, *Energia dal vento, senza le pale—Una turbina eolica senza pale potrebbe rivoluzionare lo scenario delle energie alternative* [Energy from the wind—a blade-less wind turbine could revolutionize the alternative energy scenario], Focus, 18th May 2015, www.focus.it

PREFER Project—Product Environmental Footprint Enhanced by Regions: www.lifeprefer.it

ISO—International Organization for Standardization www.iso.org

EMAS Eco-Management and Audit Scheme http://ec.europa.eu/environment/emas/index

The Copenhagen Wheel, www.superpedestrian.com

Earth Overshoot Day, www.overshootday.org

Global Footprint Network—Advancing the Science of Sustainability, www.footprintnetwork.org

Communicating the Environment Is a "Public Right and Duty"

6

Abstract

To know how to communicate an *Environmental message* with purposes of *pubic interest* implies the cooperation of "more communicators" mirroring at improving the social role of communication by reducing sensibly the risks of misunderstanding or misinterpretations as a result of "noises" of different origin. This applies both to the "external" interest addressed to a big public and both to the "internal" interest, addressed to subjects working within an organization. The particular complexity of the *Environmental matter* and the increasing demand for its information on the part of the public opinion, has created new rights and services of *Environmental Participated Communication* with the citizens and stakeholders. Communicating the *Environment* has therefore taken different forms meant to deepen the knowledge of the various themes from different points of view and allow the parties to interact and express their opinions. How can we communicate the *Environmental issues* at an institutional level? The Reporting, Information and Communication Desks, the facilitation techniques and peer review are just a few examples to ensure the efficacy of a suitable public Communication Plan. Common denominator: a cooperative approach aimed at eco-responsible decisions and behaviours concerning the local and global community.

© Springer Nature Switzerland AG 2019
M. Abbati, *Communicating the Environment to Save the Planet*,
https://doi.org/10.1007/978-3-319-76017-9_6

a   **b**

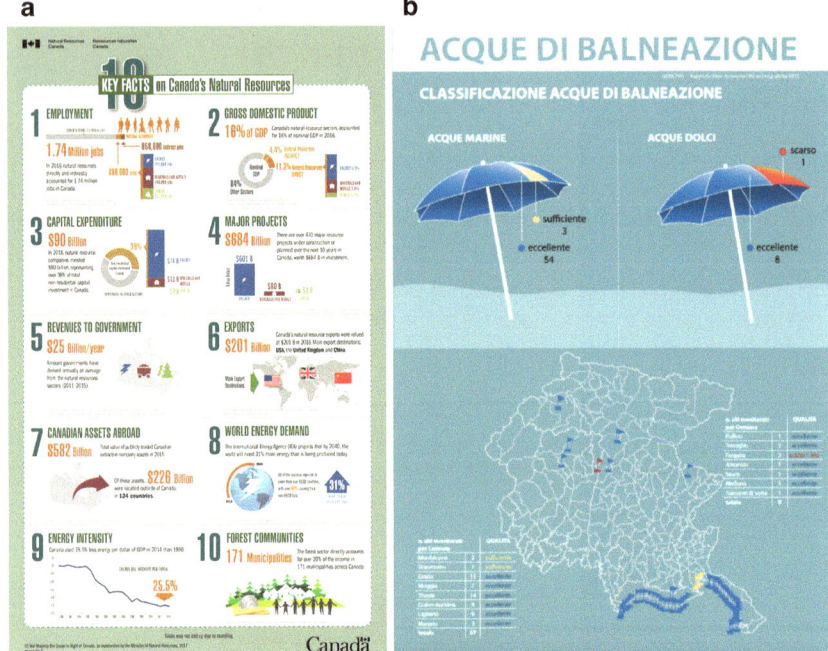

"We must remember our duty to Nature before it is too late"
Margaret Thatcher, British Politician
(Grantham, 13th October 1925 – London, 8th April 2013)
Speech at the 2nd World Climate Conference (Genève, 1990)

c

(**a**) "10 Key facts on Natural Resources" by the *Minister of Natural Resources*, Canada, @ 2017 - Natural Resources Canada/Ressources Naturelles Canada (Copyright Holder) www.nrcan.gc.ca; (**b**) *Rapporto sullo Stato dell'Ambiente ARPA Friuli-Venezia-Giulia*—[Report of Environmental Status ARPA—Italian Environmental Protection Agency], Friuli-Venezia-Giulia, Italy (Copyright holder) www.arpa.fvg.it; (**c**) "How long until it's gone", Environmental Awareness Campaign on biodegradability of solid waste in the ocean by NOAA *National Oceanic and Atmosphere Administration* USA/*Woods Hole Sea Grant*, USA (Copyright holder) www.noaa.gov

Reading Proposal: In Between Words and Images!

Transmitting an environmental message as a help for a principle of Sustainability implies the study of suitable techniques of communication underlying the fundamental elements necessary to the audience to build up their own critical point of view in the environmental questions relying upon certified sources. This operation is everything but easy. The major amount of environmental information in fact is based on scientific or statistical technical data or indicators that must be known by the citizens in order to interpret correctly "the environmental status". This is one of the greatest challenge faced everyday by the European and world public administrators which we are going to deal with in *"Chapter 6—Communicating the Environment is a public right and duty"*. Thanks to the computer graphics, a new form of communication has made its way since long time: the *info-graphic*. The written text and the visual part merge together and integrate one another thus increasing the audience involvement with the environmental message. Specific studies have shown how the average reader's attention level increases considerably in front of a picture able to communicate information (*Elaboration Likelihood Model—ELM* by *Richard E. Petty and John Cacioppo*, 1986). Visual elements represent, therefore, powers of persuasion for the diffusion of environmental data. Info-graphic is much more than a simple form of art. The sequence of images, real or stylized, accompanies the visualization of technical data making them more "pleasant" and connoting greatly their meaning. The visual impact of a beach umbrella, for instance, immediately contextualize ichnographically the environmental situation about the bathing water quality reporting, conveying and supporting the qualified data coming from the Italian *Regional Agency for Waters Protection* [Figure (b) on the cover, *Rapporto sullo Stato dell'Ambiente ARPA Friuli-Venezia-Giulia*—Report of Environmental Status ARPA, Friuli-Venezia-Giulia, Italy]. The matching among graphics, designs, chromatic effects and the use of the character-size of writing communicate efficaciously, in the space of one page, the ten key-factors of the Canadian Natural Resources management [Figure (a) on the cover, *Minister of Natural Resources*, Canada, 2016]. The info-graphic environmental sub-marine area called *"How much time before it is going to be dispersed in the Environment"* catches the audience attention making them think of the waste impact thrown into the sea, emphasized by the diameter of the circles proportional to the years of decomposition [Figure (c) on the cover, NOAA *National Oceanic and Atmosphere Administration* USA/*Woods Hole Sea Grant*, USA]. A *"written-visual narration"* is created which is easily understandable and disseminated through the Net of *Social Media*. This graphic device is able to reach millions of people and be understood and remembered. On the other hand, we know that the association between ideas and images is the basis of the mnemonic technique of the *mental maps*, exploiting the tendency of the human brain to elaborate images before words. The only pre-requisite is then always to specify the sources from which the information is taken to guarantee the credibility.

6.1 The Strength of Public Environmental Communication: A Multi-voiced Journey

The need to communicate the *Environment* to achieve targets of general interest.

"Public relations are a key component of any operation in this day of instant communications and rightly inquisitive citizens". This **Nineteenth century quotation by** *Alvin Adams* (1804–1877), founder of the Adams and Company, one of the most important *Express rail courier of the United States*, seems to be very modern and belonging to our time.

A need of "how to communicate" to the public and to employees taking different forms according to **WHO** communicates, the Contents (**WHAT**) and the Objectives (**WHY**) to be communicated. As already said the very etymology of the word "communication" includes: an exchange, a dialogue, a common and shared production. A semantic field that makes communicating a *bidirectional* and *circular process*, promoting its social function as reminded by the linguistics scholars. A *multi-voiced* and *multi-eared journey* with specific tasks for the operators. One of them is of course to foresee how the message will be decoded by the addressees in order to reduce or eliminate the "noises" prejudicing the correct interpretation; for the communicator this means that he must know how to recode the message quickly, adapting it to the answer given by the addressee.

Two factors play a relevant role if the communicator is a public authority or a body organization acting with a public function. In any case, this happens when the contents of a message pursue a general interest linked to a specific field of application, whatever the object of communication is: institutional, social, political or of solidarity. The interest of communication will be *"external"*, directed to the citizens or *"internal"*, if addressed to the operators.

Starting from the Twentieth century, after the Second World war, the *social function* of communication becomes the dynamic milestone of public relations, those between the organizations and the social actors: the people. Communicating is not a *static cornerstone* anymore, between addresser and addressee but an element constantly evolving. A *communicative process* involves also the public administrations which are supposed to share objectives, targets, action plans whose goal is to put into practice and pursue the community interest, i.e. citizens as an integral juridical part of a democratic system.

Hence, the juxtaposition of the term *"public"* given to the term *"Communication"* to connote the word that in Italian language covers **a wide range of meanings** (worldwide applicable), herewith some examples:

– *All that belongs to everybody's sphere*, becoming then of "public utility" or "of public interest" i.e. the public administrators or political powers;

- *All that is accessible by everybody* or usable by everybody i.e. public services and public spaces;
- *All that is visible*, respecting the principles of transparency and efficacy i.e. general public sphere.

Public Communication reveals its complexity. *What happens when the public communicators are supposed to deal with the Environmental matter?* Whoever communicates the *Environment* must set specific goals to answer the needs of the public he is addressed to. The **Environmental Communication** may play important roles such as:

- *informational* = in this case the information is meant to sensitize the addressee suggesting more sustainable lifestyles or consumption, respectful of the natural resources;
- *reporting* = in this case the information is meant to inform constantly the citizens or the public employees on the *Environmental status*, for example: on *Environmental impacts* of works and infrastructure, on the ecological footprint—complex indicator used to evaluate the human consumption of natural resources compared to the Earth—to regenerate them; consumptions of goods and services in a specific community;
- *participatory* = in this case the communication is meant to stimulate the participation and involvement of the citizens who take part to some decisional processes and evaluate the performance obtained by the public administration.
- *mediator* = in this case the communication is meant to act as a qualified information between the public administration and the interested parties when, for example, *Environmental emergencies* could occur such as works to drain—in such situations, in fact, it is necessary to develop *Media* campaigns that make available the *Environmental data* with legislative references of technical nature.

The function of the **public Environmental communicator** is proving to be more and more relevant in consideration of the strategic importance, given at an international level, to the *Environment Communication*.

One of them, the approval of the **Agenda 21** on the part of 178 Governments all over the world, Italy included. A *document of intent*, but also an *operative programme* of application, at a global level, of the principles of the *Sustainable Development* as a direct consequence of the **UN Conference on *Environment* and Sustainable Development** of *Rio de Janeiro* in 1992. A process strongly followed at European level, inviting the local authorities to change themselves into "*privileged Environmental communicators*" for the community that lives on the administrated territory (A21—Chap. 28). New forms of information, training, education and participation to the political decisions were adopted in order to increase the awareness on the critical *Environmental issues* and the consent for actions to face them.

Following, the other fundamental stages as the **Aarhus Convention** of 1998, the first international document stating the *right of access* to *Environmental information*. A Right strongly supported by the **European Union** that considers the *Environmental*

Communication one of the milestone of the *governance*, together with the participation and access to the *Environmental information*. The targets of safeguard, protection and improvement of the *Environmental quality*, as laid down in Article 191 of the **Treaty of the European Union Operation**, are then implemented. We cannot help reminding the *Directive 2003/4/EC* about public access to *Environmental information* and the *Decision 2005/370/EC* about the access to information, public participation in decision-making and access to justice in *Environmental matters*.

Communicating the Environment becomes an unconditioned and participated right forcing the public administrations to supply with information about the *Environment status* but also health and safety conditions of the administered territory. In this way, the citizens are allowed to participate to decisional steps of the policies on *Environmental matters*. A *right* that enjoys a legal protection being the citizens able to appeal against decisions which are not respectful of the principles of accessibility and participation just mentioned.

The *Public Environmental Communication* in some Countries, Italy included, has been evolving in the passage between the Twentieth and the Twentieth-first century, moving from a "*right*" to a "*service*" by the entry into force of *Legge Quadro* (*Framework Law*) No. 150 of 2000. A regulatory cornerstone likely to change the public administrations' approach and a starting point of the successive legislation on the matter, such as: *Legge* (Law) No. 349/86 introducing the citizen's *right* to *Environmental information*; *Legge* (Law) No. 150/2000 establishing the *URP—Uffici Relazioni con il Pubblico* (PRO—Public Relations Offices); *Legge* (Law) No. 108/2001 ratification of Convention of Aarhus; Decreto Legislativo (Legislative Decree) No. 152/2006, called "*Environmental Code*".

Communicating to the public, thus, is not "*optional, episodic, tactical*" but "*compulsory, permanent, strategic*". The communication of public administrations has become then a "*public service*", a complex organic structure of activities aiming at meeting the needs of the community, *Environment* and *Sustainable Development* included.

Getting on with the **Aalborg Commitments** implemented for the *European Conference on Sustainable Cities* in 1994 in the homonym Danish city. Evolved, 10 years later, with the **Aalborg Commitments + 10**, established by the *Sixth European Conference on Sustainable Cities* in *Dunkerque*. Ten collective commitments inspired by *Agenda 21* and subdivided by areas of competence such as: governance, local sustainability management, responsible consumption and lifestyle choices etc. More than 600 local administrations have joined the commitments and they have taken on promoting the *Sustainable Development* through concrete actions meant to convey properly the forms of non-economic reporting: The **Environmental reporting**.

Recently, the **Sustainable Development Goals** (September 2015), a series of sustainable objectives have been adopted by the *United Nations* intended to preserve the *Planet*, to defeat any form of poverty and to guarantee prosperity for everybody. A challenge that incorporates *Environmental*, social, economic aspects and *welfare*, generating a further evolution in the choices of forms and tools of *communication*.

Against this background, we have decided to deal with the mutual exchange of information between the citizens, the public administrations and their employees assimilating the term "*public administration*" to the other institutions carrying on a function of public interest. We will omit therefore any reference to the "*social sphere*"—non-governmental and semi-public organizations having some functions as public communicators on the *Environmental matter*—and to the "political sphere" of the public *Environmental Communication*.

Finally, we apologize for making the distinction between *public communication* and *political communication* that is a synonym in the Anglo-Saxon speaking world. However, during our dissertation, our *food for thoughts* can be applicable also by the communicators working in the above mentioned spheres, adapting the contents to the *context of reference*.

Green Tweets
#PublicRelations   #EnvironmentalCommunication   #EcoMessage
#SemanticField   #Noise   #SocialFunction   #PublicInterests
#EnvironmentalReporting   @Agenda21   @AarhusConvention
#EuropeanUnion#AalborgCommitments#SustainableDevelopmentGoals
#UnitedNations #PublicCommunication #Communicators

6.2   How Can the *Environmental Matter* Be Better Spread Out at the Institutional Level?

The instruments and the strategies of a public administration: a field choice.

Considering the **Techno-scientific Nature** at the basis of the "*Environmental language*", it is necessary to simplify the concepts through a clear, understandable, usable and shared language. The *communication Media* must allow the public to find out the key-concepts in order to having, at first sight, the perception of the quality of the *Environment*.

The **reporting** is then the tool to inform the citizens of the results of the *Environmental policies* promoted by the public administration, and it is often accompanied by **graphic elements** such as "*semaphores*", "*infographic*", or "*emoticons*" more or less smiling according to the target achieved, thus communicating visually *the performance*. But also *written-visual* tools may help to decode the *Environmental message* by using different characters, lists, acronyms, signs, colours, shapes etc.

The **Media tools** of communication enjoy a large number of devices for the *public Environmental communicators* interfacing more and more frequently to the **Web** and the **Social Media**, which must be carefully monitored and checked as far as the information are concerned. Nevertheless, these telematics multichannel instruments: *Internet* sites, *Intranet*, *Social Media*, hypertext etcetera have no equal for their *speed of information* and *communicability*.

It follows from that the necessity of giving the citizens the chance to know the identity of the recipients and of the information sources, in other words enabling them to "**make an aware act of communication**". It will be then their *critical capacity*, based on their cultural *background* and their level of *Environmental awareness*, to make them believe or not the information received.

The technological age, anyway, has not completely deleted the most traditional forms of *public communication* like the *interpersonal* and *papery approach*.

Information and consulting help desks are an example as an evolution of the *PRO*, *Public Relation Offices*, whose role in *Italy* was reinforced, above all at a local level, by the *Legge* (Law) No. 150 of 2000. With one or more branches in the Town hall, permanent or temporary, according to the needs, these offices were born with the precise intent to communicate with the public in a bidirectional way, giving a useful and quite often personalized support of the *Environmental information*. The *Information desks* are faithful allies of the local *governance* and inform the citizens about critical situations (e.g.: waste deposits unauthorized, electromagnetic pollution, emission of unidentified fumes into the atmosphere, etc.) and any other question on *Environmental matter*. In many cases such a kind of information acts as a means to prevent, foresee, manage and overcome an emergency situation: floods, earthquakes, avalanches, territorial pollution, etcetera.

The **Environmental Communication** **Plans** represent often an ultimate instrument allowing the public administration to manage systematically all communication actions through a specific programme to be approved each year by its bodies. A set of objectives and specific targets meant to a single purpose: communicating the citizens the *Environmental actions* promoted by that specific public administration, with respect for its values and policies. A group identity is thus created to push the communicators (employees and administrative officers) to act in a coordinated manner and participate actively in favour of the *Environment Protection* and *Sustainable Development* values.

It is very important to underline, then, that the *digital world* which we are living in, is projected to the constant research of the latest news which can make "obsolete" the ones just published a few minutes before. This digital world is giving priority to more and more synthetic and immediate forms in communicating the results of the *Environmental actions* for this public area. This trend sometimes even substitutes new types of communication and the papery reports, published *una tantum*, that risk to give an *Environmental* or *sustainable image* not updated anymore.

Anyway, the concept of *unilateral communication*, from the public authority to the citizens, is greatly surpassed, and it could go unnoticed by the addressees of the *Environmental messages*. Much better a journey which could make room to many events in order to help the citizens' decisions: seminars, conferences, round tables, forums, meetings and laboratories.

From this point of you, thanks to *ad hoc* tailor made information and communication campaigns, eco-sustainable behaviours and best practices are promoted to make the citizens be more aware. Examples of actions:

- to disseminate clearly the *Environmental message* through brochures of public utility;
- to deepen recent topics or new *Environmental services* offered to the citizenship;
- to communicate behavioural procedures or operative guides on specific themes.

A *"polyphonic choir"* whose conductors are the public communicators that bear their own responsibility to harmonise the *Environmental information* of public interest. So that to avoid any possible *"false note"*, metaphor of any form of misunderstanding, conflict and mistake, not supported by reliable sources.

Case Study in Short
United for a public communication of quality → The *Public Service Department of the Italian Government* has established its own brand name guaranteeing the citizens the reliability of the actions made by the Public Administration in the *Environment* and Communication field. The *Pubblica Amministrazione di Qualità—PAQ* (Public Administration of Quality)— **PAQ** is a new instrument of communication, a *web portal* promoting a series of activities to improve the quality of the public services. The partnership *Formez* (*Training Centre for the up-dating of Public Administrations*), *Confindustria* (Federation of Italian Employers and Italian Industrial Federation), *Cittadinanzattiva* (Active Citizenship), *Ministero di Giustizia* (Italian Ministry of Justice), *Ministero della Pubblica Istruzione* (Italian Ministry of Education), Regions and Local Administrations, allow the most important *stakeholders'* involvement on the Italian territory to highlight the "visibility" of the Institutions toward the citizens with the view of *open government*. Furthermore, this approach is achieved by the dissemination of best practices, definition of an operative know-how coordinated at European level. A system of evaluation of performances which is based also on the *benchmarking,* an Eighties–Nineties business economic *technique.* It implies the comparison between their own performances and those achieved by a similar structure used as a starting point in order to improve the quality. These are only some of the cornerstones of the project. Qualified guidelines concerning the available forms of communication are supplied with a particular overlook to the *Environmental matter*, considered a strategic subject for the public functions. Among them is to be mentioned the *Manifesto della Comunicazione Pubblica in campo Ambientale* (*Manifesto* of the Public Communication for *Environment*) carried on by the *Italian Association of Public and Institutional Communication.* This document, stating that *Communicating the Environment* is a citizen *right*, consider information and communication as precious resources to "*listen to the citizens, know their needs, support the participation as the actors of Environmental choices*". Fundamental targets that any public administration must be committed to respect and implement.

PAQ: www.qualitapa.gov.it (Web Site)

Associazione Italiana Comunicazione Pubblica e Istituzionale:
www.compubblica.it (Web Site)

Green Tweets
#Language #Environment #Communication #Sustainability #Reporting #GraphicElements #Infographics #Performances #Smileys #Web #SocialMedia #CommunicationPlan #Governance #LatestNews #Ecosustainability #Bestpractice #PublicAdministration #Benchmarking @ManifestodellaComunicazionePubblicaAmbientale

6.3   The Public *Environmental* Communication: A Citizen's Right at the Service of the Community

The "social" dimension of Communicating the *Environment*.

Visibility, transparency, dialogue, sharing, listening, respect, trust, participation, cooperation, mutuality, efficacy, these are only some of the k-words connoting *legal texts*, *public announcement*, *official speeches* or any other *medium* referring to *public communication* and that cannot be missed when facing a dissertation about *Environment* and *Sustainable Development* issues.

For *Communicating the Environment*, to achieve the public's agreement, in fact, a *dynamic approach* is assumed and open to the citizens with a continuous exchange of information, opinions and ideas.

A *"revolutionary"* process of change that moves from the mere *information* to the most authentic concept of *communication,* and it is able to give "a soul" to legal texts or other legal acts. Herewith the various steps:

- to be flexible, that is being able to adapt to the context in which we communicate;
- to be able to listen to and develop the point of view of each conversationalist, the so-called *role taking*;
- to be able to remodel the original message on the basis of the feedback receive;
- to be able to call into question.

An *Environmental Communication* chain that elevates itself from strategy to *right*, i.e. to a behavioural rule to be followed and required by all citizens. A *Status* anything but predicted and which points out the long regulatory path born and developed at European and International level.

But it requires a lot of professionalism. Getting the public of citizens involved in the debate about *Environmental themes* and Sustainable Development, in fact, needs a precise training for communicators who must know what to say, how to say it and to whom.

"Public participation is based on the belief that those who are affected by a decision have a right to be involved in the decision-making process". This is one of the

key value fixed by the *International Association for Public Participation*—IAP (Core Values 2016).

An absolute right to be informed correctly through a suitable communication on single *Environmental issues*. A right that enables people to express freely their own active contribution to the *decision-making*. In case of legal disputes which enables the citizens are allowed to intervene on *Environmental legal process* with *Environmental information* meant to inform about facts or to defend themselves individually or collectively. In many legal systems, based both on *the Civil Law* or on the *Common Law* the proposition of *class-action* to protect a group of subjects (mainly consumers and users) is set out. To them homogeneous rights to be protected are recognized. The Italian legal system, for example, introduced the collective action in 2010 with the entry into force of Article 140bis of the *Consumer Code*.

A *"hard core"* of rights on the basis of a democratic legal system which makes the *Environmental communication* a "public service" therefore an activity of public interest. This is what an *Environmental communicator* has to bear in his mind.

Now, let us think carefully about each of these *rights* and *values* in order to guarantee a correct *"adversarial procedure"* and to trigger a participative communication with the readers.

1. **The Right to Know:** states one of the most characterizing actions of the dialogue between the Public Administration and the citizens, that is the *access to information*. A transparency assurance in any democratic regime, this right is relevant for the *Environmental matter* as it was underlined by the **United States Conference on Environmental Matter and Development** [Principle No. 10 of Rio Declaration, 1992]. Also, it has been emphasized many times at the international level [e.g.: *Declaration of Bizkaia*, 1999]. Following this "wave" of shared information, in some countries a real legislation was born, the so called **sunshine laws**, in order to promote the information and the citizens' involvement into the decisional activity, once an exclusive privilege of the international *Elite*. Among the first examples, the *Freedom of Information Act* (FOIA), approved by the American Congress in 1966 and the following *Electronic Freedom of Information Amendments*, in 1996. A *vademecum* ratifying the free access to the archives, and recently, to the data bank of some key-institutions of *Environmental management* of the American territory (e.g.: *U.S. Forest Service, Bureau of Land Management, Department of Energy, EPA—Environmental Protection Agency*, etc.). To know the pollution levels in the atmosphere (air quality), in the waters and soils is a right to be claimed in the event of legal disputes or major public scandals. The American political scientist *Mark Stephan* pointed out the *"domino"* effect caused by the outraged reaction of the public opinion supported by the *Media* about the consequences of pollution of huge industrial plants on human beings and *Environment*. In these cases, the polluters admit their guilt and implement corrective and definitive actions: the *shock and shame response* approach. It is not accident the access to *Environmental information* has been an essential element in the *American Presidential Programmes*. On the national and international scene, particularly impactful was the decision to limit the values

expressed by the *Freedom of Information Act* (FOIA) during the 8 months following the terroristic attack to the *World Trade Centre* in New York, on the 11th September, 2001.

Case Study in Short
Erin Brockovich Case → the missing access to *Environmental information* is the theme which the famous American legal case is based on and whose protagonist the activist *Erin Brockovich Ellis* is, at present socially involved as *consumer advocate* for similar cases at international level. In short: lawsuit in 1993 against the *Pacific Gas & Electric Company*, energy giant, having as a subject the contamination due to hexavalent chromium in the water of the Californian town *Hinkley*. This pollution caused numerous deadly diseases such as cancer and leukaemia, found in the local population. This legal case lasted 30 years. Thanks to *Brockovich*'s investigative capacity the **Environmental data**, hidden by the company under investigation, **were made public** and showed the massive presence of the pollutant in the aquifer of *Hinkley*. So far it has been one of the greatest legal action on *Environmental matter*.

As a result, a great indignation by the whole American public opinion and sentenced to pay a compensation of $ 333 million to more than 600 inhabitants in the contaminated area. As *Erin Brockovich* reminded us: *"There are a lot of things out there that are going to harm us. But, as a consumer, we have the right to know, so we can make the different choice"*. Filmography: *Erin Brockovich*, Filmmaker: Steven Soderbergh—Cast: Julia Roberts, Albert Finney, Aaron Eckhart, 2000.

Erin Brockovich, Consumer Advocate: www.brockovich.com (Web Site)

2. **The Right to Speak:** being able to express our own opinions is on the basis of any democratic decisional process but in order to do it at its best we need a qualified and complete information based on reliable sources. Anything else belongs to the sphere of personal perception of the *Environmental* question. In the case of polluted aquifer, to remind again the theme of *Erin Brockovich*'s case, once you know the scientific definition of the pollutants and their harmful effects on human beings and eco-system, the public opinion is much more influenced by what they try on their own skin. For example:
   • to discover that relatives, friends, acquaintances are affected by serious illnesses;
   • to witness painfully their dears' premature deaths;
   • to experience the effects on their lifestyles;
   • to be the witnesses of the *Environmental impacts* originated by human-induced actions.

Hence, the importance for the public administrations of guaranteeing a suitable space for public forum, round tables, feedback gathering and proposals finalised to projects of public interests.

Case Study in Short

Aarhus Convention—Public participation at *glocal* level → at international level we have seeing for a long time the promotion of participatory actions of *governance* at the institutions and international organizations giving substance to Principle No. 10 of *Rio Declaration.* Among them we would like to remind the *Convention of Aarhus*, adopted in 1998 in the Danish Aarhus city. What makes this agreement unique is the fact that for the first time they have put on the same plane the *Environmental and the human rights.*

Article No. 1: *"In order to contribute to the protection of the right of every person of present and future generations to live in an Environment adequate to his or her health and well-being, each party shall guarantee the rights of access to information, public participation in decision making and the access to justice for Environmental issues in line with the provisions of this Convention".*

Key-words: access to the *Environmental information*, public participation and access to justice.

Added Value: right of each single citizen to get *Environmental information* including: the status of *Environment*, health and security and any other policy concerning the territory in which he lives. Within 1 month from the enquiry and without having to justify or explain it, to guarantee the citizen's direct participation.

Case Study in Short

The Swiss Case → The Article No.10g of the *Environment Protection Law* (LPAmb) allows each citizen to access the documents containing *Environmental information.* The rule of law is valid both at federal or canton level. Each single Canton, furthermore, is bound to discipline independently, with due regard to *Aarhus Convention* on the access to the *Environmental information.* The LPAmb establishes then that the Federal Council has to present the Chambers an *Environment Report* every 3 years (articles No 6 and 8). It is also envisaged the public consultation with reference to the plants subjected to the *Environment impact.* The major judicial protection is guaranteed for each citizen who was not entitled to have access to the *Environmental information* or if a right was violated.

To sum up, the request of participation to public questions concerning *Environmental issues* is growing so fast to become a real cornerstone of public management even more subjected to awareness campaigns and included among the key-themes also in the election regardless of any political identity.

It is not by chance as *Arvin Singhal*, professor at the Communication Department of the Texas University (Campus UTET, El Paso) and international expert for ***participatory communication***, is reminding us that "it is a *dynamic, interactional and transformative process of dialogue between people, groups, and institutions that enables people, both individually and collectively, to realise their full potential and be engaged in their own welfare".*

Green Tweets

#PublicComunication   #Environment   #RoleTaking   #Feedback
#DecisionMaking #CivilLaw #CommonLaw #ClassAction #RighttoKnow
#SmartLegislation #SunshineLaws @ErinBrockovich #RighttoExpress
#Medium   @AarhusConvention   #SvissLegislation   @ArvinSinghal
#ParticipatoryCommunication

6.4   Let Us Explore the Key Elements of an Effective Public Environmental Communication Plan

The "toolbox" of public eco-communication.

"The art of effective listening is essential to clear communication, and clear communication is necessary to management success". This quotation by *James Cash Penney* (1875–1971), one of the first *businessman* to manage successfully, at the beginning of the Twentieth century, a chain of supermarkets in the West of the United States, shows us a new vision of communication that puts emphasis on the recipient. An aspect anything but a clear one.

The public administration therefore has established a dialogue with the citizens, moving from an "authoritative" to a "cooperative" role meant to facilitate the participation of each member of the society in creating *a fluent, clear and participated communication*.

Now we can pretend to be administrators of a specific territory and have to draw up a suitable **Communication Plan**. *Where shall we start from? Which aspects are to be considered?* In order to answer these questions properly it is useful to fix some priorities to be analysed:

- **Being able to look "inside" yourself—Internal Communication**: A fundamental starting point is to know very well the **kind of communication** operating within those who are working inside the public administration. Each single member of the organization can contribute effectively to its behaviour and make the communication more or less strong.
- **More "heads" are better than one: the *working team* is essential** to set up an efficient Communication Plan. To ensure an adequate **formation of the Environmental information** means to know how to communicate with the whole *working team* inside the organization. This does not mean managing the work anonymously and mechanically but, on the contrary, giving a "soul" to the public administration. Password: *work the system and share the same values*.
- **Being able to look "outside"—External Communication: identifying and working the system with all the external subjects** interacting with the public

administration of reference in the communication *"production chain"*. Each single stakeholder must be involved in the Communication Plan and know his role well in advance. The system of public communication, in fact, is for its nature open to the outside, that is ready to gather all *stimuli* and needs.

- **Being able to persuade. Advertising campaigns:** if we want to reach a large public, the public administration can take advantage of real **information campaigns** meant to disseminate messages of public utility in order to catch the attention and get visibility to the institutional organization, a function comparable to the one of the commercial-advertising branch.
- **Building on solid values:** each communication action, both internal or external to the organization, must be founded **on guide values**, inspired by the **professional ethics**, the **legality** and **compliance** with the normative, the **transparency** and **coherence**.
- **Creating the identity:** the organization in setting up communicative relations (internal and external) with its public must be original and easily recognizable so that it allows its identification. An **"identity card"** of the public administration based not only on symbolic supports (logo, trademark, emblems, etc.) but it is also the result of a series of systematic strategies**.
- **Giving ourselves targets:** to better plan the organization and the actions of communication we must give ourselves **SMART goals** (Specific, Attainable, Relevant, Time-Based) which can be **checked** looking back. The creation of a protected green area in a big town where people could walk, drive bicycles, take the children to play in the open air, respecting the nature, for example, is generally perceived by the citizens as an improvement of the urban quality. Having the reforestation of a tropical area as a target, on the contrary, can arouse less interest in the community because they do not feel the same immediate benefit in *Environmental terms*. Only after a specific campaign the public opinion would get awareness about the importance of actions of compensation meant to reduce, with tree plantation, the global emission of CO_2.
- **Being able to relate to the *Media*:** the public administration has to be able to balance its own targets, that is the public's perception in order to obtain through the **Communication Plan** (*pending image*), with the *real image* created by the addressees after the decoding of the **Communicated Message** (*found image*).
- **Looking for professionalism:** the search for human resources is a delicate moment for each public administration, sometimes a critical one because of the risk of recruiting unskilled workers. The *Environmental Communication*, on the contrary, needs *specifically trained human resources* whose certified *curricula* are accompanied by relevant professional experiences.
- **Acting at several levels:** the *Environmental Communication* Plan must foresee **more levels of action** corresponding to an equal number of professional roles: **(a)** *Internal Communication*: *Intranet*; **(b)** *Communication to the Mass Media*: relationship management with external individuals; **(c)** *Communication to the citizens*: management of suitable data to communicate *Environmental messages* and information to citizenship.

- **Disseminating communication:** the Communication Plan must provide a different **form of communication** according **to the target to be achieved**. Some actions will be directed to sustain and reinforce the key values: *strategic communication; training communication; functional communication* and *marketing oriented approach*.
- **Using the right tool at the right moment:** the Communication Plan, furthermore, will be able to *specify the means of communication* for each planned action. They could be written, oral, visual, computerized and technological.
- **Interacting with the System Documents:** in order to improve the relationship between the public administration and the citizens to reinforce the transparency of the "*institutional language*", there are different, simple and immediate tools of public communication. The **Social Balance** communicates periodically and voluntarily, the administrative, financial and accounting results achieved. A *medium* of communication based on criteria, methods and qualified information. With the same targets, there is the *Environmental Balance* and other *Documents of System* under the International ISO norms or the EMAS Regulation (*Eco-Management and Audit Scheme*), drawn up by the European Union, following *Deming*'s cycle: **Plan-Do-Check-Act** (*William Edwards Deming*, 1956).
- **Spreading a good example:** in order to make the communication more effective, the Communication Plan should provide a section devoted to *best practices*, that is the experiences, procedures and **the most meaningful actions from the Environmental and Sustainable point of view** that have proved the best results. Without forgetting the participation to projects promoted by the European Union or by other International Institutions as a key-role in the exchange of ideas, best practices and professional contacts.

As our "*Priority List*" points out, the *Environmental Communication* by the public administration can become an engine for the community change and participated approach to political. A "*local*" dimension that can influence in a decisive manner also at a "*glocal*" level. The everyday "*small gestures*" have a great impact on the *Environment System*.

Green Tweets
@JamesCashPenney #ClearCommunication #Participation #Communication Plan #InternalCommunication #ExternalCommunication #Teamwork #AdvertisingCampaign #Marketing # #InformationCampaign #SocialValues #Ethics #LegalConformity #Transparency #Coherence #Identity #Targets #HumanResources #CommunicationTool #SocialBalance #MassMedia #Citizens #FormofCommunication #CommunicationStrategy #Training #DemingCycle #PlanDoCheckAct #Bestpractice #Environment #Sustainable Development

6.5   Cooperating to Communicate the Environment "In Public"

Setting up an open, constructive and civil dialogue is the "new" key for a *"green"* success.

At the beginning of the years Nineties, many projects promoted by public institution involving the citizens in the *Environmental decisional* process have spread out at the international level starting from the United States experiences: an example is EPA—*Environmental Protection Agency*.

"Communicating" can be the creation of working groups whose purpose is **to cooperate together** to collect proposals coming from the "protagonists" of the society: citizens, businessmen, city planners, researchers, consumers, trade and professional associations, etcetera. The *Citizen Advisory Committees* in fact represent a privileged moment of meeting where public and private administrations can exchange ideas and feedbacks. Last but not least, it is also an important moment to verify the efficiency of the *Environmental policies* promoted on the territory (*opinion poll function*) and to inform, train and make the public opinion aware on the main local *Environmental issues*.

A kind of *"talk show"*, moderated by qualified communicators, takes place through representatives. The *"microphone"* is given on rotation following a standard procedure meant *"to give voice"* to all speeches in order to make recommendations to suggest the public authorities in charge to decide or making laws on a specific *Environmental issue*.

"Cooperating" means to **gather spontaneously** for a common *Environmental cause.* More and more often citizens volunteering, non-governmental associations, non-profit organizations-NPO, rangers, environmentalists, country commissioners, government officials collect their strengths giving birth to *"sustainable"* groups to arise the public opinion's awareness or to convince the competent authorities to act. Actual *"Green Partnerships"* are able to affect local or national government's decisions to improve the life quality and the *Environment*.

"Cooperating" means, eventually, trying to foresee or solve critical situations at risk that can become social conflicts. They ask for **forms of collaboration promoted by the Community** living in that specific territory. Insufficient areas destined to public green might create discontent among the citizens that feel deprived of the possibility to spend their free time in a natural *Environment*. Narrow bike paths without safe barriers are often at the basis of heated arguments between motorists and cyclists.

A *cooperative approach* between citizens and public administrations can trigger effective forms of constructive communication provided that the stakeholders are involved with an open dialogue meant to solve the problems by adopting shared resolutions. That means to supply the public with the suitable means of information such as: reports, press releases, written reports, etcetera.

On that purpose new techniques of facilitation have spread out in order to create forms of *exchange of ideas* to guarantee a service of quality able to reach the intended results:

- immediate *output* or mediated *outcome* → **effectiveness**;
- with most suitable resources in terms of costs and involved communicators that is facilitators of public communication → **efficiency**.

Common denominator for the new public communicators is *to promote any possible form of cooperation.*
As far as a fair participation in the debate is concerned:

- supporting **listening moments** in adverbial procedures through *facilitation techniques* as the ones used in *peer-review* (evaluation procedure and selection of articles and scientific projects to verify the objectives and the financing suitability expected);
- the *peer-review* as a means of communication in the public debates to help the immediate contradictory to *problem solving* and looking for new ideas.

Reaching a consensus in fact is through the agreement between the two parties on a detected "common ground". A result not always taken for granted when debating about *Environment* and *Sustainable Development* where there is the high risk of clashing into specific positions determined by political and economic interests. Without forgetting the obstructive positions produced by the syndrome **NIMBY** = *Not in my backyard*—a form of inhabitants' opposition facing the project to change radically the territory where they live. It has happened recently, for example, in some urban areas of North Italy with cases of building incinerators.

Case Study in Short
Communicating the *Environment* in the "greenest city" in Europe → for some decades a town in the North Europe has been known for its "*green vocation*" besides its beautiful lake. The Municipality of *Växjö*, a town of 70,000 inhabitants in the *Småland* county, in the Southern Sweden, started in the early Years 2000, a *public management path* inspired by the principle of *Aalborg Charter*. The need to clean up the lake whose waters were polluted, gave the input for a long-term sustainable Action Plan allowing *Växjö* to be still considered and well-known as the "*greenest*" city in Europe (definition by the BBC of London). A place destined to become the first urban area to ban the use of fossil fuels within 2030. But let up step back. In 1991, Växjö states

his intention of not using fossil fuels anymore. This is the first example in the world. At the stroke of the year 2000, historical date switching between two centuries, the Municipality Council of the Swedish town, considering the urban expansion due to the presence of a university campus and the conse-quent *Environmental impacts*, undertook the adoption of an innovative *system of Environmental management*. A tool responding to the specific needs, flank-ing the traditional system of monetary reporting (*Economic Balance*) and human resources with a similar system taking into account the *natural resources*: water, soil, air, raw materials and climate stability. This is the birth of ***ecoBudget***®, a tool of territorial policy management based on a *set* of indi-cators calculated *per unit* of measurement describing the trend of the use of natural resources (*based on a year of reference*) thus guaranteeing a sustain-able management, more efficient and contrary to wastes. The resulting docu-ment, submitted to periodical approval by the public decision-makers who must approve the long and short-term targets (*master budget*), implies an *accountability* on the part of the public authorities, both internally and exter-nally toward the citizens, made aware by the trend of actions necessary to safeguard the quality of the *Environment* in which they live. Involvement, transparency and quality checks: the passwords to a system bearing fruits. The ecological commitment of the municipality administration and the whole community of *Växjö* is stronger than ever. It is not a surprise if we consider the average growth of 1100 units *per year* of residents to whom temporary pres-ences must be added, thousands of asylum seekers coming above all from Syria and Afghanistan. The 90% of the home heating (a quarter of the electri-cal energy) comes from natural vegetation and biomasses produced naturally by pinewood forests surrounding the urban area. The majority of the electrical energy used by the citizens is produced by the *water mini plants, wind energy* and *photovoltaic plants* or powered by biogas, contrary to the rest of Sweden greatly based on large hydroelectric or nuclear plants. The public transporta-tion, another impactful factor from the point of view of CO_2 emissions, uses biogas as fuel always produced by the biomass present on the territory. The policies about *Environment* are encouraging the population to use the bicycles or *pedelec* (***Pedal Electric Bike***) and electric cars, used also by the local taxi companies. The *Swedish city commitment* does not neglect any "green" detail in the lifestyle of its citizens. A virtuous example of eco-sustainable coopera-tion which has become object of studies at the international level and it shows how an urban community can be itself a *medium* of communication. The com-mitment in the sustainability direction "*makes everything be possible*" in the words of the *slogan* of the PR video (promotional) of Växjö, *the Greenest City in Europe*.

www.vaxjo.se/english (Web Site) - Växjö Anything is Possible (YouTube)

Case Study in Short

Greenwich, from Zero Meridiem to "green heart" of London → within walking distance from London metropolis a protected water reservoir of four acres (1.62 hectares) hosts an interesting "urban" biodiversity of animal species and vegetables. Utopia or realty? The *Greenwich Peninsula Ecology Park* (GPEP) is a real experience of ecologic self-management promoted by a group of voluntary citizens who have been carrying it on for 15 years, since the park was inaugurated in 2002. A marsh that surrounds two lakes, a naturalistic "paradise" for many varieties of migratory birds such as: snipes, water rails, nightingales, swifts, amphibians and insects. It ends in an alder tree wood and a field where wildflowers are growing. The main target of the project is to allow any visitor to discover and benefit of Greenwich "green heart". A *place* that has become itself a *means of communication* for its capacity of awareness of the importance in preserving an urban area where people can walk, *plunged into nature* or taking part to the events promoted by the *Conservation Volunteers*: guided tours and teaching lab for the new generations. A good example of eco-management meant to enhance responsibly the natural beauties with due regard to the biodiversity and the fight against any form of human pollution.

@GreenwichEcologyParkOfficial (Facebook), @greenwicheco (Twitter) www.tcv.org.uk (Web Site)

Case Study in Short

Agenda21   WorkingTable—*Climarchitettura*   **(Climate-architecture). Municipality and Province of Ferrara** → the local public administration of Ferrara (North Italy) is particularly aware and committed in the sustainability and *Environmental policies* and it represents a virtuous example of sustainable and participated *governance*. On the basis of the values of the *Local Agenda21* [approved at the *Agenda 21 City Forum*], in the years 2007–2008 an experimental path of reflection and public discussion on the sustainability and energetic construction issues was started. The project *Climarchitettura* (Climate-architecture) intended to spread out among builders, professionals and citizens working on the Ferrara territory the principles of the bio-construction and bio-architecture through public shared meetings (*round tables*). It gave origin to a shared path managed by *qualified Agenda 21 communicators* who, using the innovative techniques of *peer review* allowed the different stakeholders to reflect on the effectiveness of the *benchmarks*, to check the criticalities on a selected sample of buildings and plants and to work out proposals of upgrading energy to project-based field testing. *Points of strength* → the shared decision-making enabled to realize tools enhancing training paths and investments in energy saving to help more building sustain-ability, reinforcing the relationships between the Public Administration (*Municipality* and *Province*) and the citizenship. ***Added value*** → parallel activities of Work Groups made the creation of public information initiatives be possible based on the ***pilot schemes*** using the ***problem-solving*** approach in order to increase the citizens' knowledge on the critical issues considered. Specific educational paths were integrated in the *school syllabuses* for the purpose of making the new generations and their families aware on issues relating to the construction energy efficiency.

Comune di Ferrara, homepage:
www.comune.fe.it > Check > climarchitettura (Web Site)

Green Tweets
#Communication #Partnership #SmoothRunning #RuleofLaw #Fairness #Administration #PeerReview #EnergyEfficiency #Efficacy #NIMBY #Notinmybackyard #Quality #PublicService #GoodAdministration #AddedValue #ProblemSolving

6.6   Conclusions and Reflections: The "Biodiversity" of the Public Eco-communication

Communicating the *Environment*, a multidimensional tool.

The ***Environmental Communicator***'s role, as already seen in this chapter, is also to address to more recipients simultaneously and to be able to listen to their voices. If we analyse the public sphere from a holistic point of view, in fact, we realize that it does not concern only and exclusively the public administration. On the contrary, it is composed of interconnected realities to form a variegated and complex ecosys-tem. First of all, the citizens, single or in group, the associations committed into *Environmental campaigns*, the scientific and research world, the business sector, the lobbies, the journalism, the student associations and the movements contrary to the "*Environmental cause*".

Furthermore, the ***roles of Environmental Communication*** are so different inside each group of interest: journalists, writers, spokesmen, communication experts, marketing consultants, politicians, project managers and so on. We can say that nowadays the *Environment* and the *Sustainable Development* have won ground in the public debate becoming an instrument of *policy making* each time we have to make decisions even when we are unaware. From the identification of *Environmental criticalities* to the implementation of specific Action Plans to overcome them, the techniques, the forms and the means of communication available can be particularly effective to lead to changes toward sustainable lifestyles.

In the light of this, the *Environmental Communication* should become an essential prerequisite on each process brought into being by the general interest and start a dialogue between the public administrations and the citizens in order to manage the *Environmental criticalities* with awareness and responsibility. To *Communicate the Environment* also means to listen to the public *opinion feedback*, thus becoming an active subject able to express not only criticisms against the activated actions by the public sphere, but also propositions aimed at solutions and improvements for the common interest. *Communicating the Environment* means also to *be assessed objectively for the decisions* taken by the citizens and by any stakeholders as well as to be monitored by independent qualified organizations, in any phase of the policy making until the achievement of the prefixed goals.

Participation, involvement, cooperation of all social parties, here it is the essence of the public *Environmental Communication* that cannot be conceived without the development of the *e-government* **technologies** able to offer the citizens not only easily understandable and usable information but also *online* services to facilitate the democratic network process. All this converts into a best efficiency of administrative acts, best transparency of procedures, best capability of reaction to the *Environment criticalities* and a response to the citizens' requirements on the basis of the real needs of the managed territory. A communication not only as a transmission but also as a social relationship in full respect of the rights of the individual, of the protection of personal data and, in particular, the sensitive and reported data, related to minors.

The technological innovation has allowed to make giant leaps forward from this point of view; in fact, the *ICT—Information and Communication Technologies* evolution helps to cope with the complexity of the *Environmental matter* overcoming the various contradictions that distinguish it. As already reminded, some environmental questions are still characterized by a scientific uncertainty and this increases the risk of lack of interpretation. The extremely technical Language identifying some concepts linked to the *Environment* and *Sustainable Development* may decrease the public opinion interest to deepen the contents.

Building up an Environmental collective knowledge and *reach the multistakeholder deliberation* represents an effective tool of information contributing to create an *Environmental culture* in the community. The ICT can be used as user friendly tool making the difficult *Environmental concepts* or, as the German theorist of design and strategic planning *Horst Wittel* called them, the "wicked problems" accessible.

The *message understanding* is a necessary step to find out reasonable solutions to *Environmental issues* apparently impossible to solve. This is the basis of any kind of a short, medium or long public planning following a *scenario* **approach**. It is necessary to consider all the dimensions useful to represent correctly each *Environmental issue* well summed up in the *Tetrahedral FrameworkTM*, conceived by a group of researchers of "Environmental Evaluation and Nature Capital" at University of *Versailles—St Quentin en Yvelines C3ED* in the framework of certain French projects at European level. The four-dimension economic system based on information and indicators is the evolution of the traditional three-branch concept of *Environmental Sustainability*: economic, *Environmental* and social, introducing a bipartition of the "social" aspects: the political-institutional level actions and those implemented by other territorial entities. The new model has the objective to answer a simple but basic question for the *Environmental Communicators*: "*sustainability of what, for whom and why?*".

Green Tweets
#EnvironmentalCommunication   #Sustainability   #Environmental Communicator   #PublicSphere   #HolisticApproach   #Marketing #ProjectManager   #Feedback   #InformationCommunicationTechnologies #ICT   #Egovernment   #BuildingKnowledge   #SharedDecision #MultiStakeholderdDeliberation   #UserFriendly   #WickedProblems #Knowledge #Messagge #Scenario #TetrahedralFramework

Bibliography

Myria W. Allen, *Strategic Communication for Sustainable Organizations 2016: Theory and Practice*, Berlin (Germany), Springer International Publishing AG, 2015, pages: 328.

Robert Cox, Phaedra C. Pezzullo, *Environmental Communication and the Public Sphere*, Los Angeles – London – New Delhi - Singapore – Washington DC – Boston, Sage Publications Ltd, 4th edition, 2015, pages: 422.

W. Edwards Deming, Mary Walton, *The Deming management method: bestselling classic for quality management*, TarcherPerigee- Penguin Book USA, pages: 288.

Nelli Mikkola, Linda Randall, Annika Hagberg, *Green Growth in Nordic Regions, 50 ways to make it happen*, Stokholm (Sweden), NordRegio NORDREGIO Nordic Centre for Spatial Develppment, 2016, pages: 124.

Martin O'Connor, *Our Common Problems – ICT, the Prisoners' Dilemma, and the Process of working out Reasonable Solutions to impossible environmental problems*, Cahier n°00-06, Paris (France), Université de Versailles Saint Quentin en Yvelines, June 2000, pages: 26.

Martin O'Connor, "*Building Knowledge Partnership with ICT? Some lessons from GOUVERNe and VIRTUALIS*, Cahier n°06-01, Paris (France), Université de Versailles Saint Quentin en Yvelines, January 2006, pages: 20.

James Cash Penney, *My experience with the Golden Rule by James Cash Penney*, Literary Licensing, LLC 5/01/2012.

Graziella Priulla, *La comunicazione delle pubbliche amministrazioni*, Rome(Italy), Laterza e Figli, 2008, pages: 198.

Antonio Saturnino, *Le Agende21 Locali*, Quaderni Formez [The Local genda21, Formez Notebook], Rome (Italy), Formez – Dipartimento della Funzione Pubblica per l'efficienza delle amministrazioni - Area Editoria e Documentazione [Department of Public Function for the administrative efficiency – Publishing and Documentation Sector], Notebook No 17, June 2003, pages: 126.

V.A., *Eco-Courts Storie di Economia collaborativa*, a project financed by the European Commission as part of the Life + Programme, Policy and Governance, 2014, pages: 61. www.life-ecocourts.it

V.A., *The EcoBudget Guide*, ICLEI – Local Governments for Sustainability, Växjö (Sweden), Davidsons tryckeri, 2004, pages: 118.

V.A., *Metodo CLEAR – Dalla contabilità alla politica ambientale* [CLEAR Method – From the Environmental Accountability to the Environmental Policy], *City and Local Environmental Accounting and Reporting*, Life, Milan (Italy), Edizioni Ambiente, 2003, pages: 176.

Web Site List

Municipality of Växjö: www.vaxjo.com

Aarhus Convention, ec.europa.eu/environment/aarhus/

United Nations Sustainable Development Goals, www.un.org/sustainabledevelopment/sustainable-development-goal

Life IDEMS Project – Integration & Development of Environmental Management System www.idems.it

Qualita PA – Pubblica Amministrazione di Qualità (Dipartimento della Funzione Pubblica) [High Quality Public Administration (Department of Public Function)] www.qualitapa.gov.it

European Commission: Aarhus Convention http://ec.europa.eu/environment/aarhus/

Federal Council – Swiss Confederation https://www.admin.ch

Erin Brockovich, Consumer Advocate: www.brockovich.com

Eco-communication: Environmental *Management*, Big Events and Legal *Eco*-language

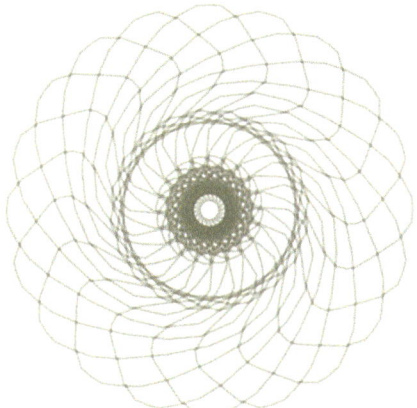

Abstract

The semantic field of the *Environment* and *Sustainable Development* needs a constant commitment of simplification, but not trivialization of complex concepts that if they are correctly understood contribute to improve the *Environmental quality* of the ecosystem. Adapting to local to promote the natural resources has been the study object for many years. Instruments like *Agenda 21* and the *Environmental Accounting* are fundamental pieces of the *Media* puzzle of the *Environmental Communication* quality, supported by *Public Administrations* to help research, discussion, thematic insights, participation, coordination and idea and information exchanges. All this is part of the *"green revolution"* process in which it is essential to build the eco-sustainable identity for each organizational reality. It is not by chance that *Environmental certification* and registration tools aim at a correct and constant communication involving each internal and external structure of the organization using the best technologies and procedures. A *systematic and holistic approach* which increases its importance when *big events* are involved whose recipients of the *Environmental message* reach relevant numerical proportions. *Will everybody be ready to accept favourably the Environment and Sustainability values and accept their rules?* Communicating the *Environment* has become also a prerogative of the *legal language. Is the legislator able to guarantee the language harmonization of the Environmental message in the present globalized society?* A feasible result on condition that we consider the *Environmental Communication* as an added value, a guiding principle, a criterion, a requisite and a right.

© Springer Nature Switzerland AG 2019
M. Abbati, *Communicating the Environment to Save the Planet*,
https://doi.org/10.1007/978-3-319-76017-9_7

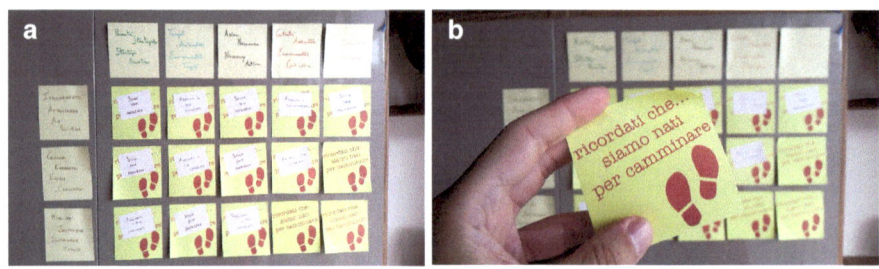

"We won't have a society if we destroy the Environment"
Margaret Mead
Anthropologist (Philadelphia, 16th December 1961 – New York 15th November 1978)

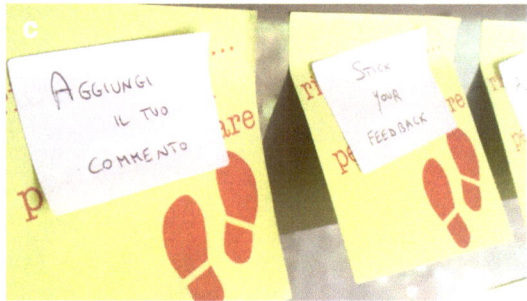

(**a**), (**b**) and (**c**) *Peer Review* Model using reminder stickers by Maurizio Abbati (Author and Copyright Holder)—© 2017 (**b**) "Remember that...we are born to walk"

Reading Proposal: Between Words and Images

The managerial activity, both at the public or private level, makes use of innovative tools of communication thanks to the information and communication technology whose prevailing element is the computer, as we are going to see in *"Chapter 7—Eco-communication: Environmental management, big events and juridical eco-language"*. Nevertheless, this is not enough to ensure the effectiveness of *"communicating the Environment"*. An example is given by the conceptual analysis drawn up by a team to facilitate the maximum participation in the decision-making phase. In order to visualize the agenda items to be discussed and interact with them, the use of **reminder stickers** is a useful *medium*. They allow to build up a conceptual map (*mind map*) where to point out easily the different environmental issues, using various colours, shapes and types of writing (italics or capital letter). This enables the *forum* of participants i.e. experts, policy decision-makers and citizens to answer in real time with very effective but short messages. The use of reminder stickers elicits various mental associations or dissociations according to the urgency or non-urgency, seriousness or non-seriousness of certain actions. At the beginning, they were used for teaching or researching purposes, especially in the English-speaking world, later on they proved to be a good allied also of the modern evaluation techniques. We are thinking about the *peer review*, the evaluation of work by one or more people of similar competence to the producers of the work (*peers*): a process meant to evaluate in details the performance of each single action in comparison to models taken as reference and based on *criteria* scientifically proved or good practices already experimented. In that way, a certified method has been drawn up such as the *Peer Review for European Sustainable Urban Development*—**PRE-SUD**, resulting from a Life Project coordinated by the Municipality Council of Newcastle (England), meant to monitor the state of progress of the environmental performances and Sustainable Development among public administrations. In the figures of the chapter cover (a), (b) and (c) we are showing an intentionally simplified example analysing, with the use of reminder stickers the efficiency of an *Environmental Action Plan* subdivided into thematic areas: *Atmospheric Pollution, Energy Consumption, Sustainable Mobility,* etcetera. The participants at a company, political or administrative meeting are then invited to reflect on the strategic priorities, the necessary actions, the environment criticalities, the targets to be achieved and the subjects to be involved. They can express their own point of view *sticking* their "environmental messages" in the various sections. The **tool flexibility** allows many approaches to the topics dealt with suitable adaptability to the chart structure. The added value to such a kind of system, generally negotiated by qualified facilitators, is also the highlighting of critical situations and environmental emergencies. This method is suitable to evaluate the citizens' satisfaction degree in relation to the actions undertaken or being implemented in order to improve or fill the gaps and inadequacies quickly.

7.1   Local Agenda 21 and Environmental Accounting: The Shared Choice to Communicate the Environment

A local *governance* process to communicate with the territory and to "report" its *Environmental performances*.

An implementing tool of the *UN Conference on Environment and Development* (Rio 1992), *Local Agenda 21*, approved by 173 Governments, is still the most important global *Action Plan* for implementing the principles of Sustainable Development as defined by the *Bruntland Report*. The heart of the project is, in fact, to encourage the birth of "Local Agenda 21" with the local authorities in the adherent Countries in order to bring the citizens closer to the government of their territories. A process that triggers a new concept of *governance,* that is a set of principles of the rules and the procedures concerning the management and government of a company, institution or a popular phenomenon.

The **dissemination of good practices** represents an important step to begin the *communication chain*. The efficiency of the communication depending, as already seen, on a suitable deepening of the issues we are communicating.

The *Environmental* and *Sustainable* **matter**, must be simplified, but not trivialized, in order to adapt itself to the addressees' language. It means to communicate clear concepts of immediate understanding, setting achievable goals easily perceived by the citizens.

Hence the necessity to rely on the reality of the territory to be governed when the actions are being planned but also on the choice of the best tools of communication. **To adapt to the local** is a basic element to communicate to establish a dialogue with the public.

All this must not interfere with the opening up to the outside world. To communicate the efficiency of the *best practices* of a different community enriches our *Environmental knowledge*, **triggers new ideas and encourage a cultural exchange**. Having "*communicators* **of** *best practices*" able to interact with the citizens is a suitable flywheel to increase the collective responsibility and reinforce, at the local level, the values of the Sustainable Development, as confirmed at the *World Summit for Sustainable Development* in Johannesburg (2002).

Local Agenda 21 is therefore a political process, completely voluntary, shared by political representatives, technical staff of the local administration, citizens and other parties concerned. *Subsidiarity, sharing and integration* are the backbone of *Agenda 21* and must influence the **communication chain**. Research, discussion and thematic insights, participation, coordination, ideas and information exchange are only some of the key-words of the *Carta di Ferrara* (Ferrara Chart, Italy), the policy paper that, in 1999, ratified the **dissemination of Local Agenda 21** on the Italian territory giving the start to the *A21 National Coordination*.

These values would risk remaining theoretical concepts if not included in a suitable planning whose targets could be really achieved by each single community. Hence, the parallel need to communicate the citizens and stakeholders the quality of the *Environmental state* where they live, a starting point of the *Local Agenda 21*. It is necessary, then, to communicate *the* **state of the Environment** to the public through a reasoned analysis based on certified, reliable indicators and shared at the national and international level.

The methodologies developed since the Years Nineties are various. Starting from a base year, chosen as a starting point, they describe the *trend*, the tendency to improvement, to stabilization or worsening of the *Environmental data*:

- **CLEAR®** *City and Local Environmental Accounting and Reporting* (2001–2003);
- **European Common Indicators (ECI)** [2001–2002];
- the territorial *management* system *eco***Budget®** [2004];
- the **Aalborg Commitments** [2004; 2010];
- the **Sustainable Development Goals** stated by the United Nations (UN).

They are only some of the main certified sources of indicators shared at European and international level.

Fundamental requisites for the public administrators increasingly engaged to maintain the credibility of their governmental programmes and to look for the highest number of citizens' consents.

A "process": **The** *Environment Accounting System* and a "product": **The Social or** *Environmental Balance* originated from the evolution of the *Environmental Communication* techniques. The latter is a voluntary tool of non-monetary accounting (*Environmental Accountability*), addressed to the inside and outside of the public administration, that *organizes, manages, communicates* information and *Environmental data*, "translating them" into **Environmental indicators**. Physical and accounting units with which the Public Administration aims to develop an important social service: "*answering to*" their citizen or privileged interlocutors of the already obtained results, the ones are being obtained and those not achieved. A kind of *audit report* that derived from an initial sharing of intents among the Public Administration, the Citizens and the *Stakeholders*. It usually follows the most shared models at international level as the one by DPSIR (*Determinants, Pressures, State, Impact and Responses*) developed by the EEA Agency (European Environment Agency) that identifies, in an easy way, the *Environmental* phenomena through their random connections. Or the *ECI—European Common Indicators*, to monitor local sustainability, promoted by the *European Commission*, largely used by public authorities throughout Europe. Or even the indicators in implementation of the *Aalborg Commitments* (*Aalborg + 10*), resulting from the European Conference of the Sustainable Cities held in the homonymous Danish city in June 2004.

The *Environmental Accounting* has taken on increasingly more importance within the Public Administration and the Private Sector: small, average, or big businesses of any kind, owing to their capacity to convey their *Environmental* and *Sustainable* commitment of the organization, a contribute to improve their image in front of their "public": citizens, users or consumers.

To be informed and involved into the *Environmental policies* means also to **improve the reputation**, that is the specific social evaluation that draws the public opinion to associate the "trademark" to quality *standards*. Let us think about *Ferrari* trademark. All over the world it identifies specific cars, with high performance, long lasting, attention to details inside and outside. Result of a long journey that led *Enzo Ferrari* to develop a unique *know how* in manufacturing engines.

The same "journey of reliability" should lead the *stakeholders* to believe the *Environmental performance* by the Public Administration as far as to compare it to a virtuous system for its commitment in the *Environmental* and *Sustainable* field. But everything contributes also ***to validate*** the actions triggering new "green" commitments and good practices that may ***influence*** the lifestyle of the community which they are addressed to.

The ***Environmental Balance*** (Fig. 7.1), from the point of view of the communication, then, follows a clear and transparent style, characterized by a *user-friendly layout*:

- short paragraphs and highlighted key-sentences;
- charts, tables and graphs;

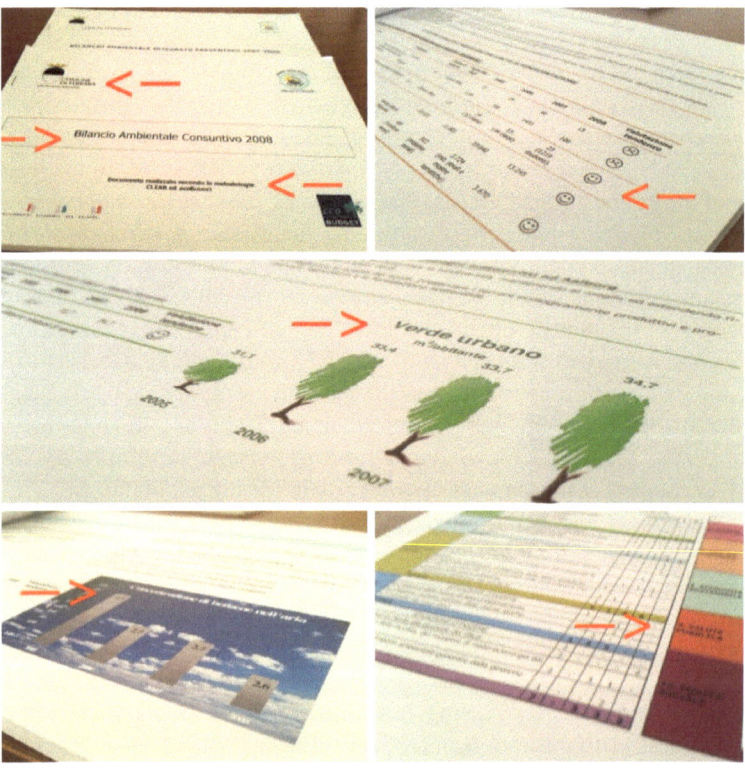

Fig. 7.1 Environmental Balances 2007–2008 following CLEAR and Eco-budget Methodologies co-drawn up by the Author—Ferrara Municipality—Italy—Photo by Maurizio Abbati (Photographer and Co-Author) © 2017

- notes and keys for technical words;
- use of colours to identify areas of competence, that is the areas which the Public Administration acts on;
- graphic signs/markings pointing out the trends in performance: i.e. the *"smiley"* borrowed from the Social Net to adapt to the new generations' language.

Can we think of a "version 2.0" of the Environmental Balance? The answer is definitively positive, thanks to *hyper-textual links, QR-code* connected with the *augmented reality,* or even *3D versions* of visualization of *Environmental data,* whose understanding can be made easier by graphs, histograms or info-graphic.

Its pragmatic, concise and functional nature makes it a communicative tool of great interest. In our opinion, it will be used for a long time to support the decision process in phase of *governance* of the local territory. An instrument of democracy and transparency able to report on the efficacy of *Environmental policies* to the audience. A role of communicative participation that can be enhanced by the new technologies of Information and Communication through the creation of *webinar,* educational and informative sessions via the web, or *question time,* live online sessions devoted to questions between the Public Administration and the citizens, in order to debate on the most urgent *Environmental issues,* close to the citizen's heart.

Here are some key items the *Environmental Balance* must contain, regardless of its form. An oriented reading called *skimming,* a speed reading technique which is to identify quickly the useful clues to give a rough idea of the document contents. It should in fact highlight straightaway ***the eco-sustainable identity of the organization*** indicating:

- Trademark and logo identifying the Organization and its partners;
- Type of Balance, budget or final;
- Year/Years of reference;
- Staff who drew it up;
- Description of the methodological aspects such as: indicators, rules of Law, regulations and other system documents.
- Areas of expertise such as: sustainable mobility, urban development, energy resources, water resources, sustainable economy, public health and so on; *plus,* detailed prepared actions for the targets and results already achieved and those non-achieved yet [in case of final balance];
- Tables with description of the single actions, indicators, unit of measurement, source of data, actors involved with the *trend* corresponding to recent years (generally from 4 to 5).
- Summary of trends and indicators, both positive, negative or stable.
- Any other reference to documents, commitments or agreements to be integrated.

Case Study in Short

Communicating the *Environment* through shared indicators → since long time new communication tools have been considered to adapt the "language" from "technical" to "political" in its etymological meaning linked to the Greek concept of "polis", that is a city-community. An accounting in which culture, knowledge, transparency and responsibility enhance the sustainability policies. These are some of the fundamental values of *CLEAR Method®— City and Local Environmental Accounting and Reporting* derived from the 3-year experience (2001–2003) of the first European Project, with Italian heart, of *Environmental accounting* applied experimentally to eight local authorities whose leader was *Ferrara* Municipality in Italy. The *CLEAR®* path is based on many levels corresponding to three basic phases. (a) **The Environmental data collection** necessary to find out the *Environmental policies*, through natural resources of the territory and among communities, *Environment* and the local economy [**Counting**]. (b) **The creation of a system ordering data** to facilitate the understanding by the addressees emphasizing, through specific physical and monetary indicators. The relationship of cause and effect with pollutants, Man's health and abnormal phenomena that happen in Nature [**Accounting**]. (c) **The communication with citizens and stakeholders** about the choices by the Public Administration and the results achieved, in compliance with the *Environmental issues*. And not only for the existing law enforcement purposes but also, and above all, voluntary [**Reporting**]. A local commitment to improve the quality of the state of the *Environment* besides the same parameters established by Law. All that leads to a document: *the CLEAR Environmental Balance®* that, like the economic balance, is drawn up "budget" and "final" and whose systems and accounting criteria are "tailor-made" compared to the reality of each single Local Public Administration. Common denominator: the subdivision in **areas of expertise** corresponding to *Environmental macro-sectors* forming the key issues of the debate of the Local Administration. For example: (A) for a Municipality: sustainable mobility, energy resources, urban development and waste; (B) for a Province: natural *Environment*, water resources, production activities, roads and transport. And then the **areas of reporting** that is the real actions taken by the Public Administration on the *Environment* and *Sustainable Development*. *CLEAR Environmental Balance®* is a "*system document*" following the same path of approval of a financial balance. The document follows the same procedures involving the same decision-making bodies (City Councils, managers, officials) whether it is examined by the same session of discussion or by another specifically reserved to it.

Life Project: Leading Partner: Comune di Ferrara - Servizio Città Sostenibile e Partecipata [Municipality of Ferrara - Department of Sustainable and Participated City]

www.ec.europa.eu/environment/life/project/Projects/ > CLEAR® (Web Site)

Green Tweets
#Environment #SustainableDevelopment @LocalAgenda21 @Bruntland Report #Governance #BestPractice #CommunicationChain @Cartadi Ferrara (Ferrara Charter) #StateoftheEnviroment #CLEAR #CityandLocalEnvironmentalAccountingandReporting #European CommonIndicators #ECI #EcoBudget #AalborgCommitments #SustainableDevelopmentGoals #EnvironmentalAccountingSystem #EnvironmentalBalance #SocialBalance #EnvironmemtalIndicators #Stakeholders #Ecoidentity #Brand #Trademark #Logo

7.2   When the Environmental Management Systems (EMS) Are Turning "Green"

The *Environmental Communication*, both internal and external, is a factor of vital importance to ensure the success of an *Environmental Management System, how can we make it more efficient?*

"It is not enough to do your best; you must know what to do, and then do best", used to say *W. Edwards Deming*, an American engineer, essayist, teacher and consultant of business management and manager. In this keen observation lies the heart of the "continual improvement" principle, the cornerstone of all the ***Environmental Management Systems*** (EMS). *But what is it exactly?* What we are referring to are *detailed organisation charts*, the result of Community rules such as: *EMAS Eco-Management and Audit Scheme Registration*, founded on a *European rule,* or ISO 14001:2015 Certification, the first international standard in the world for the *Environment* meant to manage each single activity of any organization producing *Environmental impacts*. This applies to any "green" product or service. In this case, the public administrations; the companies; the associations, the cultural or tourist facilities; the banks and other institutes must follow a "logical path" synthesized in the sequence **Plan-Do-Check-Act** defined as ***Deming's* Cycle** named after its creator, whose reflections above we are indebted.

This implies a strong network of contacts between the internal and external parties to the organization, the *stakeholders*, in which the ***role of communication*** is of ***vital importance***. And this is worth for each phase of the process:

- the identification and evaluation of the critical issues;
- priorities-setting;
- planning: actions, roles and responsibilities;
- the follow-up phase of the effectiveness of its action plan (preventing, detecting and correcting errors concerning the *Environmental issues* and verify the effectiveness of the cycle as a whole).

In this cyclical dimension, the communication is an essential step to "blend" the involved subjects as far as the data and the information they carry. It is no coincidence that the communication is always a key-element in any *Environmental Management System*. Despite considering it as an "expected" path, the **Communication Plan** is in reality a kind of "glue" among the various steps of the *Environmental certification/registration* (*Kevin J. Sobnosky*). The success of the *Environmental performances* is naturally followed by the improvement and increase of the productive reality or service provision. The first *Guidelines on Environmental Management and Sustainable Development*, published by the *United Nations* in 1992 considered ***education, information and Environment*** as key-elements.

In which way does the communication work in an Environmental Management System? First of all, we must point out that the communication elements take on different forms inside the *Deming's Cycle*. They often depend on the type of organization the *System* works in. A manufacturing company is sometimes a complex reality formed by different structures dealing with business management and administration plus different production divisions, each of them characterized by a subdivision into departments. In order to apply an *EMAS* it will be necessary to ensure ***the interaction among all the internal structures*** respecting the organization *hierarchical chart* from executives to employees and *vice versa* or, similarly, from a public administration to the citizens/users and *vice versa.*

Hence, the need to manage a *vertical Environmental Communication* addressed to all the structures from the bottom upwards and downwards. The spokesmen can be all the representatives of a professional group: the executives, secretaries, consultants, employees and so on. We must go further by implementing another kind of communication working inside each production structure among: technical staff, managers and workers, that is *the horizontal communication*. Without forgetting the need of ***"interacting" with external parties*** who the organization comes into contact with such as: consultants, contractors, suppliers, transporters, distributors, clients, citizens and users requiring a devoted suitable "channel" of communication.

Beyond such, it is interesting to point out how many different forms of communication may coexist in the same *Environmental Management System* and they should be foreseen and planned *a priori* in order to achieve the targets of *Environmental quality*. The *Environmental issues*, in fact, are an integral part of the business management, both public and private, from the family business to the multinationals. Nowadays, it is essential to assess as our production chain or services affect the *Environment* and how we can remedy to potential negative effects that this entails (*compensatory actions*).

We need to pinpoint exactly the *Environmental criticalities*, and find the most suitable actions and measures to overcome them: corrective and preventive actions or non-conformity. Finally, last but not least, the ability to communicate the public the achieved improvements, in terms of performances achieved and those being programmed for the future scenarios. A suitable flux of information (*transparency*) may affect positively the public opinion even in the presence of activities such as building sites, renovation works, closure of the road and rail networks causing more or less temporary inconvenience. As already mentioned, the **Aarhus Convention of 1998** introduces for the first time ***the right to information*** and so the citizens and the users/consumers are now more careful of the *Environmental messages* and are looking for reliable information to keep up constantly. An important and determining way to the success of a project or to the *Environmental policy*.

All these **communication paths** require **personalized contents** according to the goals pursued. The first essential core of *Environmental information* to be conveyed refers to regulations and standards at the base of the *Environmental Management System* and what concrete actions must be taken in order to respect them. This last aspect focusing on "how can I do" can be particularly useful in the ***System implementation phase***, considering the rapid evolution of the *Environmental matters*. Just to make an example, regulatory measures to amend the *"**Testo Unico Italiano sull'Ambiente**" (**Italian Consolidated Law on Environment**, Legal Decree. No. 152/2006, as subsequently amended and supplemented) are numerous. The term "*consolidated law*" in reality is not technically correct because the legal text does not regulate the whole *Environmental matters* but it submits their application to other external laws. Nevertheless, this type of communication is useful also in the **production phase, service provision, staff management and financial resource management.**

Another key-information not to be neglected is about the ***monitoring of their performances*** that is generally conveyed on vertical form, supplying the tools to verify the state of the art in the path for the *Environmental certification* or for maintaining it. The *certifications/registrations* have usually a 3-year duration and therefore they have to be renewed prior ***audit*** by an impartial external Body. The *audits* are implemented on:

1. the number of violations or the exceeding of the legal limits imposed by law such as: the level of air pollutants, water pollutants and soil pollutants;
2. the achieved or being planned targets for the reduction or the pollution prevention;
3. the level of programme, project and planning implementation and their available *budget* to achieve *Environmental objectives* of energy efficiency, waste reduction, dissemination of the sustainable tourism and so on;
4. the potential *Environmental risk* arising out the harmful use of highly polluting substances or, even, the number of incidents occurred and their impact on the *Environment*.

These are only some of the possible items on which the form of communication is applied.

After analysing the form of communication and its contents, now we turn to the ***media of communication***. In a *EMS* there are a lot of *Media* subjected to the free choice by the organization. From the mere exchange of information during a meeting to other forms of communication such as the telephone, email, videoconference, Internet, Intranet and so on. The verbal communication seems to be the most efficient however, but it is difficult to use in our busy *routine*.

From our personal experience, we know that it is always useful to plan regular meetings on a monthly or weekly basis. This helps to maintain "live" the object of the *team work*, that is the implementation of the *Environmental Management System*. It needs frequent *audit* procedures (internal audits) to put into practice the key principles of the *continual improvement*. Finally, the ***informal communication***, helps to increase the professional "feeling" among colleagues.

The **communication via telephone** is effective (only minor to the face-to-face meeting) because it allows an immediate exchange of information, ideas and points of view. The use of **sms or emails** is particularly useful to collect information, and reasoned data being also a *"written proof"* of the actions.

Videoconferences are highly functional even though infrequently used. Here the technology enables to combine the advantages of a telephone call with a valuable opportunity of "decoding" the messages from the *facial expressions* and the *body language*. The same happens for the **virtual conferences** and the **webinars** (neologism formed by the words web + seminar) guaranteed by the Internet network. The web, both internal (*Intranet*) or public (*Internet*), has become one of the main sources of information thanks to the *speedy in disseminating the messages and the easiness to update data*.

Furthermore, the choice of the tool/instrument must be contextualized on the basis of the characteristics and needs of each single organization. It is, therefore, important to draft a plan enabling the choice of the most effective *Media* at our disposal. In the case of *EMS* this step is facilitated. Most of the *Media* enjoy already an **Environmental Policy**, that is a programme document of the organization laying down the objectives and the general principles of action for the *Environment*. Really, the implementation of the *Environmental* Policy should be the object of a more articulated plan arousing awareness. Operational meetings, round tables, *webinars*, internal audits, newsletters, emails are relevant documents to prevent misunderstandings and oversights. One of the highest risk in the internal and external audit is the demonstration of a superficial reading of the *Environmental Policy* which may lead the speakers to rely on not updated texts especially if they consider the hard copies hung in the bulletin board.

The core of the **spill reporting** about the *Environmental issues* is perhaps one of the most efficient tool to communicate the message. It is a quick communicative form which foresees an answer within the 24 h, based on a concrete case and spread out through emails or faxes.

Each **Environmental Management System** (*EMS*), in order to be maintained, needs to communicate at many levels through several *Media*. A suitable Plan must be prepared not to undermine the foundations of *Deming's Cycle* within complete information. To avoid such a kind of "collapse", an *ISO 14001* standard of *Environmental Management System*, fixing the certifiable requisites by an accredited *Certification Organism*, offers the opportunity of:

1. revising your own *Communication Plan* emphasizing strengths and weaknesses;
2. enabling the setting up of corrective actions necessary to overcome the criticalities, with a shared and certified approach.

Specific indicators to be included in the **System documents**: *Surveillance Plan*; *Environmental Impact-Aspect Table, Procedure of Communication, Environmental Balance*, and so on, allow an objective **constant monitoring** able to stimulate key-questions:

(a) *What do we need to communicate?*
(b) *Whom do we have to communicate it?*

(c) *How do we have to communicate it?*
(d) *How can we check the correct understanding of what has been communicated?*

The first step to make is certainly the ***Mapping***, preparing a list of possible communicative paths and their potential contents. A kind of *"inventory"* of *Environmental Communication* in which all the subjects involved in the "chain" of transmission of the *Environmental message* are specified, with the indication of their respective roles according to their competences. Herewith some as an example:

- Business Manager;
- Environment Service Manager;
- Coordinator;
- Environmental Manager;
- Energy Manager;
- Environmental Engineer;
- Environmental Lawyer.

The next phase of the *"**Documentation**"* is the second step. We are led to winder how to use the means of communication identified in the first step, at their best. A real schedule, numbered in progression from the most suitable tool for each situation (*scenario*) to the least flexible. Herewith some simplified examples:

- *Is more effective that the Production Manager communicates personally with his colleague of the Packaging Sector/Department—face-to-face meeting—or via telephone, email, fax, videoconference, etcetera?*
- *In case of emergency: who is the responsible of the communication? What is the best means to answer quickly?*
- *We have to be informed and to know "who does what", to be verified at the internal audit.*

The third step is ***"Creating a unique work tool"*** including a detailed description of each step of communication of the *Environmental Message* identifying the involved subjects, the contents to be conveyed and the way to convey them. This approach allows the increase of the awareness of the *Environmental Management System* guaranteeing a constant flux of information and data addressed to the targeted public.

In the stage of practical application, the crucial role of the *Communication Plan* in disseminating the concept of **Sustainable Development** should not be underestimated. Thus, being in a true sense more *"educational"* in the case of a Public Administration. To introduce the citizens, the commitment for the *Environmental policies* means to influence the individual behaviours for the improvement of the *Environmental quality* and consequently of their territory health. The evolution of *ISO 14001 standards*, in its updated formulation (2015), gives more freedom to different kind of communication with written, oral or partially written procedures, provided they are effective.

We are given below an example of managerial *medium* of an **Environmental Communication Plan** taken from the author's experience in the area of *Environmental Management in the Public Administration*.

Communication Plan				
Addresser (from):	Addressee (to):	Document:	In charge	Media
(1) T.S. ISO 14001 →← Environment Service →←	Environment Service →← T.S. ISO 14001 →←	→ **EMS Reporting** → **EMS Violations** → **EMS Emergencies** **Energetic Consumptions** ← **Communications** ← **Environmental data** ← **Information** Procedures: →**SP Recycling Management** →**SP Emergencies** →**SP Non Conformity** → **SP Corrective Actions, Preventive actions** ← **Comments** ← **Changes** ← **Updates**	Office/Manager Name and Surname Business/Trade Name Contact	📞 ✉ 🖨 💻 Intranet ▦ *e*
(2) T.S. ISO 14001 →← PRO →←	PRO →← T.S. ISO 14001 →←	←**EMS Communications** ← **EMS reporting** ← **Required information on EMS** →**Answers to Requested Information** → **EMS information** Procedures: → **SP Communication** ← **Comments** ← **Changes** ← **Updates**	Office/Manager Name and Surname Business/Trade Name Contact	📞 ✉ 🖨 💻 Intranet ▦ *e*
(3) T.S. ISO14001 →← Emergency Service →←	Emergency Service →← T.S. ISO 14001 →←	→**Reporting** **EMS Emergencies** →**EMS Information** ←**Environmental Data and Information** Procedures: →**SP Emergencies** ← **Comments** ← **Changes** ← **Updates**	Office/Manager Name and Surname Business/Trade Name Contact	📞 ✉ 🖨 💻 Intranet ▦ *e*
(4) T.S ISO14001 →← Energy Service →←	Energy Service →← T.S. ISO 14001 →←	→ **EMS Reporting** → **EMS Information** ←**EMS Data** **energetic consumption** Procedures: →**SP Energy** →**SP WEEE** ← **Comments** ← **Changes** ← **Updates**	Office/Manager Name and Surname Business/Trade Name Contact	📞 ✉ 🖨 💻 Intranet ▦ *e*

Table-Chart "Communication Plan" by Maurizio Abbati (author and Copyright Holder) © 2017

KEY ➜ = incoming communication ← = ongoing communication

T.S. Technical Secretariat, *S.P.* System Procedure, *OP* Operational Procedure, *WEEE* Waste Electric and Electronic Equipment, *P.R.O.* Public Relation Office, *E.M.S.* Environmental Management System

>> Communication Path

(1) ISO 14001 Technical Secretariat → Environment Service/Environment Service → Technical Secretariat ISO 14001

(2) ISO 14001 Technical Secretariat → P.R.O. Public relation Office/P.R.O. → ISO 14001 Technical Secretariat

(3) ISO 14001 Technical Secretariat → Emergency Service/Emergency Service → ISO 14001 Technical Secretariat

(4) ISO 14001 Technical Secretariat → Energy Service/Energy Service → Technical Secretariat ISO 14001

Etcetera, etcetera

The table above, deliberately simplified, is a rough indication at the base of the *Environmental Communication Management* tool on which are emphasized: the **subjects involved**, their **roles,** their **contacts**, the **medium** or **Media** to be applied in order to convey the *Environmental messages* and the updated **System Reference Documents** that cannot be dismissed. The "*communication path*", furthermore, is meant to clarify an essential element for the success of any *Environment-Security Management System*. The integration of the *Environmental Management System* in the organizational processes, on which the *ISO 14001 standards* (2015 version) placed special emphasis, is based on the principle of "**high level structure**".

The **flow chart** can be produced also in a discursive text and be included inside a **System Procedure** (SP) as in the following example:

"The ISO 14001 Technical Secretariat (T.S. ISO 14001) uses emails to report the integration of new documents in the Intranet area to send communications of general interest referring to EMS (Environmental Management System). In some cases, the T.S. choses to send mails to the whole personnel, using the "internal communication channel" centrally managed by the Personnel Department or whether to more restricted groups of addressees […]. In this latter case, the T.S uses its own Internet mailbox: xxx14001@yyy.com. Through this device the Documents of the System are made circulating for their approval and so the communication of the Work Group meetings and the Operational Management are disseminated".

Case Study in Short

IDEMS Project → in the Public Management the *"Communication and Information"* spreads the cross-cutting aspects on different levels:

- the *strategic*, to answer the needs and issues by the citizens;
- the *operational*, where are to be found concrete results through an efficient action plan;
- the *organizational*, resulting by the balance between the public policies addressed to citizens and the institutional targets.

This is the efficient organigram identified in the Project Manual **Life IDEMS** (*Integration and Development of Environmental Management System*, co-author: Maurizio Abbati) that, between 2005 and 2008, investigated with hands on experience, in four EU Countries (Germany, Greece, Italy and Sweden), on the correlation between the *EMS* and the main *Environment Accounting Systems* (EAS) such as *CLEAR* and *Eco-budget*. The model defined by **11 Points**—whose *"Communication and Information"* is Item 10.

List of Partners: *Ravenna Municipality (Italy, beneficiary); Ferrara Municipality (Italy); Mantua Municipality (Italy); Amaroussion Municipality (Greece); Dresden Municipality (Germany); Heidelberg Municipality (Germany); Växjö Municipality (Sweden).*

IDEMS was awarded in 2010 by the European Commission among one of the best Life projects.

Prior actions in an Organization are those meant to:

1. guarantee the access to Environmental Information, both written or conveyed by images, sound or any other *multimedia* form;
2. carry out an efficient *Environmental Communication* giving the possibility to deepen each information, measured according to the recipients' interests and competencies.

Life IDEMS Project - Integration and Development of Environmental Management System (Co-Author: Maurizio Abbati)

www.idems.it (Web Site)

Case Study in Short

The interconnected *Data-Base* → The search of new *means of communication* together with the increasing security need in the filing, management and data transmission, has pushed the scholars to improve the knowledge of such a kind of complex and evolving matter: the ***Environmental Legislation***. Among them we want to remind *Cloud Computing* available, *on demand*, on *Internet* or *Intranet*. This is the main *mission* of the ***ilregistrolegislativosga.com*** *(thelegislativeregisterems.com)*, a start-up project, created by the author, *Maurizio Abbati*, on the basis of his long-standing research *when working as an Environmental Project Manager* at *Public Administrations* (*Ferrara* and *Ravenna*) (Fig. 7.2). This online *Data-Base* is not a mere list of laws, regulations, procedures and agreements but an interactive and interconnected tool. It allows the various stakeholders to exchange data and information *in real time*, ensuring thus their up-dating. This Document of System, in fact, records the *state of the Environment* highlighting graphically and through acoustic signals the due, ongoing and concluded actions. The document structure is subdivided into thematic areas: *Air Pollution, Water Pollution, Energy Consumptions, Noise Pollution, Sustainable Mobility, Green Area Management*, and so on. These are identified by visual icons to look for topics of interest quickly and to send straightforward comments, alerts or emergency requests through a suitable messaging addressed to the interested recipients. In this way, the Data-Base becomes a new *means of communication* appropriate to the maintenance of the *Environmental Management System*, in order to make shared decisions with the partners involved. Besides, it is a virtual platform able to create communication processes such as *webinars, peer review evaluations* or *online debates*, eliciting ideas at long distances. A sustainable *chain of management*, that is "*dematerialised*", because there is no paper consumption.

Ilregistrolegislativosga.com

www.maurizioabbati.it (Web Site)

Fig. 7.2 Presentation and Installation USB of the *ilregistrolegislativosga* © 2017 Maurizio Abbati (Author and Copyright Holder). www.maurizioabbati.it

Green Tweets

#EnvironmentalManagementSystem   #ISOStandards   #EMAS   #Eco-ManagementandAuditScheme   #Education   #Information   #Environment   #CommunicationProcess   #Media   #InformalCommunication   #EnvironmentalPolicy   #Indicators   #Mapping   #Documentation   #EnvironmentalCommunicationPlan #HighLevelStructure #ilregistroleg-islativoSGA (**thelegalregisterEMS**) *#LegalCompliance #PeerReview*

7.3   The Big Events: The Multi-faced Challenge to Communicate the Environment

When the *Environmental Communication* "*thinks big*" it must test new forms to make the *Environmental Message* attractive and effective. How to make it properly?

The *Environmental Communication*, as already said, is the result of a complex process between the *addresser*, who creates and sends the message, and the *addressee*, who gets the message and decodes it. This path is influenced by many factors that may help or not its understanding, according to the sources of "*noise*" it meets "*along the way*". This seems clear, for instance, when we communicate through the *Net* and addressing to a virtual audience composed of millions of *Internet* users, the "*Internauts*". *What happens if, on the contrary, we want to communicate the Environment "live" to thousands of people gathered in a big event?*

First of all, what is a big event? Usually we mean **temporary bigger events**, **of high international importance**, whose visibility is to attract a big multitude of visitors from different Countries, enhancing the host city and the Country concerned. They may deal with *global issues*, but also *local*, exploiting *a multimedia approach*, particularly *advanced* and *spectacular*, emphasized by new structures, technological installations and buildings that require substantial resources, investments and actions. The final target is to increase the awareness of the whole community of people and citizens on the topics covered, thus increasing their cultural heritage.

All this generates a series of criticalities and challenges from the *Environmental point of view* that must be taken into account *a priori* by the organizers. We cannot forget the *Environmental impacts* produced by:

- the work sites building;
- the necessary works for the event;
- the consequences of a massive influx of visitors during the performance or during the phase of management of the structures and facilities, once the event ended.

The territory concerned could get worse by *"side effects"*, from all the *Environmental points of view*, such as: increased production of waste, increased mobility, increased water and energy consumption; whatever the event (sports event, concert, festival, big party or a big exhibition). The capacity to counterbalance the sustainable actions before, during and after the big event is vital and it influences its success significantly. Various studies on the matter have been evaluating the benefits of such an event in sustainable terms: *Environmental*, social, economic and touristic. In this process of *strategic planning* technically called *"legacy"*, communication plays a key role to make it as much as possible *eco-friendly*.

But why should we communicate the Environment in a social moment of general thoughtlessness? Couldn't it have the opposite effect: that is a departure from the Environmental good practices because of excessive communicative "fussiness"? These are only some of the questions the *event managers* wonder about in the phase of planning a large event having nothing to do with the *Environmental issues*. Let us think about a *rock concert*. This kind of events are held in the open air into contact with Nature. *But is this enough to catch the audience on the Environment cause?* The point of the matter seems to be another. *Should we consider the Environmental Communication to make the big event less impacting or consider the event itself as a medium of Environmental Communication?* The emotions felt by the participants, while the events are in progress, are so strong that last for a very long time (Klöckner 2015). So this may cause a *"domino effect"*, a *spill over effect*, useful to raise awareness among the people to follow eco-compatible styles of life (Zeelenberg 2008). This is something that goes far beyond the mere rules of conduct linked to the event.

Is the audience of a big event receptive enough to get an Environmental message? Some organizers could say: *"No, they are not"*. On a closer reflection, nevertheless, the management of an event able to gather thousands of people in a specific place justifies the need of communicating *Environmental rules*. We mean in particular those strictly referred to *security, becoming more and more severe due to serious news stories and the threat of terrorism*. On the other hand, some big events traditionally are considered a privileged stage for debating on social and *Environmental issues* being, sometimes, the *mission* of the event (Sharpe 2011). Generally, who participates to the event is someone well off (considered the high-priced tickets) and thus with economical resources sufficient enough to invest in sustainable actions and *"clean"* technology.

In the light of the above, a spontaneous question arises: *how can we convey a message in a context of such general euphoria and adrenaline?* The *sense of humour* is suggested by the doctrine as a winning recipe (Eisend 2009). Recent studies have pointed out how humour determines a positive attitude in the addressees of the *Environmental message*, besides improving the image of public figures and trademarks. But it is less effective as far as the credibility of the sources of information and *Environmental data* are concerned (Banas et al. 2011). Therefore, its use is not recommended by the other part of the doctrine. The **debate is still open**. Furthermore, it would be necessary evaluating case by case and finding out the most suitable *medium* to catch the public attention. This should be done through an effective *Communication Plan* without missing the generational and cultural *target* in order *to adapt the linguistic code* to the audience of reference. The approach should avoid the creation of barrier between communicators and addressees in order to facilitate participation and cooperation between the two parties and guarantee a high degree of understanding and eco-awareness. The *Environmental message* should not include, then, imperative tenses or other words that could make it as imposed by a culturally or hierarchically superior authority.

The element of Sustainability is often chosen as a prerequisite for the selection of applications of the city and the Country hosting the event. It is also the main object of *audit* and studies on *Environmental impacts* in each phase of the *planning*. Since the eco-sustainable management of an event can enhance the commercial image of the organizers having the possibility to access to natural protected areas, more and more event managers are interested in *Environmental issues*. An example, a drastic reduction in waste means less sweeping costs of the areas involved, *a posteriori*.

The ***Environmental Communication*** is, then, the ***driving force*** of each single action or sustainable process and technological, social and infrastructural innovation accompanying the big event in phase of preparation, development and report *ex post*.

The big event may be the *medium*, or better the system of *Media*, aimed at communicating the *Environment*. Let us think about **Milan EXPO 2015,** the international exhibition entitled: "***Feed the Planet, Energy for Life***", that hosted more than 20 million visitors from May until October 2015 in Milan.

Communicating the Environment at Milan EXPO 2015 has been an opportunity to ***inform***, to ***train*** and ***sensitise***, through the ***ideas exchange*** and ***good practices***. All the volunteers working in the exhibition, more than 13,000, attended awareness meetings organized by experts of the sector aimed at putting into practice correct actions from the point of view of the Environment inside the vast exhibition areas of about 1,100,000 m². They were always ready to act as sustainable "*tutors*" *for the visitors*, with particular reference to recycling of bio-plastics and traditional plastics. The various *Media* used were addressed to the big public: seven informational videos, broadcasted by the 44 EXPO *totems* and a video about the recycling of the cardboard packing for the children. Furthermore, a kind of ***communication-reporting*** was able to communicate the result of the virtuous environmental policies. The ***Environmental Counter*** made the exhibition easy to understand, thanks to a suitable graphic and musical accompaniment and this allowed the visitors to be periodically informed, every fortnight, on the *trend* of recycling and waste.

Besides that, there were forms of "***experimental communication***" like the installation of 30 benches realized with recycled packaging, tangible and visual metaphor of the *Circular Economy*. Finally, the *Recycling Tube*, an interactive video installation, designed by the Agency *Piano B* and by the architectural *Principioattivo* simulated the path of the packaging waste, EXPO waste included, from the recycling bins to the recycling start, to the transformation in a new product as *secondary raw material*.

Milan EXPO 2015 was the right place to compare with ***innovative forms*** of ***Environmental Communication*** interpreted in different **Pavilions** representing single Countries and the international community. The peculiarity of their own architectural structure was able to be easily assembled and disassembled to ensure a new social and *Environmental use* after the exhibition.

An example the ***Azerbaijan Pavilion*** (Fig. 7.3), considered by many one of the best for its form and innovation. Designed by the Italian Architecture Firm *Simmetrico*, founded by *Daniele Zambelli*, its objective was to represent the *Environment* and the Nature of the Caucasian, whose biodiversity is unique. The

Fig. 7.3 Milan EXPO 2015 www.expo2015.org—Azerbaijan Pavilion: "Azerbaijan, Treasure of Biodiversity"—Designer: Studio Simmetrico www.simmetrico.it—Photos by Maurizio Abbati (Photographer and Copyright Holder) © 2015

structure, totally extended on three floors, more than 1000 m², "communicates" *Environmental values* for the **predominant use of wood**, shaped in a sinuous structure and modelled in thin sheets, technically called *louvers*, with the function to protect the structure micro-climate, thus rationing the energetic consumptions. An architectural metaphor of the Azerbaijan wind and the cultures of its communities. A perfect framework of the three biospheres devoted to Geography and Nature, to Biodiversity and the relationship between tradition and innovation.

Last generation **audio–visual Media**, in the indoor areas, enabled the visitors to interact tangibly with technology through **3D projections, augmented reality, multimedia walls, light sculptures,** and **iconographic structures** symbolizing natural elements like air, water, soil, pomegranate plant (national fruit) and the silhouette of an eagle, symbolising both the natural world and the shape of the Country. Everything was united by *robotic tulip fields* able to interact with the public by hands movement. Just to remind us that also our human body is a producer of energy that, if stored correctly, is a real source of *sustainable and renewable energy*.

The **innovative Environmental Communication** could reach a visual impact particularly efficient through the clever **mix** of **visual effects** and of the **semantic strength of words,** that is **visual and non-visual communication.** That was the challenge of **Spain Pavilion** (Fig. 7.4), designed by the firm *B720 Arquitectos* of *Barcelona*, led by *Fermín Vázquez*, that was notable for two fundamental aspects, under the communicative aspect. First of all, for the architectural structure inspired by a greenhouse, subdivided into two spaces to symbolise the tradition and innovation; completely built with natural materials: wood, cork and esparto grass (common plant species in the whole Mediterranean area).

Secondly, the journey into *"the language of the taste"* through the creation of more *Environments* in which every element: walls, ceilings, paper sheets, waterfalls, lights and videos were communicating sets of words belonging to the semantic area of *Environment, Food, Tradition and Culture*, translated into the various Indo-European languages. The words are *trait d'union* between the visitor and the *Environment* around becoming *Media* to elicit idea exchanges, information and competencies. In other words, a **communication chain** made even more original by the rooms covered with *interactive dishes* able to create, both individually and synergistically, images of ingredients and recipes of the Mediterranean tradition. As a Result of a sustainable management enhancing the *zero kilometre products* in harmony with Nature.

Many Pavilions aimed at communicating the *Environment* through their technological structures allowing the creation of real *vertical vegetable gardens challenging* the laws of gravity. Among them the **U.S.A Pavilion** (Fig. 7.5), designed by the team of *James Beard Foundation* and the *International Culinary Center*. Exploiting a kind of hydroponic farming mounted on movable panels, able to follow the solar radiation, they realized a **vertical vegetable garden** cultivated in fruits and vegetables and powered by a system of collection and recycling of rainwater. An architectural element that besides offering an example of urban farming, played the role of *"catch phrase"* to encourage the public to visit the indoor space aimed at giving more information about the nutritional properties of fruit and vegetables and the

Fig. 7.4 Milan EXPO 2015 www.expo2015.org—Spain Pavilion: "The Voyage of Food"—Designers: Fermín Vázquez Arquitectos www.b720.com/es—Photos by Maurizio Abbati (Photographer and Copyright Holder) © 2015

biological and sustainable techniques of cultivation with technological installations, fountains with digital curtain and 3D screens.

At *Milan EXPO 2015* the eco-communicative experimentation was not limited to big structures. Even the tiniest element, coming from an ordinary object, was designed to become a *medium* of communication thus overcoming the frontiers of the traditional tools of communication.

The *Principality of Monaco Pavilion* (Fig. 7.6) was a perfect example of *integration between traditional and experimental Media*. **The whole exhibition area,** designed by the German Arch Firm *Facts and Fictions*, was conceived as a "*warehouse of ideas*" with constant references to the sea which the Monegasque territory is closely linked to. A journey among visual images, sounds, words, symbols and interactive installations. The common denominator of the thematic *Environments* was the "*wooden crates*", visual metaphor of the eco-sustainable commitment travelling

Fig. 7.5 Milan EXPO 2015 www.expo2015.org—U.S.A. Pavilion—American Food 2.0: United to Feed the Planet—Designers: Biber Architects www.biber.co in cooperation with GLA Genius Loci Architettura S.r.l. www.gla.it—Photos by Maurizio Abbati (Photographer and Copyright Holder) © 2015

around the world. Real shipping crates for the freight transport transformed into *means of communication* through the text prints, info-graphic, video screens, LED lightings special effects and video projections. Being able to interact with the visitor by giving information on different key topics about sustainability. From the *Environment protection* to the sustainable fishing, from the deforestation to the increasing jellyfish colonies in the seas; from the oceans acidification to the cooperation projects promoted by the *Prince Albert II of Monaco Foundation* in *Madagascar, Burkina Faso and Mongolia*. All that framed by a structure, designed by the Italian architect *Enrico Pollini*, derived from the ***recovery of 19 containers*** for sea transport, to stress the subject of "*crate*" as reusing and recycling material. Finally, the *vegetable roof* subdivided according to vegetal species, from the high trunk woody perennials, to the low perennials, bio-indicators of the typical Mediterranean woodland with some *hibiscus*, exotic element, source of an eco-system nourishment to be preserved.

Fig. 7.6   Milan EXPO 2015 www.expo2015.org—Principality of Monaco Pavilion: Pavillion Monaco—Designer: Enrico Pollini, architect @monacopavilion (Facebook)—Photos by Maurizio Abbati (Photographer and Copyright Holder) © 2015

Green Tweets
#EnvironmentalCommunication   #MultimediaApproach   #Environmental Message   #SenseofHumour   #EXPOMilan2015   #Information   #Training #Awareness #ExhangeofIdeas #BestPractice #NewFormsofCommunication #CommunicationChain

7.4   Legal Language and Environmental Communication: Utopia or Reality?

The legal language as a *medium* of *Environmental Communication* needs a precise terminology and information quality, but *how can the legislator communicate effectively? And to which extent?*

The *Environmental Communication*, as seen in our journey so far, must follow *guide values* and *principles* encoded in policy documents, procedures and legislation linked to international *standards* or of legal nature. Communicating the law is in fact a duty and a commitment. For this latter reason, the legislator is more and more often called to deal with *Environmental* and *Sustainable* issues penetrating into linguistic-semantic areas particularly complex and originally out of the *"legal system"* and lawyers' training. In doing so the legislator must refer to an international dimension and not only local. Inside of the European Union (EU), for example, the national legal systems are, in many cases, subdivided on more local levels: (1) Regions, Provinces, Municipalities and Metropolitan Areas in Italy; (2) *Regiones Autónomas*, in Spain; (3) *Länder*, in Germany; (4) *Régions*, in France etcetera, and they are coordinated with **the European legislation** that is with:

– decisions;
– regulations;
– guidelines and opinions;

and with the **international legislation**, whose sources are from:

– agreements or treaties;
– customs;
– unilateral acts;
– international organizations' acts.

The *"Environmental-legal Communicator"* is generally identified with the subject who draws up documents. The **legislator** therefore, when *drafting*, has to make the legislative act understandable and being known and accessible as much as possible. He should look after both the linguistic aspects and the structural ones without forgetting the institutional role that is officially typical of the *legislative communication*. He [the legislator] must follow the different *linguistic registers* linked to the ranks of the regulation source: the same words, in fact, may have different meanings if used in a *constitutional Charter* or for *a simple Act* (*primary source*) or an *administrative Regulation* (*secondary source*). The **Environmental matter,** anyway, requires **further challenges**. The *Environmental language* makes frequently use of **technical formulae** that may be rather difficult to be understood by an average reader since they belong to a **different linguistic code** and the legislator is supposed to mediate this difficulty. The **multidisciplinary Environmental matters** provide for engineering and architectural terms as well as economic, legal and social words. All that means a **greater commitment** to the *legislator* in order to ensure *clarity* and *precision*, characteristics of the typical nature of any piece of legislation.

The good understanding of a legal act depends on many factors such as: sociological, psychological, pragmatic and not strictly linked to the consulting of legal texts. Some factor examples: the knowledge and cultural *background*; major and minor willingness to deepen the topic; the values; the traditions and the real possibility to put into practice the contents of legal obligations. *"There is not democracy if the laws are hung so high that they cannot be read"* used to say, with an acute vision, the philosopher *Georg Wilhelm Friedrich Hegel*, a member of the German idealism.

First of all, the legislator has to find the tools conveying effectively the contents of the legislative acts with the final purpose to comply with their regulations and avoid the penal or administrative sanctions. A role of **communicative mediation** nowadays considered a value of democracy. On the other hand, the final result is not far from the one already dealt with in other chapters of the *Environmental Communication*; that is the addressees' attention and the influence on their behaviours towards eco-compatible objectives, reinforced by a binding legislative apparatus (*Command and Control approach*).

Making the legislative text be understood is a prerequisite for its spreading out and *"existence"*, with consequences not to be undervalued. Since the end of the Years Nineties the jurisprudence has stated, in many European Countries, how the knowledge of the legislative text is an essential condition to enforce the rights. To such an extent as to represent, in case of serious ambiguity, a serious threat to the principle of law accessibility expressed by the well-known Latin *brocards* [legal maxim] *"ignorantia legis non excusat"*, ignoring the law is not excuse. The same maxim is to be found, for example, in French language: accessibility and intelligibility of the law, in the French Constitutional Jurisprudence, in the interpretation of the Spanish Constitutional Court and in the Italian Constitutional Court.

The **legislator**, therefore, is searching for a linguistic *medium* useful for the understanding a specific text, avoiding a confused and ambiguous vocabulary, and thus turning himself into a *"qualified Environmental Communicator"*. Hence, the sections devoted to "definitions" generally put before a sequence of various articles of the law. Otherwise explanatory cards accompanying the laws; or even disclosures; or the drafting of coordinated texts to facilitate the understanding of key-concepts, essential to ensure homogeneity in the application of the legal requirements. An *explanatory approach* that cannot ignore the technology information (IT) *medium* of the *Environmental legal language*, used both for creating the text and for making the contents more suitable and convincing. As well as the hyper textual links to further legislative, jurisprudential and technical references, or of any other nature necessary to a correct interpretation of the legal requirements.

The *"Eco-communicator Legislator"* is given a great responsibility that, in some cases, takes the shape of a partnership among various legislators, working in synergy of communication operators. An example: the European **INTERREG IIIB LexALP Project—Legal Language Harmonisation System for Environment and Spatial Planning within the Multilingual Alps** (January 2005–February 2008). A virtuous example of interregional cooperation promoted by international, national and local public bodies aimed at **harmonizing legal terminology** used in the **Alpine Convention**. An International Treaty undersigned (in September 1991) by the Countries concerned with the Alps Ranges: Austria, France, Germany, Italy, Liechtenstein, Principality of Monaco, Slovenia and Switzerland, as well as by the European Union to manage, in a sustainable manner, the territory to protect the Alpine eco-system.

In these places further problems of communication, owing to the different linguistic groups such as: French, Italian, Slovenian and German areas, arise beyond the understanding of technical terminology and their different legal systems.

Primary objective of the three-year project was to enable the understanding of the Convention Implementing Protocol: "*Regional and Sustainable Development planning*". A group of legal experts drew up a collection of legal materials (*Corpus*), and a terminology database (*Term Bank*) and legislation (*Bibliography database—Bib Base*) subdivided into five thematic areas: (1) Protection of Nature and landscape; (2) Transports; (3) Regional economic development; (4) Rural areas; (5) Urbanized areas. The purpose is to find out a unique system of classification able to adapt to legal texts belonging to different systems.

Certainly, an interesting type of *Environmental Communication* whose purpose is to reorganise the legislation and promoted by a group of legal experts, called *Harmonization Group,* responsible of the lexical validation in the various languages concerned. The *Informative System LexAlp* makes the *lexical Data-Bank* available online on the website of the *EURAC—European Academy of Bozen* (Italy). This Data-Bank is the result of a team work focused on legal concepts linked to a sustainable local planning respectful of the *Environment*. Finally, the lexical cards, subjected to a continuous updating, represent a precious shared *medium* of a qualified legal-environmental information, for a fair and correct communication (***the terminological accuracy***).

Consequently, the necessity to pursue the ***legal certainty***, a principle by which any single individual is allowed to assess and predict the legal consequences of his conduct according to the current legislation. This is to apply both to real, certain and traceable sources, and to clear terms. The evolution of the IT technologies has introduced *non-conventional forms of written communication*. We cannot help considering the ***smileys****,* that is the small faces used in the *Social Media*; or the use of the computer "*snail*", the well-known @, sometimes used instead of "a" to catch the public attention above all in the advertising and publishing industry. Here are some new Italian spellings: @*mbiente* or ☺*mbiente,* standing for "*Environment*" (Italian: *ambiente*); or English spellings: Clim@te Ch@nge or Clim☺te Ch☺nge, standing for "*Climate Change*" but they are not compatible with the Legal "ecosystem".

Clearness and *certainty* are the essential elements of any legal document. These key principles once were considered of secondary importance. On the contrary, sociological studies have proved how the influence of Media is great for the legislator's and the judicial authority's choices, during the administrative procedures and the criminal proceedings. An irrefutable evidence is the great deal of attention devoted to the judicial reporting in the newspapers. This includes an increasing number of cases involving *Environmental* and Sustainable issues. Hence the need of a right balance among ***freedom of expression****,* the ***right to information*** and ***the data processing***. Particularly the need of protecting the *honour*, the *professional confidentiality* (unless it undermines the right to transparency of the public administration), as well as the *right to privacy* and "*sensitivity*" of the personal data of the subjects involved, unless they are public officials or civil servants. The need to protect the "weakest" and vulnerable subjects, that is the *minors,* having the *right to anonymity*. This limit is applied to the *Media* of communication, *Social Network* included, as a guarantee to preserve the physical, psychological and emotional integrity and social life (*Carta di Treviso*, Protocol signed by the *Italian Association* *of Journalists* and by the *Italian National Press Federation* to regulate the relationship between Media and minors, on the 5th October, 1990). Such delicate issues involve also the information referring to the *Environment* conveyed by a journalist or a Net communicator as we have already considered in Chaps. 2 and 3 of this manual.

The very sensitive crux of the **Environmental Legal Communication** has raised a legal debate in order to find the **best solutions**. An example:

1. *Media* able to enhance the communicative role of the act as a synthesis;
2. sheets of complex Laws, unofficial, but drawn up by experts;
3. the predisposition of quality educational materials for the laws regulating technical matters;
4. *online* juridical database integrated and connected to links of external in-depth analysis able to convey *Environmental messages*;
5. technological tools able to link the binding legislation apparatus to *Environmental volunteering tools* of product and process like *eco-labels*, *Environmental certifications* and *registrations*.

The **Environmental-legal Communicator** must be aware of the recipients, anyway, that is the citizens or qualified groups like institutions, foundations, professional associations and professionals to adapt the *"linguistic code"* of articles of Law to the real addressees.

Among the possible existing solutions, we would like to underline the project **ECOLEX,** an international information service about the *Environmental legislation*, as a result of the cooperation among FAO—*Food and Agriculture Organization of the United Nations*; IUCN—*International Union for Conservation of Nature* and UNEP—*United Nations Environmental Programme*. A tool conceived to strengthen the global understanding skill of the *Environmental legislation*. The data collected by the three international organizations are now available on an interactive *online platform.*

The *Database* includes information about the *acts of Law*, the most important *international treaties* and the apparatus of *non-binding regulations* which derive from the evolution of the social and economic dimension of each single Country. We are referring, for instance, to the *codes of self-regulation* of businesses or other organizations, to the *codes of various professional associations*, to the *basic rules* of the *international commercial relationships* (*Lex Mercatoria*) and finally to some *collections of principles* and *rules* drawn up by international, governmental and non-governmental organizations. What is covered by the concept of *Soft Law*. A para-legal apparatus making a substantial part of the regulations always more often negotiated by international bilateral or multilateral agreements linked to the national legislations, the literary production of the legal doctrine as well as to the judicial judgment courts making the *case-law.*

A synthetic abstract of each trusted source document is then available to highlight the key-concept and links to the full text in order to guarantee a deepen study to professionals. The requirements of certainty, legal clarity and precision already mentioned, are then fulfilled. A condition at all obvious in the legal-*Environmental sector*. The regulatory apparatus complexity and the jurisprudence often prevent the research of legal materials through the search engines.

The coverage of the Net allows the *online c*ommunication to reach the most vulnerable areas of the Planet especially from the legal, social and economic point of view. We mean for example those Countries affected by wars or political, economic crisis or subjected to terrorist invasions where the access to any kind information, *Environmental* included, is forbidden by the very strict censorship involving also the *World Wide Web* (Internet). That is what *ECOLEX* means, resulting a precious *medium* to increase the awareness of the *Environmental criticalities* and international regulations to be enforced in those areas with serious legislative gaps.

Undoubtedly, the collection of data is meant to have a rapid access to information in respect to their completeness and multilingualism. The information, in fact, is conveyed in English, French and Spanish, the three most spoken languages in the world. The efficiency, affordability, flexibility and interoperability of this *medium* allow the possibility to meet the information requirements of the public opinion in a very easy, fast way and with savings of human resources. All this, in view of putting together the parties concerned to trigger a *communication chain* that does not know any boundaries, both national or cultural or linguistic. These are the main strengths of this project and we think they should be the starting point to create *Media* in the future, aimed at communicating the eco-legislation.

Green Tweets
#Environment   #Law   #Legislator   #Legislation   #LegalAct   #Eco-LegalCommunicator   #Transparency   #Precision   #Multidisciplinarity   #Values   #Principles   #CommunicationMediation   #RuleofLaw   #FreedomofExpression #RighttoInformation #DataProcessing #SoftLaw

7.5   Conclusions and Reflections: The Eco-diversity of Environmental Communication, a Future Target or a Present Need?

To communicate eco-sustainable values makes us *eco-communicators* who act responsibly.

The variety of *Environmental management tools*, both of voluntary or juridical nature, as already mentioned in the previous chapter, allows a lot of forms of *Environmental Communication* to adapt to the managerial and legislative *mission* and thus meeting the addressees' requirements. The common denominator of this "*Media diversity*" is the protection of the *Environment* in all its forms, with a view of *a holistic vision of the Eco-system Earth*.

The *Encyclical "Laudato Si"* drawn up by *H.H. Pope Francis* is totally devoted to *Environment* and it reminds us to counteract the disastrous consequences due to the unchecked activity of exploitation of the Nature by the human beings and therefore the need to change our behaviours. The pastoral and non-scientific approach does not diminish the importance of the principles expressed. This document, in fact, is full of inspiration for the *Environmental Communicators* and a guideline to start an efficient route to the *Environment* and *Sustainable Development*.

First of all, the spirit of participation to the *Environmental* cause which no one can escape. The contribution of each of us, in fact, is essential to achieve the target of the *"ecological global conversion"* which the Pope refers to. An approach of very constructive cooperation we are invited for *"a conversation which includes everyone, since the Environmental challenge we are undergoing, and its human roots, concern and affect us all".*

A sustainable commitment underlined in the text by the sharp juxtaposition of the lexical terms referred to *Environment* belonging to the *positive semantic area* such as: development, life, progress, growth, joy, truth, authenticity, respect, safety, utility, wellness, vital strength, fullness of values, change, good, equity and eco-centrism; and those words related to the Man's bad deeds not caring the *Environment* of the *Planet Earth* he lives, on the contrary, belong to the *negative semantic area* such as: destruction, degradation, risk, repercussion, deterioration, dysfunction, suffering, problem, crisis, anthropocentrism, worries, pain, pollution, loss, in-equity, weakness and evil.

Secondly, the *strong sense of responsibility* must accompany every action by virtue of sharing the same eco-system with other living creatures that inhabit it. The technological innovation and the progress, be considered as such, must be at the service of *Environmental Communicators* who have the task to point out the strong bond that links the human being to his surroundings. As the *Pope's Encyclical* reminds us the term *Environment* incorporates the concept of eco-system in which the Nature is strictly linked to Man, by a *symbiotic relationship*, in its most literal sense, the one derived from the ancient Greek σύν "with, together" and βιόω "to live".

Thirdly, the importance of the *dialogue and a communicative approach* oriented toward a complete information, a listening, a pro-active debate, a sharing with all the social parties of the territory management policies in a view of the utmost clarity of contents and transparency in the *decision making*.

Finally, the *educational character*, more and more relevant in the *Environmental Communication* as a source for any human change and as a metaphor of social progress but also *Environmental* and *Sustainable*. As mentioned during our *"experimental and practical journey"*, without a specific basic training, the *Environmental Communicator* will find greater difficulty in convincing the addressees of the *Environmental* message to change their minds and their lifestyle turning to good practices.

Can the Environmental Communication be claimed as a fundamental right in the modern society? The answer is positive. To *Communicate the Environment* means the *right to freely express our thinking*, being guaranteed since the *Universal*

Declaration of Human Rights (article No. 19). Naturally, the condition to express our thoughts derives directly from the free access to information which enables a correct *chain of communication*.

The achievement of these rights has allowed worldwide the fight against the poverty and has insured suitable forms of *Environmental protection*. This happened in 2014 in *Bangladesh*, one of the poorest Country in the world, subjected to natural disasters. The population and some *Environmental movements* forced the introduction of a new *Information Law* binding the local administrations and the governmental authorities to report on their *Environmental local management*, with the utmost transparency.

Communicating the Environment then acts as: value, guiding principle, criterion, requisite, right according to the social, economic and legal dimension. The purpose of our dissertation on the "*Communicating the Environment to Save the Planet*" underlines how the *Environmental message* is able to affect our lifestyle turning us into responsible "*Environmental Communicators*" using a lot of *Media* addressed to different types of recipients.

Green Tweets
*#EnvironmentalCommunication   @LaudatoSiEncyclical   #PapaFrancesco
@PopeFrancis   #Environment   #SustainableDevelopment
#EcologicalGlobalConversion   #semantics   #EnvironmentalCommunicator
#DecisionMaking #FreedomofSpeech #UniversalDeclarationofHumanRights
#CommunicationChain   #CommunicatingtheEnvironment   #Values
#GuidingPrinciple #Criterion #Requisite #Right*

Bibliography

Allen Myria, *Strategic Communication for Sustainable Organizations: Theory and Practice*, Berlin (Germany), Springer, 2016, pages: 308.

Atkin David J., Lin Carolyn A., *Communication Technology and Social Change – Theory and Implications*, Abingdon-on-Thames (UK), Routledge Communication Series, 2006, pages: 350.

Banas J., Dunbar N. E., Liu S.-J., & Rodriguez D., *A Review of Humor in Educational Settings: Four Decades of Research*, Communication Education, 60, 2011, pages: 115 – 144.

Bonetti Enrico, Raffaele Cercola, Francesco Izzo, Barbara Masiello, *Eventi e strategie di marketing territoriale – gli attori, i processi e la creazione di valore* [Events and Strategies of territorial marketing – the actors, the processes and the value creation], Milan (Italy), Franco Angeli Management, New Edition, 2017, pages: 321.

Carretero González Cristina, *El Derecho en los medio de comunicación* [The Right of Mass Media], Cizur Menor (Navarra, Spain), Thomson Reuters Aranzadi, Universidad Pontificia Comillas ICAI – ICADE, 2013, pages: 290.

Compton Paul, Dimitri Devuyst, Luc Hens, Bhaskar Nath, *Environmental Management in Practive: Vol 1 – Instruments for Environmental Management*, Abingdon-on-Thames (UK), UNESCO – Routledge, 2002, pages: 507.

Cox Robert, Pezzullo Phaedra C., *Environmental Communication and the Public Sphere*, Los Angeles – London – New Delhi - Singapore – Washington DC – Boston, Sage Publications Ltd, 4th edition, 2015, pages: 422.

Creedy Allen, Dictus Jan, *Towards Enviromental Sustainability – Report of the Peer review of the city of Helsinki*, Publication of the Municipality of Helsinki, Helsinki Environment Centre 5/2009, pages: 57.

Frey Marco, Iraldo Fabio, *Il management dell'ambiente e della sostenibilità oltre i confini aziendali: dalle strategie d'impresa alla governance nei sistemi produttivi territoriali* [The environmental and sustainable management beyond the corporate boundaries: from the corporate strategies to the governannce of the local production systems], Milan (Italy), Franco Angeli, 2008, pages: 375.

Klöckner Christian A., *The Psychology of Pro-Environmental Communication: beyond standard information strategies*, Basingstoke (UK), Palgrave Macmillan, 2015, pages: 271.

Martin Eisend, *A meta-analysis of humor in advertising*, Frankfurt (Germany), Journal of the Academy of Marketing Science, *37*(2), Europa-Universität Viadrina Frankfurt, 2009, pages: 191-203.

Nyklicek Ivan, Vingerhoets Ad, Zeelenberg Marcel, *Emotion, Regulation and Well-Being*, Berlin (Germany), Springer-Verlag, 2010, pages: 331.

Antonio Saturnino, *Le Agende21 Locali*, Quaderni Formez [The Local genda21, Formez Notebook], Rome (Italy), Formez – Dipartimento della Funzione Pubblica per l'efficienza delle amministrazioni - Area Editoria e Documentazione [Department of Public Function for the administrative efficiency – Publishing and Documentation Sector], Notebook No 17, June 2003, pages: 126.

Sharpe William F., *William F. Sharpe: Selected Works*, New Jersey, London, Geneva, Hong Kong, Taipei, Beijing, Shanghai, Tianjin and Chennai, World Scientific Pub Co Inc, World Scientific-Nobel Laureate Series, 2011, pages: 692.

Sobnosky Kevin J., *Effective Communication in Environmental Management*, Environmental Quality Management, Hoboken (New Jersey, USA), Wiley Periodicals, Autumn (Fall) 2001, 28th September 2001, pages: 47-56.

V.A., *The Expo we learned*, Ministero dell'Ambiente e della Tutela del Territorio e del Mare [Italian Ministry of Environment, Land and Sea Protection], 2016, pages: 92.

V.A., *Eco-Courts Storie di Economia collaborativa*, a project financed by the European Commission as part of the Life + Programme, Policy and Governance, 2014, pages: 61. www.life-ecocourts.it

Zeelenberg R & Other Authors, Perceptual processing affects conceptual processing (Scientific Article), Cogn Sci. 2008.

Web Site List

Aarhus Convention, ec.europa.eu/environment/aarhus/

European Commission: Aarhus Convention http://ec.europa.eu/environment/aarhus/

ECOLEX Project – The gateway to environmental law www.ecolex.org

Life IDEMS Project – Integration & Development of Environmental Management System www.idems.it

LexAlp Project: http://lexalp.eurac.edu/index_en.htm

Life pre-SUD Project: http://ec.europa.eu/environment/life/

The Rio Declaration on Environment and Development – Agenda 21 (1992), http://sustainabledevelopment.un.org

United Nations Sustainable Development Goals, www.un.org/sustainabledevelopment/sustainable-development-goal

PART II

INTERVIEWS WITH PROFESSIONALS (IWP)

Mounir Bouchenaki Algerian archaeologist, former manager of the Cultural Heritage Division at UNESCO (*United Nations Educational, Scientific and Cultural Organization*) from 1990 to 2000. He was then appointed Manager of World Heritage Centre, from 1998 to 2000 and from 2000 to 2006 as General Director of Culture. He has been General Director of the International *Study Centre for the Restauration and Conservation of the Cultural Heritage* (ICCROM). Since January 2012 he has been in charge as Special Counsellor of the General Director of UNESCO and ICCROM. He has been managing the *Arab Regional Centre for World Heritage*, based in Manama on the Bahreïn territory.

▶ **Question** Monsieur, la signification du *patrimoine mondial de l'humanité* au cœur de tous les efforts promus par l'UNESCO, au niveau international, va inclure des atouts culturels ainsi que naturels en soulignant un lien solide entre l'archéologie, l'art et l'écosystème de quoi le genre humain fait partie. Non sans raison, les premières formes d'expression artistique humaines, dans la préhistoire, ont souvent représenté des figures inspirées par la nature (ex.: *Cueva de Altamira*, Espagne; *Grotte dei Balzi Rossi*, Italie; *Lascaux et Chauvet*, France; *Wadi* In Djeran, Algerie; *Tadrart Acacus*, Sahara—Afrique du Nord; *Serra de Capivara*, Brésil nord-ouest; *Bhimbetka*, Inde, etc.). Depuis la fin des années '90, de plus, on a commencé à parler de tourisme archéologique (au sens où nous l'entendons aujourd'hui), un sujet qui couvre bien plus de points qu'on peut imaginer, un concordant également sur un élément significatif du Développement Durable: la nécessité de préserver le patrimoine éco-culturel pour les générations futures. Selon votre

Mounir Bouchenaki
@UNESCO (Twitter)
https://en.unesco.org (Web Site)

See English version on p. 347

© Springer Nature Switzerland AG 2019
M. Abbati, *Communicating the Environment to Save the Planet*,
https://doi.org/10.1007/978-3-319-76017-9_8

expérience depuis plusieurs années dans le domaine de protection et préservation du patrimoine mondial de l'humanité, comment on peut communiquer efficacement l'importance de garantir une gestion durable des sites archéologiques ou d'autres *endroits du cœur* par rapport au tourisme et aux écosystèmes qui les accueillent?

▶ **Answer** Il s'agit là d'une des questions les plus importantes que le Comité du Patrimoine Mondial aborde lors de chacune de ses sessions, en insistant sur le fait que l'inscription d'un site culturel ou naturel sur la **Liste du Patrimoine Mondial** ne constitue pas une fin en soi, mais qu'il s'agit de considérer cette étape comme un **premier pas** en vue d'une **gestion rationnelle et intégrée des sites inscrits**.

Les Orientations pour la mise en œuvre de la **Convention** sont d'ailleurs très explicites à ce sujet: "Pour cela il faut se référer aux paragraphes 108 et 109 de ces Orientations spécifiant que:

> "Chaque bien proposé pour inscription devra avoir un **plan de gestion adapté** ou tout autre **système de gestion documenté** qui devra spécifier la manière dont la **valeur universelle** exceptionnelle du bien devrait être conservée, de préférence par des moyens participatifs. Le but d'un système de gestion est d'assurer la **protection efficace** du bien proposé".

Comme le rappelait récemment *M. Kishore Rao*, Directeur du Centre du Patrimoine Mondial dans un "Avant–Propos" d'une publication récente de l'UNESCO consacrée précisément à la question de "**Gérer le Patrimoine Mondial Naturel**", "*Avec plus de mille sites naturels et culturels déjà inscrits sur la Liste, le défi actuel pour la Convention est d'assurer que les valeurs pour lesquelles les sites ont été classés soient **maintenues dans le contexte d'un monde globalisé et en rapide évolution**. La Convention ne se limite pas qu'à reconnaître et glorifier ces lieux exceptionnels: en les inscrivant sur la Liste, les Etats parties à la Convention prennent l'engagement de les protéger pour les générations actuelles et futures. Afin de maintenir les valeurs et l'intégrité de ces sites, les Etats parties doivent assurer une gestion aux plus hauts niveaux envisageables*".

Cette question est devenue primordiale au fil des années et l'acceptation des dossiers de classement est liée à l'exigence de fournir un plan de gestion détaillé pour les sites proposés.

De ce fait, divers programmes de **sensibilisation** ont été développés dans toutes les régions du monde en commençant par les responsables des sites et en associant les communautés locales et les jeunes. C'est ainsi qu'ils ont été successivement créés "**les journées du patrimoine**", les "**journées portes ouvertes**", les ateliers organisés au sein des administrations publiques de même que certains concours définissant les "meilleures pratiques" et les récompensant comme c'est le cas en Asie avec le "*Cultural Heritage Award*". On peut rappeler également le **concours national de vidéos** sur *YouTube* organisé aux **Etats-Unis** à partir de la question suivante: "Pourquoi selon vous les sites du patrimoine mondial des Etats-Unis sont-ils importants pour le reste du monde?".

Sur le plan de la communication, des accords ont été conclus avec les grandes chaînes de télévision (comme par exemple: *Arte, BBC, NHK, TV5, History Channel*, etc.) afin de contribuer à la prise de conscience des différents acteurs impliqués dans la gestion des sites.

La relation entre gestion des sites du patrimoine mondial et Développement Durable a pris de plus en plus d'importance depuis que des études spécialisées en Economie de la Culture ont montré le rôle peu connu jusqu'alors que le patrimoine culturel joue pour l'amélioration des conditions de vie des populations qui vivent à proximité. Plusieurs réunions d'experts portant sur le thème "Patrimoine mondial et Développement Durable" ont recommandé la mise au point d'**indicateurs** pour évaluer comment la **conservation** et la **gestion** du **patrimoine** peuvent contribuer au **Développement Durable**.

L'exemple le plus frappant est celui du **développement touristique** dans les sites du patrimoine mondial que l'on voit représentés sur la plupart des guides et brochures publiés par l'industrie du Tourisme.

L'Italie qui est l'une des grandes destinations touristiques en Europe a vu se développer deux institutions à cet égard, la première à Florence, ville emblème du patrimoine mondial italien, avec la **Fondation *Romualdo Del Bianco*** qui cherche à promouvoir un tourisme culturel où dialogue et échange peuvent être à la base d'un nouvel humanisme. La seconde qui a pris depuis plus de vingt ans le titre de "***Borsa del Mediterraneo per il Turismo Archeologico***" et qui a depuis dépassé les limites de la Méditerranée, est un forum unique en son genre où se retrouvent pour dialoguer, à la fin d'Octobre de chaque année, les spécialistes de l'archéologie et les représentants de l'industrie touristique.

Les organes internationaux spécialisés, tel **l'Organisation Mondiale du Tourisme** (OMT dont le siège est à Madrid) et l'**UNESCO** (dont le siège est à Paris) n'ont pas manqué de s'intéresser de près, depuis plus d'une dizaine d'années, au développement du tourisme dans les sites patrimoniaux. Il est à rappeler que les organisations saoudites ont organisé la première **Conférence mondiale** au **Cambodge** sur le thème "**Culture et Tourisme**", ayant pour lieu symbolique de cette rencontre la petite ville de *Siem Reap*, près du prestigieux site d'*Angkor*.

Ci-après un extrait du Communiqué de presse de l'OMT:

"Plus de 900 participants, au nombre desquels plus de 45 ministres et vice-ministres du tourisme et de la culture, des experts internationaux, orateurs et invités en provenance de 100 pays étaient rassemblés pour la Conférence mondiale de l'OMT et de l'UNESCO sur le tourisme et la culture à Siem Reap (Cambodge), dans le but d'explorer et de promouvoir de nouveaux modèles de partenariat entre le tourisme et la culture (4–6 février 2015)."

▶ **Question** "L'image utilise des langages que la raison des mots ne connait pas…" se plaise à dire Frédéric Lambert, professeur de Sémiotique de l'image et des Medias auprès de l'Institut Français de Presse (IFP)—Université Paris 2—Panthéon Assas. Est-ce que vous partagez cette conviction? Selon votre expérience auprès de l'UNESCO, une campagne de sensibilisation adressée aux touristes, en marquant le fort lien entre le patrimoine culturel et celui naturel, peut-il utiliser un langage visuel plutôt qu'écrit pour faire mieux passer la nuance culturelle-environnementale?

▶ **Answer** *Le professeur Frédéric Lambert a bien résumé l'importance de l'image* dans le monde contemporain où le support visuel des nouveaux *Médias* depuis l'apparition de la télévision, jusqu'aux plus récentes inventions des tablettes et autres téléphones où **l'image, partout présente, a complètement transformé notre perception de la réalité**.

*"L'image ne vaut que pour autant qu'elle soit **capable de modifier notre pensée**, c'est-à-dire de **renouveler** notre propre **langage** et notre **connaissance** du monde"* répond *Georges Didi-Huberman* au professeur Frédéric Lambert (Cité dans "La condition des images", dans Mediamorphoses, 2008)

Je voudrais rappeler à ce sujet une expérience vécue lors du lancement par l'UNESCO de la **Campagne Internationale de sauvegarde** de la **Vieille Ville de Sanaa** au **Yémen** avec M. *Marco Livadiotti*, alors responsable d'une agence de tourisme italienne implantée dans ce pays.

La situation sécuritaire était alors tout à fait normale, même si dans la population yéménite, le port de mitraillettes *Kalashnikov* était une vieille habitude. M. *Marco Livadiotti* a loué une ancienne demeure traditionnelle, qui avait abrité l'Ambassade et le Consulat des Etats-Unis.

Sans en modifier l'architecture traditionnelle typique de la ville, M. *Livadiotti* l'a transformée en hôtel et a organisé des visites touristiques de Milan, précédées d'un atelier de sensibilisation et d'information à partir de photos et de films, parmi lesquels celui réalisé par le fameux metteur en scène *Pier Paolo Pasolini*, sur l'histoire et la culture yéménite.

Il s'agissait alors d'une expérience tout à fait originale et c'est à partir du documentaire, intitulé *"le mura di Sanaa"* (Les murs de Sanaa)—tourné un dimanche, deux scènes qui faisaient partie du *Décaméron* de **Pier Paolo Pasolini**, faisaient aussi appel à l'UNESCO, que la ville de *Sanaa* fut mieux connue en Occident et qu'un tourisme culturel avait commencé à se développer.

Hélas, aujourd'hui, et après un conflit qui touche l'ensemble du Yémen depuis le 25 Mars 2015, cette expérience originale où le rôle de l'image avait joué une place considérable dans la sensibilisation et l'appréciation d'un patrimoine unique est complètement oublié, et ce ne sont plus qu'images de destruction de sites culturels et de drames où les populations civiles paient le prix fort qui endeuillent le quotidien et nourrissent les *Médias*.

J'ai eu la surprise de revoir Monsieur *Marco Livadiotti* à *Sanaa*, lors de la Mission que j'ai effectuée en Juin 2014 pour préparer un rapport sur les restaurations du plafond de la Grande Mosquée de *Sanaa*, effectuée par le Département de restauration de l'Université de Venise.

▶ **Question** Le rôle des *Médias* par rapport à la communication est, depuis plusieurs années, objet d'étude par des anthropologues, des sociologues, des sémiologues, etc. qui se divisent entre ceux qui considèrent un *medium* précieux pour augmenter des connaissances et ceux qui relèvent leur tendance à perturber la qualité des informations. Comment vous considérer le rôle des nouveaux *Médias*, et notamment des *Social Media*, dans la diffusion des valeurs clé comme l'importance de la culture, d'une éducation interdisciplinaire, de la préservation du patrimoine symbole de la diversité culturelle à respecter?

▶ **Answer** À la lumière de cette question qui se pose à tous les chercheurs et les responsables du patrimoine culturel, nous devons nous demander si le passage au tout numérique qui vient transformer notre approche au matériel et à l'immatériel va offrir les mêmes garanties scientifiques que les méthodes traditionnelles utilisées depuis plusieurs siècles.

Il est évident que les **nouvelles technologies** nous ouvrent des **possibilités immenses** notamment dans le domaine de la **restitution virtuelle** mais elles posent en même temps des **questions graves** liées aux **notions d'intégrité** et **d'authenticité**.

Les récentes expériences de **relevés en 3D** effectuées par l'architecte *Yves Ubelman* (Société ICONEM) notamment sur le site de Palmyre dévasté par les attaques des groupes extrémistes de *Daech* ont démontré de quelle manière et à peu de frais, les responsables pourront mettre en exergue ce patrimoine bâti de valeur exceptionnelle.

Depuis lors les **reconstitutions virtuelles** se sont développées et l'on a même vu qu'à Londres, sur la place de Trafalgar, puis à Rome au Colisée, certains des monuments détruits, tels l'Arc de Triomphe du site ou le Temple de Bel reconstruits à l'identique en matériau composite!

D'autre part, devenu culturel, le patrimoine est depuis quelques années soumis au développement de produits dérivés, cédéroms et dévédéroms en tête, destinés à préparer, accompagner ou approfondir une éventuelle visite. Ceci revient à dire que l'on procède à ce stade à une **dématérialisation de la trace patrimoniale**. On passe alors du monde réel au monde virtuel.

La visite virtuelle telle qu'elle est proposée par exemple au *Musée d'Olympie* en *Grèce* présente une avancée considérable dans la compréhension du site par tous les visiteurs et non pas seulement des spécialistes en archéologie. Elle donne à voir une dimension historique restituée dans une représentation spatiale dont la réalité est toute proche et que l'on peut ensuite faire revivre par son imaginaire.

Certes "l'imagination, comme le rapporte une Prépublication de l'Université de Caen Basse-Normandie sur *La mise en valeur du patrimoine culturel par les nouvelles technologies*, est un **outil précieux** mais qui pose des limites sur lesquels intervient la réalité virtuelle. En déambulant virtuellement dans un espace reconstitué, le visiteur doit pouvoir sentir que, ce que l'on considère aujourd'hui être un patrimoine a vécu, traversé le temps, son usure et ses affres, a subi des destructions, des incendies, des guerres, autant que des aménagements, des modifications, qu'il convient de nommer modernisations". (ERSAM—Sources Anciennes, Multimédias et publics pluriels. Directeur *Philippe Fleury*).

De nos jours, les "Applications pour smartphone, tablettes tactiles, QR codes, Visio guides, consoles de jeux mobiles" sont la panacée de tous les jeunes mais aussi des moins jeunes qui s'initient à ces nouveaux outils. Depuis quelques années, ces nouvelles technologies sont utilisées par les institutions de recherches de même que les institutions en charge des musées et des monuments et sites historiques.

Cette arrivée des technologies numériques est particulièrement prisée par le public qui trouve là un moyen plus facile à suivre et à manipuler que la traditionnelle guide imprimée. Au cours de mes voyages l'on voit se développer, grâce aux coûts de plus en plus réduits des produits informatiques, et dans toutes les régions du monde, de nombreuses initiatives qui permettent d'allier l'information et la connaissance du visiteur qui profite ainsi des attraits du ludique et de l'interactif.

Il ne me semble pas que de tels progrès techniques soient en contradiction avec l'importance accordée au patrimoine culturel, matériel ou immatériel, de même qu'à la promotion des expressions de la diversité culturelle, telle que préconisée par la récente Convention de l'UNESCO approuvé par une quasi-unanimité des Etats— Membres de l'UNESCO.

▶ **Question** *"Tourism is an immensely popular global social phenomenon [...] part of 'human exploratory behaviour' that serves as a diversion from the ordinary and helps to make life more interesting and 'worth living'"* (source: *Cameron Walker* and *Neil Carr*: *Tourism and Archaeology: Sustainable Meeting Grounds*, *Routledge– Taylor & Francis*, New York, 2016), comment pousser une *audience* aussi importante comme la population mondiale à mettre en place des actions respectueuses du patrimoine humaine afin d'éviter la récurrence des actions qui détruisent, en quelques secondes, le contexte socio-culturelle et environnementale (notamment: la destruction des Bouddhas de *Bâmiyân*—Afghanistan, en mars 2001, ou la destruction du site archéologique de Palmyre—Syrie, en 2015)? Quelles actions à poursuivre au niveau institutionnel, culturel, social?

▶ **Answer** Selon le **Secrétaire Général de l'Organisation Mondiale du Tourisme**, **M. Taleb Rifai**, "Le tourisme, c'est plus d'un milliard de personnes qui franchissent les frontières internationales chaque année; aussi le tourisme offre-t-il des possibilités immenses de **développement socio-économique** dans les destinations du monde entier. Le **tourisme culturel** a démontré sa capacité à accroître la **compétitivité**, créer des possibilités d'emploi, enrayer l'exode rural, générer des revenus pouvant être destinés à la préservation et alimenter un sentiment de fierté et l'estime de soi au sein des communautés réceptrices. Toutefois, si l'on veut promouvoir efficacement et sauvegarder le patrimoine dont dépend justement le tourisme culturel, il est crucial de s'inscrire dans une **démarche durable** mobilisant les différents acteurs.

La **Directrice générale de l'UNESCO**, *Irina Bokova*, a déclaré quant à elle: *"La* **culture** *forge notre identité et favorise le respect et la tolérance entre les peuples. Elle permet aussi de créer des millions d'emplois et d'améliorer la vie des gens, elle un* **moyen de renforcer la compréhension mutuelle***. La sauvegarde du patrimoine culturel doit avancer avec le* **tourisme durable***... Cet idéal guide nos efforts pour promouvoir la culture en tant que moteur et catalyseur du Développement Durable"*.

Malheureusement et alors que les chiffres du tourisme augmentent année après année, **des conflits éclatent et éloignent pour longtemps les destinations touristiques** des pays qui en subissent les conséquences. Tous ces conflits que l'on voit se développer depuis la fin du XXème siècle et au début de ce XXIème siècle visent de plus en plus souvent des symboles de la culture afin de détruire et la mémoire et l'identité des peuples concernés.

Les conflits, s'ils n'ont plus le caractère international dévastateur des deux grandes Guerres Mondiales, n'en ont pas moins continué d'éclater çà et là, depuis une trentaine d'années, avec en particulier la guerre entre l'Irak et l'Iran, la guerre civile au Liban, puis l'éclatement de la Yougoslavie, jusqu'au fameux "Printemps Arabe" des années 2011–2012, suivis de sanglants soulèvements puis de conflits internes en Egypte, en Irak, en Libye, au Mali, en Syrie, en Tunisie, et au Yémen.

A cet égard, à l'UNESCO où j'ai été chargé de suivre, avec mes collègues, pendant les 30 dernières années, tous ces conflits, l'une des expériences les plus frustrantes que j'ai vécues durant le conflit en Afghanistan a été la destruction des Bouddhas de Bamyan en Mars 2001.

Ce type de destruction a fait réagir la **communauté internationale** qui a eu un **sursaut unanime de condamnation**, faisant appel non seulement aux textes normatifs de l'UNESCO relatifs à la protection du Patrimoine Culturel, mais aussi à ceux de la Cour Pénale Internationale, à partir du moment où ces destructions en série ont été qualifiées par le Secrétaire Général de l'ONU, *Ban Ki Moon*, comme par la Directrice générale de l'UNESCO, *Irina Bokova,* de **"crimes de guerre"** et de **"crimes contre l'Humanité"**.

Interview No. 2: Mariaelena Camerini

IWP 2

Mariaelena Camerini lives and works in Bologna and Bolzano (North Italy). She graduated in Architecture at the University of Ferrara (near Bologna), completing her academic studies at the *Katholieke Universiteit* of *Luven*, Belgium, at the *Department of Architecture and Human Settlements*. Freelance, registered at the Association of Architects of Bologna, she carries out her activity at the Architecture Firm *Camerini*, as designer and site manager. Communication and management are an integral part of her professional commitment.

Premise to Questions 1: The *Environmental matter* is, for its nature, very complex since it includes different subjects thus overcoming the traditional subdivision of the knowledge. For this reason, everybody who communicates the *Environment* should do it responsibly whatever the means of communication. Words, images, sounds, films, symbols, graphic representations, designs, etc. As many communication scholars remind us each action is able to influence, inform and educate a specific "public" on condition to involve it emotionally pushing it to change its lifestyle in order to help the *Sustainable Development*. "Architecture is an art fact, a phenomenon arousing emotions […] Architecture is meant to 'move'" used to say *LeCorbusier*, a real revolutionist together with *Frank Lloyd Wright* for including Nature into his buildings.

Mariaelena Camerini
www.studiocamerini.it (Web Site)

See English version on p. 350

© Springer Nature Switzerland AG 2019
M. Abbati, *Communicating the Environment to Save the Planet*,
https://doi.org/10.1007/978-3-319-76017-9_9

▶ **Question**   Condivide questa affermazione? Secondo la sua esperienza di architetto, cosa significa realizzare un progetto di bioarchitettura *ovvero di architettura sostenibile*? Quali aspetti non possono mancare in sede di progettazione?

▶ **Answer**   Ho in mente un'immagine in bianco e nero: è il 1924; davanti ad una finestra lunga undici metri, *Le Corbusier* osserva il Lago Lemano che irrompe in tutto il suo splendore nella "Petit Maison"; il maestro ha appena ultimato la costruzione di questa casa per i suoi anziani genitori a *Corseaux*. La finestra è progettata per illuminare magistralmente gli ambienti interni, ma anche perché nei giorni d' inverno, la natura ed il paesaggio entrino nel tepore delle mura domestiche. Sì. L'architettura è per commuovere e in questo modo può proiettarci in un percorso creativo, esecutivo e di utilizzo del costruito, volto alla consapevolezza e al rispetto per l'Ambiente, in una sola parola alla sostenibilità. Bisogna intendersi sul significato di Sostenibilità in architettura, poiché troppo spesso questo termine è usato impropriamente per definire manufatti che vengono presentati come ecologici, ma che di fatto non lo sono. "Sostenibilità" è un'intenzione, un approccio e allo stesso tempo un'azione mirata che si traduce in un metodo. Fare architettura sostenibile vuole dire "progettare allo specchio": da una parte la matita, dall' altra il pianeta presente e futuro; stante il fatto che le risorse che ci offre la natura sono limitate, si tratta di dare risposte alle esigenze e confrontarle con l'impatto che esse hanno e avranno sull'Ambiente.

In questa ottica, la progettazione sostenibile si compone di alcuni aspetti fondamentali:

- Un atteggiamento olistico, cioè uno sguardo rivolto all'architettura in relazione al contesto fisico
- (quindi lo studio del verde, delle alberature, del paesaggio, dei percorsi carrabili e ciclabili) e al contesto sociale e psicosensoriale (i comportamenti, le percezioni, le emozioni).
- Un minimalismo di fondo che, a parità di confort, riduca le dimensioni e gli sprechi.
- Un approccio bioclimatico (la scelta dell'orientamento e della forma in funzione del soleggiamento, ombreggiamento e ventilazione naturale).
- La scelta di un sistema costruttivo attento alla vita delle singole componenti e al loro potenziale di riuso.
- Lo studio dell'efficienza degli impianti per ridurre i consumi a parità di prestazione.
- Il ricorso all' uso di fonti energetiche rinnovabili.

Quindi nel progetto non possono mancare materiali locali, rinnovabili, prodotti in maniera ecologica, e che in fase di manutenzione, smontaggio e smaltimento, siano facilmente separabili per poter poi essere riciclati. Allo stesso tempo vanno impiegate lavorazioni non inquinanti, non tossiche, che limitino il consumo di energia.

▶ **Question**   La Natura e le sue forme sono state spesso fonti di ispirazione dell'Architettura anche se ci sono state correnti più propense a separare l'umano dal naturale (es.: *Semper* ispirato da Goethe e dai filosofi tedeschi). Con la nascita dell'ecologismo moderno (tradizionalmente ricollegato alla pubblicazione del libro "Silent Spring" della biologa americana Rachel Carson), si è dato un nuovo impulso alle visioni architettoniche compatibili con l'Ambiente e/o che ne richiamino le forme ed i meccanismi. Le crisi energetiche, la definizione ufficiale di Sviluppo Sostenibile (Rapporto Bruntland, 1987) e Conferenze Internazionali dedicate all'Ambiente (a partire da quella di Rio nel 1992) hanno, infine, innescato un processo irreversibile di sensibilizzazione mondiale sui principali temi ambientali anche per il modo di pensare sia in fase di progettazione urbana che di riqualificazione di aree industriali dismesse. Tanto da arrivare a realizzare, a livello mondiale, progetti eco-sostenibili (es.: Bosco Verticale, Studio Boeri; Molti dei Padiglioni EXPO, ecc.) o che si ispirano essi stessi alle forme della Natura (es.: Aqua Tower di Jeanne Gang, Chicago). Sulla base della sua esperienza professionale, il progetto architettonico può comunicare uno o più messaggi ambientali non verbali, attraverso quello che rappresenta o i materiali con cui è stato fatto? O attraverso le sue stesse forme? Ci può fornire qualche esempio pratico?

▶ **Answer**   Dagli anni '90 ad oggi, molti grandi architetti si sono mossi nella ricerca di un linguaggio formale che diventasse sostanza della progettazione ambientale.

Penso all' opera di Renzo Piano che da sempre si è impegnato su questo fronte, e al suo centro culturale *Tjibau*, costruito nel 1998 in Nuova Caledonia: "scafi arcaici che contengono spazi di tecnologia moderna". Le tecniche di ventilazione passiva dell'edificio si affianca all' uso di materiali e segni mutuati dalla tradizione locale, dando vita ad un unicum di forma, funzione e tecnologia, improntato alla Sostenibilità.

Vorrei citare anche *Heliotrope*, un edificio residenziale cilindrico che ruota su sé stesso di due gradi ogni dieci minuti inseguendo il percorso del sole per catturarne l'energia attraverso un impianto fotovoltaico posto sulla sommità. Partendo dal fiore di Girasole, l'architetto *Rolph Disch*, che lo ha progettato, ha creato un'architettura la cui forma è materia prima di *comunicazione ambientale*.

Ma se osserviamo la prassi più diffusa del costruire, viene a mancare la ricerca di un linguaggio architettonico che sostanzi il concetto di Sostenibilità, e a fronte di prestigiose certificazioni energetiche o di tecnologie all' avanguardia per il contenimento energetico, l'alfabeto delle forme si basa su modelli tradizionali.

Se, come sosteneva *Bruno Zevi*, l'architettura è una scultura che include l'uomo, la sua componente allusiva, metaforica e simbolica è comunicazione allo stato puro.

In questa ottica la forma e l'immagine sono formidabili "materiali" da costruzione; come le pareti formano uno spazio, così l'immagine di un edificio ed i messaggi che ne derivano possono plasmare emozioni, riflessioni, o dar vita a stimoli e comportamenti orientati alla consapevolezza ecologica.

Pregare in una chiesa la cui struttura è realizzata in grandi tubi di cartone, o studiare in una scuola realizzata con la stessa tecnologia, penso alla "Cattedrale di

Cartone" o alla *"Hualin School"* del premio Pritzker per l'architettura *Shigeru Ban*, sono stati gesti quotidiani per alcune comunità terremotate della Nuova Zelanda e della Cina, ma i materiali e le tecnologie impiegate elevano un monito globale sul tema della Sostenibilità e della salvaguardia dell'Ambiente.

Vivere in una casa le cui strutture portanti sono formate dall' assemblaggio di pneumatici riempiti di terra, come nel caso delle "Earthship" di *Michael Reynolds*, diviene paradigma del concetto di riuso e di limitazione degli sprechi.

Sono due esempi emblematici del concetto di Comunicazione e Sostenibilità in architettura.

Le Corbusier considerava "sole, spazio e alberi" come "materiali fondamentali per la creazione urbanistica".

Anche un vocabolario architettonico che includa l'elemento naturalistico quale materiale fondante del costruito, evoca e invoca il senso di rispetto per l'Ambiente, oltre ad assolvere a funzioni pratiche.

Il *Bosco verticale* di *Boeri* che lei citava, con le sue facciate verdi favorisce la rigenerazione dell'Ambiente e della biodiversità urbana, determina la creazione di un particolare microclima, la produzione di ossigeno e l'assorbimento di anidride carbonica, ma parlando di *comunicazione ambientale*, trasmette un messaggio netto: dobbiamo promuovere, a tutti i livelli, un corretto rapporto tra Uomo, Edificio e Natura.

▶ **Question**   Se le dessero l'incarico di realizzare un edificio ad uso abitativo, secondo i canoni della bioarchitettura, che si integri con il paesaggio circostante (sia esso una foresta, o un'area costiera del mediterraneo, o un ambiente appenninico, alpino o insulare) come lo immaginerebbe? Quali sarebbero gli obiettivi prioritari da raggiungere? Quali le problematiche da risolvere? Quale *messaggio ambientale* vorrebbe che comunicasse?

▶ **Answer**   Se mi dessero l'incarico di costruire secondo i canoni della bioarchitettura, penserei a piccole unità, che si nascondano nell'Ambiente naturale che le ospita e ne ricevano l'abbraccio: architettura organica, silenziosa, minimalista, che incoraggi e promuova uno stile di vita semplice e sano. Utilizzerei materiali e tecniche locali: il concetto di Sostenibilità passa anche per il rispetto di assetti sociali ed economici legati alle realtà del territorio. Darei molto spazio agli elementi naturali costitutivi del luogo, siano essi sabbia, verde o roccia, ma soprattutto cercherei di allontanare le automobili; progetterei parcheggi distanti dagli alloggi quanto basta per ricordarci che la vita è cammino e movimento, e che l'apparente scomodità ha in serbo grandi risorse: tempo per pensare, rituali salutari, minimo impatto sulla natura. E poi cercherei per quanto possibile, di rispettare i principi che ho citato all' inizio di questa intervista.

Ma le devo dire la verità. Se mi dessero l'incarico di costruire secondo i canoni della bioarchitettura proporrei, per prima cosa, di riutilizzare edifici già esistenti: ha mai pensato a quanto sia vasto il patrimonio architettonico, talora anche di grande interesse storico o paesaggistico, che versa in stato di abbandono?

Interview No. 3: Francesca Carminati

Francesca Carminati Freelance, she graduated in Conservation of Cultural Heritage. Being an archaeologist she attended the Master in "Preventive Archaeology and Management of Archaeological risks" at *Luiss Business School*. On the basis of her expertise in archaeological excavations, on behalf of Superintendence and Universities, she focused then on the promotion and communication of the Cultural Heritage, becoming an "Expert on the management and organisation of Cultural Events".

Member of the Scientific Committee for the study on the *San Gennaro's Treasury*, she worked hard to publish "*Le Dieci Meraviglie del Tesoro di San Gennaro*" (The ten wonders of *Saint Gennaro's* Treasury) and, later on, the official catalogue on the exhibition "*Il Tesoro di Napoli*" (The treasure of Naples), held at the *Fondazione Roma* (Rome Foundation), in 2014. Acting as Project Manager for the Italian Heritage Award in 2013, for four years she had been founding and managing a Publishing House, specialising in the publishing sector. Through her professional engagement, addressed to the publishing and the cultural event management, she is still tutoring emerging authors, aiming at reinterpreting her "cultural experience" to promote and highlight the cultural heritage and its players.

▶ **Question** "Comunicare" esprime, già nella sua etimologia, un'incredibile valenza intuitiva. La "Comunicazione" viene considerata da molti come un perfetto esempio di espressione sociale che si perfeziona però solo quando il messaggio da veicolare raggiunge i suoi destinatari e ne è compreso, diventando così patrimonio comune per la costruzione di una discussione, di un sapere, di una cultura. Condivide la funzione "sociale" del *comunicare*?

Francesca Carminati
@FrancyCarmi (Twitter)
Francesca Carminati (LinkedIn)

See English version on p. 352

© Springer Nature Switzerland AG 2019
M. Abbati, *Communicating the Environment to Save the Planet*,
https://doi.org/10.1007/978-3-319-76017-9_10

▶ **Answer** Comunicare oggi è fondamentale, ancora più di quanto lo sia mai stato. Ma la cosa più importante è "comunicare bene". La divulgazione oggi è notevolmente cambiata, grazie alle nuove tecnologie. Chiunque può portare all'attenzione generale ciò che vuole trasmettere. Forse non siamo tutti poi così consapevoli del fatto che ogni nostro gesto, parola o azione "comunica" sempre agli altri qualcosa, nel bene e nel male. Se il reperire costantemente informazioni permette di conoscere ed adattarsi ai cambiamenti dell'Ambiente che ci circonda, stando al passo con la società che evolve, nello stesso tempo oggi, grazie ai nuovi sistemi comunicativi, rischiamo di essere sopraffatti da una grande quantità di informazioni che, se comunicate in maniera errata, possono creare grande confusione. Sappiamo però che ciò che non si comunica rischia di "non esistere", ed è quindi necessario riuscire a trasmettere ciò che si vuole, nel modo giusto. Chi comunica deve essere abile e pronto ad adattare la propria comunicazione al *contesto comunicativo*; così come chi recepisce deve essere in grado di distinguere tra la buona comunicazione e la cattiva comunicazione.

Partendo dal presupposto che la Comunicazione è fondamentale in qualsiasi ambito della vita nel momento in cui è necessario interagire con altri individui, ci sono due aspetti da distinguere dal punto di vista professionale. Il primo riguarda la Comunicazione all'interno del sistema sociale di un gruppo di lavoro. Una buona Comunicazione può favorire una buona organizzazione dell'attività e di conseguenza il raggiungimento degli obiettivi prefissi, così come una buona Comunicazione tra le persone può favorire un clima idoneo per un piacevole ambiente lavorativo. Il secondo aspetto è legato alla Comunicazione verso l'esterno, che proprio per la facilità che abbiamo oggi di informare, grazie ai tanti canali a disposizione, deve essere ancora più mirata, incisiva ed estesa. Devo dire che non mancano oggi gli strumenti per permettere questo, ma l'offerta è talmente ampia che riuscire a distinguersi non è facile.

▶ **Question** Quanto influisce la Comunicazione nella sua professione? Adotta "buone pratiche" ambientali per sensibilizzare il pubblico dei lettori?

▶ **Answer** Quando mi sono occupata della produzione dei libri ho sempre adottato una tipologia di carta realizzata con legno che proviene da foreste gestite responsabilmente, che rispettano cioè determinati standard ambientali, certificate dalla FSC (*Forest Stewardship Council*), il cui logo compare in genere nei libri. Inoltre per una scelta estetica, ma anche appunto "ambientale", ho sempre evitato di realizzare copertine plastificate. Inoltre, in occasione di eventi da me curati, cerco di far capire come il nuovo modo di comunicare possa aiutare anche l'Ambiente attraverso l'invio di soli inviti digitali, senza la necessità della loro stampa cartacea.

Ci sono poi le buone pratiche interne che ho cercato sempre di comunicare ai nostri lettori, descrivendole in una apposita pagina *Internet* con una sezione dedicata ai consigli e alle proposte. Venivano descritte semplici iniziative quali: evitare il più possibile la stampa cartacea di e-mail e l'uso di carta riciclata per bozze e comunicazioni interne; ed altre pratiche di buona condotta quali: non accendere le luci durante il giorno, usare un'illuminazione a basso consumo, avere macchine impostate con risparmio energetico automatico, usare l'aria condizionata soltanto nelle ore più calde privilegiando nel resto della giornata l'areazione naturale. Un obiettivo che perseguirò nei prossimi mesi nel seguire nuove pubblicazioni e nuove realtà editoriali sarà poi quello di continuare ad insistere per una distribuzione "green", realizzata cioè con mezzi di trasporto sostenibili.

▶ **Question** Quali strumenti utilizza per verificare l'efficacia della Comunicazione, sia essa visiva, verbale, per segni, semiotica, iconografica, ecc.?

▶ **Answer** Gli strumenti che ho sempre avuto a disposizione per monitorare e valutare la comunicazione sono diversi. Puntando molto sulla realizzazione di eventi legati ai libri, un indicatore che ritengo rilevante è l'affluenza di persone agli eventi e il numero di libri venduti durante la presentazione. Importante è poi l'interesse in generale dei *Media*, dai giornali di settore, e non solo, ai blog, monitorando quindi l'uscita di articoli o recensioni dei libri e degli eventi da me curati, così come l'attenzione sui *Social Network*, che permettono, attraverso strumenti quali *"like"* *"visualizzazioni"* *"retweet"* ed altro, di capire l'*appeal* delle diverse strategie di comunicazione. A questi vanno aggiunti gli strumenti che permettono di verificare l'accesso all'eventuale *sito Internet* aziendale. La richiesta delle librerie e il conseguente volume di vendita, è un'ulteriore indicazione di una buona comunicazione.

▶ **Question** Il "Comunicatore ambientale" è una nuova figura professionale che si occupa di tutti i temi cosiddetti 'green', riferiti dunque al concetto di Sostenibilità Ambientale. Che contributo può dare, secondo lei, l'editoria nella diffusione di messaggi di Sostenibilità Ambientale? L'organizzazione strutturale di un libro, compresa la parte grafica, può realmente influenzare le scelte dei lettori verso il mondo *green*?

▶ **Answer** Purtroppo oggi non basta pubblicare libri su determinati argomenti per sensibilizzare le persone. Ciò che è fondamentale è la campagna di comunicazione che viene realizzata intorno al prodotto. L'editoria, da parte sua, condividendo con il lettore le proprie buone pratiche ambientali e nello stesso tempo l'attenzione a una produzione sostenibile a livello aziendale, può aiutare a diffondere il giusto messaggio "green".

Nelle scelte che noi facciamo tutti i giorni, abbiamo sempre qualcosa che ci attira più di altro. Su questo argomento gli esperti di *marketing*, e non solo, hanno fatto studi di tutti i generi ed è assolutamente vero che determinate immagini grafiche, o effetti cromatici, possono stimolare le nostre emozioni. Il colore, per esempio, è una sensazione che viene percepita dal cervello e che ha effetti sul nostro organismo, e soprattutto sul nostro atteggiamento psicologico. Si pensi al verde, colore distensivo e riposante per gli occhi nonché rilassante e rassicurante. Sempre abbinato, non a caso, a tutto ciò che è riferito all'Ambiente, al naturale. Così come iconograficamente alcune immagini e simboli che ripropongono determinati soggetti, per esempio gli alberi, sono sempre recepite come legate ad argomenti "green". Assodato, quindi, che le emozioni hanno molto a che vedere con la vendita di un prodotto, è sicuramente possibile indirizzare, attraverso l'uso di elementi visivi, la volontà di acquisto di un cliente verso un certo prodotto piuttosto che un altro.

▶ **Question** Quali elementi contribuiscono a darle fiducia sull'autorevolezza del *messaggio ambientale*, sia esso veicolato da una campagna pubblicitaria, da uno *spot* ovvero da un'etichetta ecologica di prodotto o di processo?

▶ **Answer** Sicuramente le certificazioni sono importanti per constatare la serietà e veridicità di un messaggio. Per quanto riguarda *spot* e campagne pubblicitarie, queste influenzano certamente il mittente e quindi l'eventuale istituzione, pubblica o privata (quale che sia: società ambientalista, ESCO *Energy Service Company* o altro), che divulga il messaggio.

▶ **Question** Quali sono, secondo lei, gli sviluppi della Comunicazione in generale nell'editoria? L'apporto di *Internet* e degli strumenti "*Social*" può essere un valore aggiunto per una comunicazione di qualità?

▶ **Answer** Il mondo dell'editoria è ormai da alcuni anni in difficoltà a causa di diversi fattori ed è quindi ancora più importante una strategia comunicativa incisiva che possa riuscire a raggiungere un pubblico sempre più ampio e soprattutto conquistarlo. Portare il lettore ad interessarsi ad un libro, in particolare per le piccole case editrici che non annoverano tra i loro autori personaggi già conosciuti al grande pubblico, diventa la sfida più grande. L'offerta oggi è tale che non basta pubblicare opere di qualità, sia per quanto riguarda i contenuti che per la scelta della grafica e dei materiali con cui si realizza il prodotto libro, perché tutto dipende da come si "comunica".

È innegabile quanto oggi influisca *Internet* nella comunicazione e quanto sia importante l'utilizzo dei "*Social Network*" per far conoscere un nuovo prodotto, stimolare la curiosità e soprattutto per riuscire a raggiungere un pubblico sempre più vasto. In una società dove l'interesse culturale è oggi sempre più relegato ad esperienze dirette, i consumatori cercano esperienze immateriali e non prodotti; in questo caso è la società che aiuta la comunicazione ha trovare la giusta via. Realizzare eventi che non siano solo presentazioni di libri, ma che coinvolgano il lettore da altri punti di vista, può essere oggi il modo per suscitare quell'interesse che porta alla piena soddisfazione nella condivisione emotiva dell'evento.

▶ **Question** Quali sono le sfide per il futuro nel settore dell'editoria?

▶ **Answer** La sfida più grande per ogni casa editrice è quella di farsi conoscere e raggiungere un pubblico sempre più ampio. Ritengo che portare avanti la qualità dei prodotti e degli eventi sia un tratto distintivo essenziale, in un mercato ormai oggi saturo e sempre più agguerrito.

Portare avanti la comunicazione nel migliore dei modi è un impegno ulteriore che non si può né si deve trascurare. L'uso delle nuove tecnologie continuerà ad essere imprescindibile, così come l'organizzazione di eventi culturali che possano anche promuovere una conoscenza partecipata del lettore. Considerata oggi l'enorme offerta sul mercato librario e in generale culturale, bisogna valutare ogni libro come un prodotto a sé che va comunicato in modo differente; per questo sarà sempre più importante individuare il pubblico a cui ci si rivolge: se all'intero mercato o al pubblico nel suo complesso, se ad uno specifico segmento di mercato o di pubblico, o a più segmenti di mercato o pubblico. Occorrerà, di conseguenza, progettare diverse strategie di comunicazione per ogni segmento. Infine, fondamentale continuerà ad essere il monitoraggio ed il controllo dei risultati per capire l'efficacia delle strategie intraprese.

Interview No. 4: Cristina Carretero González

IWP 4

Cristina Carretero González Adjunct Professor of Procedural Law; Coordinator of the Investigation Group: Law and Juridical Language at the Faculty of Law (ICADE), *Universidad Pontificia Comillas*, Madrid, Spain. Graduated in Law at the *Universidad Complutense* de Madrid, *Cristina Carretero González* has been working as a Professor of *Oratoria y Redacción jurídica* (Oratory and Legal drafting) for the *Máster Universitario de Acceso a la Abogacia* (University Master to the access of the Legal Profession) at the prestigious *Universidad Pontificia Comillas* since 1998, focusing on some legal key issues, including the process of modernization of the legal language. She held many workshops and round tables, at national and international level. She is the author, of several academic books, including: "*El Derecho en los Medios de Comunicación*" (The Law in the Means of Communication) and the short Guide: "*Lenguaje claro. Comprender y hacernos entender*" (Clear Language. Understanding and Making us been understood).

▶ **Question** La comunicación legislativa se asocia generalmente a la comunicación institucional, teniendo en cuenta el grado de conocimiento y de comprensión del acto jurídico; se valora así el matiz lingüístico y estructural que va a revestir importancia vital con respecto a la legislación medioambiental en la que el legislador debe regular materias técnicas de otros sectores, ¿Cómo se pueden comunicar efectivamente los contenidos de una ley facilitando la información sobre los principios claves del Medioambiente? ¿Cómo verificar su validez en términos de claridad y calidad?

Cristina Carretero González
@criscarretero1r (Twitter)
www.comillas.edutero1 (Web Site)
Cristina Carretero González (LinkedIn)

See English version on p. 355

© Springer Nature Switzerland AG 2019
M. Abbati, *Communicating the Environment to Save the Planet*,
https://doi.org/10.1007/978-3-319-76017-9_11

▶ **Answer** Es fundamental que la técnica legislativa sea correcta y adecuada. Dentro de esa buena técnica legislativa, un factor fundamental es la claridad de la norma.

Para verificar su claridad y calidad bastaría con tener un protocolo de redacción legislativa en que el parámetro de la claridad estuviera regulado y que contemplara cuestiones esenciales en el lenguaje claro o la lectura fácil o sencilla.

▶ **Question** ¿Para realmente entregar el mensaje "Green" de una ley, pueden ser útiles notas explicativas, esquemas, síntesis? ¿Qué valor se atribuye a esta documentación explicativa? ¿Cómo verificar su autenticidad (no influenciada por intereses particulares)?

▶ **Answer** Me parece que la pregunta da en el clavo. Creo que un buen mensaje, si se quiere hacer llegar a una generalidad de personas, resulta más eficaz cuanto más explicativo es. Y lo explicativo para unos no es lo mismo que para otros. Es decir, hay personas que comprenden mucho mejor con un esquema; otras prefieren un resumen y otras, extensas explicaciones detalladas. Pero hay otras personas que encuentran más completo un mensaje en el que hay imágenes, fotografías o iconos. Creo que la variedad comunicativa conseguirá que todos o la mayoría de los destinatarios comprendan sin problema el mensaje. En mi opinión, el valor de su autenticidad, vendrá asociado a la institución o persona que lo lance.

▶ **Question** ¿Los contenidos legislativos pueden afectar de manera notable los hábitos de los ciudadanos para favorecer un comportamiento más sostenible? ¿Qué necesitan para alcanzar este objetivo? ¿Puedes usted proponer algunos ejemplos concretos a nivel local, nacional o comunitario?

▶ **Answer** Esta pregunta se me escapa de los datos que conozco. Hipotéticamente, entiendo que todo contenido normativo que resulte obligatorio afectará a los hábitos adquiridos. Los que no resulten obligatorios sino meras recomendaciones, pueden tener una buena acogida siempre que se haga una buena campaña fundamentada, clara y apoyada en datos convincentes. Por ejemplo, cuando nos han invitado a reciclar el vidrio o pilas, sin multarnos, por lo contrario, estoy convencida de que cada día vamos adquiriendo más conciencia y se tiende a hacerlo con más naturalidad. Ahora bien, si el lugar al que los ciudadanos tenemos que desplazarnos para hacerlo está lejos de nuestras zonas de movimiento habitual, es posible que solo en ocasiones se haga y en otras no.

▶ **Question** ¿El uso de medios de comunicación disponibles sobre la Red puede aumentar la comprensión y sensibilización de la opinión pública sobre el derecho medioambiental? ¿Qué papel pueden desempeñar las Redes Sociales en el proceso de explicación de la legislación medioambiental? ¿Qué hay de las imágenes visuales y todas las posibles aplicaciones multimedia?

▶ **Answer** Sí. Creo que los medios en redes pueden aumentar la comprensión y sensibilización sobre la importancia de cuidar nuestro Medio Ambiente. Son medios cercanos y rápidos de fácil acceso a todos los usuarios de móvil y si se hace con mensajes claros y breves, aún mejor. Hoy día, tengo la impresión de que son más numerosos los lectores de mensajes breves y claros.

Creo que el papel de las redes es muy importante. Es un hecho que las instituciones públicas, por ejemplo, los ministerios, tienen sus cuentas en redes y se consultan habitualmente.

Las imágenes y aplicaciones multimedia, como decía anteriormente, creo que completan los mensajes consiguiendo que lleguen a más personas que si la información es únicamente en letra.

▶ **Question** Al lado del sistema legislativo, llamado "*command & control*", hay otro que persigue los mismos objetivos de comunicación a través instrumentos voluntarios pero vinculantes por quienes sean miembros (por ejemplo: sistemas de gestión medioambiental, *ecolabel*, certificaciones de eficiencias energéticas, etc.), ¿habría algunas posibilidades de integrarlos en la legislación medioambiental para consolidar la comunicación ecológica?

▶ **Answer** No conozco bien estos sistemas referidos, pero entiendo que, si se integran en la legislación medioambiental, en el caso de que no lo estén, pueden completar las políticas medioambientales. Para consolidar la "comunicación ecológica" habría que estudiar la mejor forma de difundirla a través de las normas.

▶ **Question** ¿Qué tienes que cambiar/mejorar en el sistema jurídico actual para implementar el "sector" de la comunicación de cuestiones de Medio Ambiente?

▶ **Answer** Creo que la base radica, como indiqué previamente, en la mejora de la técnica normativa para que el mensaje que se quiere transmitir sea de lectura sencilla y comprensible y su claridad derive en confianza y en cumplimiento casi natural por los ciudadanos.

Interview No. 5: Alice Comble

Alice Comble   Engineer of Telecommunication, she is the Founder and Project Manager of *GreenMinded*, a start-up developed within the *Euratechnologies* incubator programme of *Lille* (France), a pole of excellence dedicated to the technologies if information and communication. *GreenMinded* was ranked in first place in the METHA European Competition, Europe 2016 (*Maîtrise de l'Energie dans les Transports et l'Habitation*) organized by the *Ecoles de Mines* on the occasion of the award ceremony held in the EVER, (International Exhibition of Environmentally Friendly Vehicles and Renewable Energy), at the *Grimaldi Forum* (Principality of Monaco), in April 2016.

This is the project: the creation of an "intelligent" network of urban containers for fag ends and chewing-gums able to communicate information to the users; raising their awareness on the *Environmental abandonment* without rules of this type of waste through *ad hoc* messages to modify their habits; collect useful data at statistical level allowing the recipients to send their feedbacks thus creating a community of *GreenMinders* united by the philosophy: "*I reduce, I reuse, I recycle*". The containers, in fact, allow a separate collection for recycling use and a reward system for those who act in an eco-compatible way.

Alice Comble
@GreenMinders (Twitter)
www.facebook.com/GreenMinders (Facebook)
www.greenminded.fr (Web Site)
Alice Comble (LinkedIn)

See English version on p. 356

© Springer Nature Switzerland AG 2019
M. Abbati, *Communicating the Environment to Save the Planet*,
https://doi.org/10.1007/978-3-319-76017-9_12

▶ **Question** Madame, votre projet réinvente la collecte des déchets en créant un réseau de communication aux différents niveaux, pouvez-vous nous expliquer les atouts principaux visant à communiquer le *message environnemental* (focalisé sur l'importance du recyclage des petits déchets comme les mégots) et à sensibiliser, au même temps, l'opinion publique?

▶ **Answer** L'idée de communiquer autour du recyclage tout en proposant une solution concrète et ludique est à mes yeux nécessaire car aujourd'hui, il a été prouvé que les solutions répressives et moralisatrices (amendes, interdictions, etc.) ne permettent pas de changer les comportements sur le long terme. Si nous voulons changer les habitudes des citoyens et que demain il soit "anormal" de ne pas recycler, il faut en faire un automatisme, une mode. Nous vivons dans une société qui nous a fait prendre pour habitude un tas de comportements qui ne sont pas basés sur le "bon sens humain". Et il est temps que la mentalité basée sur le "consomme et jette" devienne "consomme et valorise".

▶ **Question** Quelles sont, selon votre expérience, les piliers de la communication du point de vue du contenu du message par rapport aux destinataires ? Quels rôles vont jouer les soi-disant nouveaux *Médias* (ex. : Internet, Social Media, etc.) ? Rendent-ils plus efficace la *communication environnementale*?

▶ **Answer** Aux yeux de *GreenMinded*, les piliers d'une communication reposent sur un doux mélange entre le sérieux et la légèreté. Nous voulons donner des informations factuelles mais choquantes sans pour autant être dans la moralisation. Les nouveaux *Médias* (*Internet*, *Réseaux Sociaux*, etc.) rendent effectivement plus efficaces la transmission des informations à travers le monde. Et donc la transmission d'éléments afin d'alarmer sur la situation environnementale à travers le monde. Internet est donc certainement responsable de l'éveil collectif que la planète est en train de vivre sur la situation très difficile au niveau écologique que vit l'ensemble des êtres vivants.

▶ **Question** "You can have the greatest idea in the world, but if you can't communicate your ideas, it doesn't matter" se plaisait à dire Steve Jobs, Co-fondateur, directeur général et président du conseil d'administration d'Apple Inc., êtes-vous d'accord avec cette affirmation?

▶ **Answer** Oui je suis plutôt d'accord avec cette affirmation dans la mesure où il est effectivement très difficile de mettre en œuvre un projet ou une idée s'il n'y a pas de communication et donc de collaboration.

Et quand bien même, certaines personnes, grâce à l'arrivée d'Internet ont cette capacité à avoir une idée et de la mettre en œuvre par eux même, et ce, notamment grâce à l'arrivée d'Internet.

Mais finalement, si ces personnes-là ne finissent pas par parler de l'outil qu'elles ont mis en place, les probabilités pour qu'il tombe dans l'oubli sont très grandes.

▶ **Question** La communication implique aussi un échange des connaissances, votre projet GreenMinded, quelles connaissances va-t-il transmettre ?

▶ **Answer** Les informations transmises par *GreenMinded* sont des informations concernant le véritable impact écologique des petits déchets (dont les mégots de cigarettes) et les actions qui peuvent être mises en place pour palier à cette pollution. Nous intervenons ainsi dans les universités et entreprises afin de sensibiliser à l'impact des mégots. Pour cela nous mettons en place des affiches aux points stratégiques, utilisons des "Green Nudges" (un coup de pouce pour atteindre un style de vie plus durable) autour des points de recyclage. Nous faisons en sorte de communiquer d'une façon *"fun"* (amusant) et non pas répressive ou moralisatrice. Nous essayons au maximum d'utiliser le second degré et de donner des chiffres qui sont facilement visualisables par tout le monde (ex : "un mégot pollue 500L d'eau, c'est-à-dire l'équivalent de 3 baignoires" ou encore "chaque minute en France, assez de mégots sont jetés pour atteindre le sommet du Mont Blanc."

De plus, nos bornes sont équipées de différents capteurs permettant de superviser le taux de réponses aux questions ainsi que le profil de la personne ayant répondu. Nous sommes également en mesure de savoir si la personne ayant répondu est fumeuse ou non, ce qui nous permet de recouper avec le reste de la population grâce au travail de statisticiens. Nous distribuons également gratuitement des cendriers de poche et organisons des collectes de déchets.

Interview No. 6: Edoardo Croci

IWP 6

Edoardo Croci Professor at Bocconi University of Milan (Italy), he is the Director of the *Centre for Economics and Energy and Environmental Policy* (IEFE) and he is the coordinator of the *Green Economy Observatory* in Italy. As a former Councillor for Mobility, Transport and Environment of the City of Milan with Mayor *Letizia Moratti*, he has effectively reduced the urban traffic and the pollution levels in Milan area, introducing the free tickets on public means of transport for children up to ten and promoting the sustainable mobility through the *road pricing* and the bike sharing (*bikeMi* project). His many *Environmental* and sustainable activities were performed during his career as a former President of *ARPA* (Regional Agency for Environmental Protection of *Lombardia*) and president of *MilanoSiMuove* (Milan moves); as a promoter of 5 referendum for *Environment* and Life Quality approved in 2011 and as the founder of *Milano Ambiente Foundation*; as a national Councillor of *Italia Nostra* and as a founder of the Association for Culture and Leisure (*ACTL*) and finally as the editor of the online newspaper *Cartalibera*, set up with *Egidio Sterpa*, editor of the Smart City & MobilityLab magazine and the *Corriere della Sera* newspaper.

▶ **Question** Recenti studi sullo stato dell'arte della *Green Economy*, ovvero della *Circular Economy*, delineano un'ottima posizione dell'Italia per quanto riguarda la performance economica, ma un modesto risultato in termini di percezione del valore del "marchio verde" italiano, ancora largamente sottovalutato a livello internazionale [rif.: Rapporto del Global Green Economy Index™ (GGEI), 2016].

Edoardo Croci
@Edoardo_Croci (Twitter)
www.edoardocroci.it (Web Site)
Edoardo Croci (LinkedIn)

See English version on p. 357

© Springer Nature Switzerland AG 2019
M. Abbati, *Communicating the Environment to Save the Planet*,
https://doi.org/10.1007/978-3-319-76017-9_13

Prof. Croci, questi esiti, apparentemente in contraddizione, potrebbero essere il risultato di un insufficiente apparato di informazione e comunicazione a livello nazionale ed europeo? Quali azioni, o forze trainanti (*driving forces*), sarebbero necessarie per migliorare la qualità della *comunicazione ambientale* e difenderci dal *Greenwashing*?

▶ **Answer** Tutta una serie di studi recenti confermano che l'Italia ha una posizione rilevante nel campo della Green Economy. In particolare ci sono alcuni settori come quello delle Rinnovabili e quello del Riciclo dei Rifiuti, che hanno una dimensione economica di circa 10 miliardi di euro ciascuno. Un risultato importante, raggiunto in alcuni casi anche grazie ad un sistema di incentivi pubblici (es.: settore delle energie rinnovabili) o di tassazione ambientale (es.: settore dei rifiuti). L'Italia ha raggiunto, dunque, una posizione di *leadership* nella *Green Economy* rispetto ad alcuni Paesi della stessa Unione Europea e a livello internazionale. D'altronde questo essere avanti è anche dovuto al fatto che l'Italia è povera di materie prime e relativamente povera di combustibili fossili. Il nostro Paese deve inoltre tener conto di un quadro internazionale che spinge in questa direzione: prima con il Protocollo di Kyoto e poi con l'Accordo di Parigi alla conferenza internazionale sul clima di Parigi (COP21) nel 2015.

Non a caso l'Italia, in termini di efficienza energetica, è storicamente uno dei Paesi *leader* nel mondo; lo stesso vale pe la *resource efficiency* e quindi, in generale, la capacità di utilizzare le materie prime in modo efficiente. E questo è un vantaggio anche in relazione alle nuove strategie di *Circular Economy*, che si stanno sviluppando a livello internazionale ed europeo.

Alcuni nostri settori produttivi industriali strategici ne hanno così tratto vantaggio.

Da parte del cittadino e del consumatore, tuttavia, spesso non vi è piena consapevolezza di questo aspetto. Secondo l'indagine annuale di *Eurobarometro*, promossa dalla Commissione Europea, gli italiani primeggiano in termini di sensibilità ambientale guardando alle dichiarazioni di principio, ma quanto a comportamenti precipitano invece nella classifica. Sul fronte dell'educazione ambientale a partire dalle scuole, della *comunicazione ambientale* e dell'informazione ambientale ci sono ancora dei passi avanti da compiere. In parte questa è una responsabilità pubblica, in parte è una responsabilità delle imprese e dei cittadini.

▶ **Question** Quali potrebbero essere le possibili azioni, o soluzioni, per migliorare la consapevolezza dell'Informazione e della *comunicazione ambientale*, all'interno dei confini nazionali, ma anche a livello dell'Unione Europea?

▶ **Answer** Con riferimento al **Settore Pubblico**, lo *strumento di comunicazione* più comune è il **reporting ambientale** relativo a diverse scale territoriali. In Italia si comincia a parlare di **Relazione sullo Stato dell'Ambiente**, dagli Anni '80 con l'istituzione del Ministero dell'Ambiente (1986); la relazione ha avuto una periodicità variabile rispetto alla previsione di legge. Sono nate poi varie forme di *reporting ambientale*, piuttosto diversificate. Ciò è avvenuto anche grazie alla nascita del sistema agenziale che oggi vede al centro l'ISPRA—Istituto Superiore per la Protezione e la Ricerca Ambientale, a livello nazionale, e le ARPA—Agenzie Regionali per la Protezione dell'Ambiente, a livello regionale. Questo sistema ha giocato un ruolo importante sia nel monitoraggio ambientale che nella progressiva condivisione di una metodologia omogenea nel rilevare e rappresentare informazioni ambientali legate al territorio. Oggi ci sono nuove tendenze nelle forme di comunicazione, nel senso che rispetto alla Relazione sullo Stato dell'Ambiente, la disponibilità di strumenti *web*, in continuo aggiornamento, fa sì che ci sia la possibilità di alimentare un flusso continuo di dati ambientali, frutto di monitoraggio continuo. In qualche modo, dunque, si va oltre l'idea di una relazione, elaborata *una tantum* che registra la situazione in un determinato momento storico (tipicamente una volta all'anno), soggetto a rapida obsolescenza.

Ecco, dunque, l'esigenza di creare strumenti sempre più sintetici e di facile lettura. Come catasti *online* facilmente accessibili. Qualcosa che non sia più rivolto ai soli addetti ai lavori ma grazie a cui il cittadino possa facilmente comprendere lo stato e l'evoluzione della qualità ambientale di un territorio attraverso informazioni geo-referenziate e con il supporto di grafici semplici da interpretare.

Mentre gli strumenti a disposizione esistono, devo dire che il cittadino medio italiano non è così informato sullo stato di salute del proprio territorio. La diffusione dell'informazione ambientale è spesso gestita da associazioni ambientaliste piuttosto che dal sistema pubblico agenziale che viene ancora visto come un sistema di supporto prevalentemente tecnico, rivolto alle sole amministrazioni pubbliche. È necessario, dunque, rafforzare il suo ruolo di "comunicatore pubblico" anche nei confronti del cittadino. Una funzione chiave che è stata sottolineata dalla recentissima riforma delle Agenzie Ambientali con il Sistema Nazionale a rete per la protezione dell'Ambiente che si sta realizzando in Italia, a partire dal 2016 [Rif.: Legge 28 giugno 2016 n°132].

Il secondo aspetto è la ***comunicazione ambientale*** che viene effettuata dalle ***imprese*** nella logica della **responsabilità sociale**, ma anche del **marketing**, legata dunque al miglioramento della posizione sul mercato. Un approccio competitivo strettamente legato alle dinamiche del mercato, con i dipendenti, con gli *stakeholder* e con il territorio. Da questo punto di vista, l'Istituto di Economia e Politica dell'Energia e dell'Ambiente (IEFE) presso l'Università Bocconi di Milano, ha realizzato, qualche anno fa, uno studio specifico sulla *comunicazione ambientale* in Italia, da cui si evincono interessanti indicazioni. Emerge, innanzitutto, che questo tipo di comunicazione, sia istituzionale che di prodotto, ha un *trend* positivo di crescita nel nostro Paese, anche in termini di investimenti da parte del sistema delle imprese. Si tratta poi di una comunicazione che, nella maggior parte dei casi, si pone come *claim* principale il "vantare" le qualità del prodotto o del servizio in termini di *performance* ambientale. Il termine "*claim*" nel linguaggio pubblicitario identifica appunto l'affermazione delle caratteristiche distintive del prodotto che offrono un beneficio ai consumatori. Altro elemento molto rilevante è l'economicità del prodotto o del servizio: avere *performance* ambientali migliori significa beneficiare di risparmi nei consumi. Per esempio, consumare meno carburante o consumare meno energia è un risultato facilmente riscontrabile dall'opinione pubblica e quindi efficace dal punto di vista comunicativo. Dal nostro studio, emerge tuttavia un limite che persiste tutt'ora. Nella maggior parte dei casi i *claims* non vengono accompagnati da certificazioni riconosciute a livello nazionale o internazionale che rendano i dati oggettivamente documentati e verificati attraverso meccanismi di certificazione di soggetti terzi indipendenti. Rientra tra le responsabilità dell'Autorità Garante della Concorrenza e del Mercato (AGCM) sanzionare ogni forma di pubblicità ingannevole. In questi ultimi anni sono stati più di uno i casi apparsi sulle prime pagine dei giornali, riferiti alle presunte prestazioni ambientali di prodotti e servizi, poi rivelatesi non rispondenti al vero. Più in generale vi è dunque un rischio reale di *greenwashing*. Spesso infatti la descrizione delle caratteristiche ecologiche di un prodotto e/o di un servizio provengano direttamente dalle imprese produttrici/erogatrici senza che vi sia stata alcuna verifica esterna, *super partes* ed indipendente. Occorre, dunque, continuare ad **investire sulla qualità** dell'informazione e della *comunicazione ambientale* se vogliamo che la loro **credibilità** migliori, per comunicare in maniera **sempre più efficace** a vantaggio della corretta informazione dei cittadini.

Le imprese devono rendersi conto che la *comunicazione ambientale* non può essere superficiale ed incompleta ma che debba essere certificata da soggetti terzi, autorevoli ed indipendenti per essere credibile. Senza dimenticare i meccanismi già ben definiti all'interno di sistemi codificati: come la Dichiarazione Ambientale richiesta da EMAS (*Eco-management and Audit Scheme*, sistema di gestione ambientale volontario, conforme all'omonimo Regolamento Europeo), o aspetti legati alle norme internazionali ISO 14001 o dichiarazioni legate al prodotto mediante il ricorso a marchi ambientali a livello europeo (Ecolabel, dichiarazioni ambientali di prodotto, etichette energetiche, ecc.).

▶ **Question** Ritiene ancora validi gli strumenti come il Bilancio Ambientale e Sociale o i documenti creati nell'ambito di Sistemi di Gestione Ambientale?

▶ **Answer** Su questo tema si è assistito in Italia ad una grande evoluzione nel tempo. Questo vale sia per gli strumenti a disposizione delle amministrazioni pubbliche che per quelli in uso dal settore delle imprese. Sono stati fatti molti passi avanti nel codificare questi strumenti. Oggi si parla sempre di più di Bilanci di Sostenibilità e più recentemente di reporting on finanziario (ambientale e sociale), integrato nel bilancio economico di un'impresa. È importante ricorrere a forme di *reporting* codificato. Ricordiamo, innanzitutto, le linee guida del **GRI—*Global Reporting Initiative***. Il *reporting* rimane, dunque, una forma in continua evoluzione di comunicazione qualificata (e quindi verificata) delle performance ambientali.

▶ **Question** Ritiene che questi strumenti di rendicontazione ambientale debbano diventare cogenti per la pubblica amministrazione o per le imprese per garantire maggiore trasparenza ed un migliore accesso all'informazione ambientale? Prevede nuove proposte di legge a riguardo?

▶ **Answer** Per un lungo periodo, in Italia, si è cercato di codificare in via legislativa la *contabilità ambientale*. In realtà credo che con gli strumenti a disposizione di oggi e gli sviluppi che ci sono stati a livello internazionale, non sia tanto necessario avere degli strumenti che obblighino per legge le amministrazioni pubbliche, così come le imprese, ad adottarli. La loro natura di impegni volontari consente infatti di essere percepiti come fattori premianti, sia dal settore pubblico che da quello privato. A condizione naturalmente di aderire a protocolli di tipo internazionale (es.: norme ISO, EMAS, Ecolabel, ecc.). Ci stiamo riferendo, in ogni caso, ad una materia in continuo divenire. Nel 2015, per esempio, sono stati approvati da parte delle Nazioni Unite i *Sustainable Development Goals*. È evidente che questi Obiettivi di Sostenibilità rappresentano il nuovo *vademecum*, a livello nazionale ed internazionale, per adottare le politiche di Sostenibilità più adeguate. Ma questi *goals* offrono anche nuovi strumenti di *reporting* perché ad essi sono associati *set* di indicatori codificati a livello internazionale e, soprattutto, confrontabili. Molte imprese, così, oltre agli Stati, hanno così deciso di conformarsi agli stessi Obiettivi di Sviluppo Sostenibile. Al contrario, il grande rischio di avere un sistema cogente è quello che diventi già vecchio prima di partire. In un certo senso anche la stessa *contabilità ambientale*, così come concepita qualche anno fa, appare superata in conseguenza della mutata scena internazionale che richiede nuove forme di *reporting* e rinnovati parametri per stabilire gli indicatori di *performance*, nonché strumenti diversi di diffusione, come impone l'uso crescente del *web*. Siamo sicuramente di fronte ad un processo continuo di innovazione che si perseguire meglio se resta volontario.

L'Agenda 21 globale, e poi quella Locale, frutto della prima Conferenza di Rio del 1992 sullo Sviluppo Sostenibile, ha fin da subito riscosso un grande successo, soprattutto in Europa. In Italia è attivo un **Coordinamento Nazionale Agende 21 Locali** di cui io stesso ho fatto parte quando ero amministratore a Milano, ricoprendo il ruolo di Vice-Presidente. A livello di coordinamento italiano ci si è però resi conto che non si può solo promuovere Agenda 21 come strumento di partecipazione e *reporting*. Oggi ad esempio il Coordinamento promuove i PAES (Piani di azione per l'energia sostenibile) previsti dal Patto dei Sindaci promosso dalla Commissione Europea. Il percorso di Agenda21 è dunque anch'esso in continua evoluzione. Più in generale si deve parlare di un processo che fa sì che ci siano strumenti di confronto tra i vari livelli all'interno delle amministrazioni pubbliche nonché tra il livello nazionale e quello internazionale.

▶ **Question** La Rete (ovvero il Web) è sempre più usata come strumento di informazione e *comunicazione ambientale*, a cui si aggiunge il "boom" esponenziale dell'uso dei *Social Media*. Secondo la sua opinione, il loro utilizzo può migliorare la qualità della *comunicazione ambientale* o al contrario costituirne un problema in considerazione della mancanza di un'adeguata regolamentazione globale del *Web*?

▶ **Answer** Non c'è dubbio che la *comunicazione ambientale*, ma in genere ogni forma comunicativa, punti sempre più alla diffusione dei messaggi via *web* o attraverso i *Social Media*, a discapito della carta stampata, ma anche della stessa televisione. Questo cambia la forma ed il linguaggio con cui vengono veicolati i messaggi in una dimensione poco controllata. Si pensi ai rapporti *peer to peer* che si instaurano in Rete tra persone a volte difficilmente identificabili e per questo non affidabili. Questo fa sì che si renda anche difficile sanzionare i comportamenti scorretti come le situazioni di comunicazione mendace o non fondata su fonti autorevoli. Le recenti elezioni presidenziali americane hanno dimostrato la volatilità e infondatezza di molte informazioni comunicate dai *Media*, tramite *Internet*. E non esiste, a tutt'oggi, un'autorità mondiale che possa realmente garantire gli internauti sulla qualità di quello che circola in rete o sui *Social Media*. Questo apre un campo estremamente complesso e dibattuto.

Credo, dunque, che ci debba essere una qualche forma di regolamentazione, quanto meno in termini di verifica delle affermazioni pubblicate *online*, soprattutto in campo ambientale. Indicare sempre e comunque la fonte su cui si basano le informazioni dovrebbe essere obbligatorio, lasciando poi al giudizio del singolo cittadino internauta di trarre le conclusioni per dare più o meno credito ai loro contenuti.

Si rischia altrimenti di far sì che i *Social Media* diventino la fonte, ovvero l'"editore" di loro stessi. Questo comporta una deresponsabilizzazione rispetto al *messaggio ambientale* inaccettabile. Molto spesso infatti alcune piattaforme *online* consentono di raccogliere in bacheche virtuali informazioni non verificate, con il pretesto di non essere né una redazione giornalistica né una istituzione pubblica. Ben venga, dunque, il mantenimento del dovuto ruolo di responsabilità del direttore di redazione e dell'editore, o di ogni altro soggetto dotato dell'autorità di controllare la qualità dell'informazione. Senza ricorrere, naturalmente, a nessun tipo di censura, soffocante, che ne vieti l'utilizzo a priori. L'ideale sarebbe quello di raggiungere un giusto equilibrio dei Media *online* volto a responsabilizzare il ruolo del "comunicatore" cui è richiesto il compito di fornire al destinatario della comunicazione i giusti "attrezzi" per valutare l'attendibilità o meno di quanto riceve.

Lo scenario della *comunicazione ambientale* sta cambiando e vi è sempre il rischio di trovarci di fronte a quello che nel mondo anglosassone viene definito come "*rubbish*" o "*trash*". Fenomeni "virali" come le recenti "secchiate d'acqua" si sono trasformati da un gesto simpatico, originariamente promosso per una giusta causa, in una moda "virale" che ha portato a snaturare i suoi contenuti originali. Altre azioni o progetti promossi sulla Rete, invece, si fondano su basi più solide e scientifiche. Pensiamo ai film di *Al Gore* e di *Di Caprio* che trattano temi ambientali. Naturalmente ognuno ha diritto di esprimersi, e lo può fare con un "linguaggio" accattivante a seconda del *target* di audience a cui è indirizzato. L'importante è sempre verificare il contesto, la fonte e la "filiera" delle informazioni veicolate.

▶ **Question** Come vede il panorama italiano ed internazionale con riferimento alla *comunicazione ambientale*? Quali azioni sono necessarie per sensibilizzare l'opinione pubblica?

▶ **Answer** Potrebbe sembrare paradossale ma più che sensibilizzare i cittadini ciò che è più importante fare è sensibilizzare la nostra classe politica. Come dimostrano i risultati di un sondaggio qualificato a livello europeo, come *Eurobarometro*, ci accorgiamo che gli Italiani sono in realtà molto sensibili ai temi ambientali, almeno a parole. Certo, sul fronte dei *comportamenti di acquisto* questo non è sempre evidente anche se, negli ultimi anni, alcuni settori come il biologico stanno esplodendo a dimostrazione del fatto che vi è una crescente sensibilità tra i consumatori. Il punto chiave nei confronti dei cittadini è l'**attendibilità**, la **veridicità** e la **credibilità** delle informazioni ambientali comunicate, aspetti di cui si dovranno fare carico le amministrazioni pubbliche e le imprese. L'ulteriore aspetto su cui investire è quello delle **politiche ambientali italiane** che risultano indietro rispetto a quelle di altri Paesi dell'Unione Europea. Questo è dovuto ad una forma di timore infondato (ancora presente nel pensiero politico italiano), in base al quale chi promuove azioni ambientali in grado di cambiare lo stile di vita dei cittadini potrebbe risultare "impopolare", creando un effetto *boomerang* penalizzante per il proprio consenso. Questo "preconcetto" tuttavia è sempre meno vero in quanto gli Italiani si stanno dimostrando, al contrario, molto aperti al cambiamento del loro stile di vita a condizione che alla modifica dei loro comportamenti corrisponda un *feedback* informativo che dimostri i vantaggi reali in termini di qualità di vita, risultanti dalle loro *eco*-azioni. Solo così, infatti, si riuscirà ad esempio a convincere un crescente numero di cittadini a rinunciare all'uso dell'auto a vantaggio dell'utilizzo del mezzo pubblico. Occorre fornire le prove del miglioramento della qualità ambientale (e della salute) conseguente alla riduzione della congestione del traffico. Comunicando i dati che dimostrano il miglioramento della qualità dell'aria o del riutilizzo degli spazi urbani, tradizionalmente destinati alle quattro ruote, a favore di altri usi sociali e ricreativi. Io stesso sono stato testimone di un'iniziativa referendaria che ha avuto luogo a Milano dove ho presieduto un Comitato Cittadino ("Milanosimuove") promotore di alcuni *quesiti* su temi ambientali. Nonostante la difficoltà iniziale del Comitato promotore nel raccogliere le firme necessarie (più di 15.000) coinvolgendo tutte le parti interessate (es.: cittadini, associazioni ambientaliste, vip, esponenti politici, esponenti della cultura e della scienza) e di farle autenticare (seguendo una procedura burocratica che andrebbe snellita e adeguata all'era informatica), nel giugno 2011 si è raggiunto il *quorum* richiesto per la validità di tutti i referendum proposti che sono stati approvati da larghissima maggioranza. Un'iniziativa, divenuta una *best practice*, nata e cresciuta all'interno della comunità dei cittadini che ha dato (e continua a trasmettere) un *messaggio ambientale* forte all'amministrazione comunale milanese, influenzando la *governance* territoriale delle ultime legislature, indipendentemente dal loro colore politico.

Interview No. 7: Barbara Frateschi Moreno

IWP 7

Barbara Frateschi Moreno Born in Turin, since her childhood she has been keen on painting. Successively, she moves with her family to Paris, Rome, Rabat, Florence, where she attended classical studies. Then, she moves to New York, where she attended courses of drawing and painting, getting in touch with the informal art and the American artistic *Environment*. In the years Eighties, she moved to Geneva (Switzerland) to stay permanently. After attending the *Ecole des Arts Décoratifs*, thanks to her mentors: *Gwyneth Barth-White*, *Jean-Marie Borgeaud*, *Claude Mura*, Barbara starts her career as a painter. Since 2000, she has been working at the University of Geneva, without neglecting his artistic activity with even more intensity. Her art is nourished by sign, emotion, light and colour. She has taken part successfully to a series of personal exhibitions and group exhibitions. Since 2010, she has presented her artworks at exhibition of great impact such as *Geneva*, *Venice*, *Milan*, *Sanremo*, *Florence*, *Innsbrück*, Nice and the Principality of Monaco. Her paintings can be found both in art galleries or private collections. Being of Italian-Switzerland nationality, she shares her time in between Geneva, Rome and the Western *Liguria* where she has her art laboratory in *Ospedaletti* (near Sanremo), close to the sea, a place dear to her husband, the Ambassador *Maurizio Moreno*, former President of the International Institute of Humanitarian Law in Sanremo.

▶ **Question** *Umberto Eco* nella sua *Opera Aperta* (1962) affermava che "l'opera non smette mai di essere modificata dal suo lettore e dal suo spettatore" consentendo così una molteplicità di interpretazioni frutto di una dialettica tra la forma ed il movimento dell'interpretazione. Signora Barbara condivide il pensiero del noto semiologo? La sua produzione artistica è strettamente legata agli elementi

Barbara Frateschi Moreno
www.barbarafrateschi.com (Web Site)
Barbara Frateschi (LinkedIn)

See English version on p. 361

© Springer Nature Switzerland AG 2019
M. Abbati, *Communicating the Environment to Save the Planet*,
https://doi.org/10.1007/978-3-319-76017-9_14

263

della natura ispirandosi spesso alla biodiversità del *Mediterraneo* che comunica attraverso sensazioni di colore e luce. Ritiene che un'opera pittorica possa influenzare la sensibilità ecologica dello spettatore contribuendo a generare una sua personale "crescita" verso uno stile di vita più sostenibile o più attento alla Sostenibilità?

▶ **Answer** Condivido il pensiero di *Umberto Eco* se si considera che un'opera sia espressione di un pensiero, di uno stato d'animo, di una sensibilità, su un tema che vuole essere presentato e condiviso dallo spettatore il quale con occhio e mente guidati da emozioni, consapevolezze, maturità diverse può giungere ad una diversa lettura delle forme di quella originaria sentita dall'autore.

Un'opera d'arte può sensibilizzare maggiormente lo spettatore sulla grandezza del nostro patrimonio naturale. È possibile proporgli una riflessione, un'interrogazione sulla possibilità di un suo cambiamento di stile di vita, indirizzandolo verso una consapevolezza ecologica che aiuti a contribuire alla salvaguardia dell'Ambiente. Un Ambiente curato non può che incidere sullo sviluppo della persona in maniera positiva e la relazione con un Ambiente sano e armonioso non può che aiutare l'uomo nel suo comportamento di vita mentre un Ambiente degradato provoca il deterioramento della qualità dell'abitante.

Voglio credere che se un'opera riesce a stimolare i sensi emotivi, visivi e perché no olfattivi del suo spettatore forse può anche pervenire a stimolare il desiderio di preservare l'immenso tesoro che la natura ci offre con uno stile di vita più sostenibile

▶ **Question** "L'apparizione di un'immagine, a prescindere dalla sua 'potenza' e dalla sua efficacia, ci 'investe' quindi ci sveste", afferma *Georges Didi-Huberman*, storico dell'arte, filosofo e docente presso l'*Ecole des Hautes Etudes en Sciences Sociales* di Parigi. L'affermazione ben descrive uno degli elementi chiave del processo di *comunicazione ambientale* ovvero quello della consapevolezza o nuova conoscenza che il *messaggio ambientale* inviato dal comunicatore genera nel ricevente. Se dovesse trasmettere attraverso un'opera pittorica un messaggio che sensibilizzi il pubblico al rispetto per l'Ambiente che cosa rappresenterebbe?

▶ **Answer** Probabilmente rappresenterei, come ho già proposto in alcune mie opere, il mare e l'aria, elementi essenziali di vita che tra le risorse naturali fondamentali devono essere l'oggetto di un più grande rispetto. Il linguaggio pittorico si è sempre prestato particolarmente bene a raffigurare il mare che si offre all'artista con una ricca paletta di colori, invitandolo a evidenziarne la forza e aiutandolo a esaltarne la bellezza mentre l'aria con leggere velature sovrapposte e trasparenti che ne rilevano la purezza gioca rispecchiandosi sulla superficie dell'acqua. Il mare è un immenso patrimonio ed è nostro dovere preservarlo e, tramandarlo alle generazioni future come un tesoro di rara bellezza.

▶ **Question** Può descrivere alcune opere di sua produzione o che l'hanno colpita in particolar modo, ispirate ai valori ambientali o sostenibili, offrendoci una proposta di "lettura" che sottolinei la loro potenziale capacità di influenzare il pensiero dello spettatore, contribuendo ad una modifica della percezione del mondo naturale.

▶ **Answer** L'arte contemporanea ha diversi modi per tentare di sensibilizzare lo spettatore al tema ambientale. La *Land Art* famosa intorno agli anni '70 conciliava Arte e Ambiente, portando la natura stessa ad essere partecipe dell'opera. Questa, generalmente di dimensioni gigantesche, era sovente situata in un sito isolato e di difficile accesso, si valeva di un paesaggio come soggetto e utilizzava gli elementi naturali come materiale.

Tali costruzioni per lo più concettuali, meditative, erano volontariamente integrate dall'artista al luogo che le ospitava, modificandolo sostanzialmente e a volte anche in modo definitivo. Le opere per lo più efemere si valevano dell'arte fotografica per immortalare la loro esistenza. *Robert Smithson*, artista americano, ne è stato una dei grandi esponenti. Una tra le sue opere maggiori è la "*Spiral Jetty*", situata sul "*Great Salt Lake*", nello Stato americano dello *Utah*.

Smithson fece deversare tonnellate di pietre basaltiche, di composizione vulcanica e di colore nero, creando un molo circolare di circa mezzo chilometro, con motivo spiralico. Ma la natura con il tempo, prende il sopravvento e ne diviene il vero artista. L'opera inghiottita dopo pochi anni dalla sua realizzazione dalle acque altamente saline del lago, riaffiora saltuariamente ben trent'anni più tardi, secondo il livello dell'acqua. Ma il sale e le alghe avevano portato avanti il lavoro artistico: il molo cristallizzato dal sale da nero è ora bianco, le alghe, solo organismo vivente nel lago così altamente salato, colorano l'acqua di rosa e per qualche sublime e naturale trasformazione il colore dell'acqua passa da rosa a verde. *Smithson* ha attirato gli sguardi da tutte le parti del mondo su questo sito inospitale e non facilmente raggiungibile che l'*Uomo* aveva abbandonato dopo un fatuo tentativo di sfruttarlo industrialmente.

La *Natura* può improvvisarsi artista sfoggiando le sue immense capacità di adattamento ma se noi spezziamo e distruggiamo un Ecosistema vecchio di milioni d'anni la nostra salvezza e quella dei nostri figli è definitivamente in pericolo.

Fig. IWP 7.1 © Barbara Frateschi Moreno: *Les danseuses* (The dancers), acrylic painting on canvas 70 × 50

Fig. IWP 7.2 © Barbara Frateschi Moreno: Acquatica Emozione (Aquatic Emotion), acrylic painting on canvas 80 × 80

Fig. IWP 7.3 © Barbara Frateschi Moreno: Nostalgia (Nostalgia), acrylic painting on canvas 80 × 60

Interview No. 8: Maurizio Giani

IWP 8

Maurizio Giani CEO since the 2nd January 2016 of *Waste Recycling* S.p.A., a company specialized in the treatment and disposal of Industrial Waste, has had a great experience in the sector "waste management" (since 1992). He has been managing the corporate business, since its birth, with special regard to: the beginning of separate-collection facilities, the chemical laboratories, the biological and chemical treatment installation, the evaporation plant, the inertisation system, the hazardous waste treatment and storage. In this context, he has been dealing with the corporate art department, *SCART* whose communication project is *Waste Recycling*. His enthusiasm in carrying out this successful project made him a firm believer in the strong *Environmental connotation* the *design objects*, made out of recycled materials, give to the public.

▶ **Question** Dott. Giani, oltre a dirigere una delle aziende toscane *leader* nel trattamento e nello smaltimento dei rifiuti industriali, lei cura, da diciotto anni, il progetto di comunicazione *SCART®* che nasce da un'intuizione alquanto creativa. L'idea che da anonimi rifiuti industriali potessero nascere delle opere esclusive, espressione di arte e design (es.: componenti di arredo, abiti, strumenti musicali, sculture, mosaici, dipinti, ecc.). E per poterlo realizzare al meglio ha ideato progetti artistici in grado di coinvolgere creativi italiani e stranieri, e anche alcuni docenti e studenti delle Accademie di Belle Arti di Firenze e Bologna. Che cosa ha ispirato questo progetto di forte valenza ambientale? Qual è stata la risposta da

Maurizio Giani
www.w-r.it (Web Site)
www.scartline.it (Web Site)
@Scartline (Twitter)
Maurizio Giani (LinkedIn)

See English version on p. 362

© Springer Nature Switzerland AG 2019
M. Abbati, *Communicating the Environment to Save the Planet*,
https://doi.org/10.1007/978-3-319-76017-9_15

269

parte dei giovani artisti all'idea di dare nuova vita a materiale considerato un "rifiuto"? Cosa è emerso dalle loro creazioni dal punto di vista comunicativo (si sono ispirati a forme della natura; hanno evidenziato problematiche ambientali come inquinamento, cambiamenti climatici, corretta gestione dei rifiuti; hanno voluto inviare un messaggio educativo-informativo volto a valorizzare il riciclaggio, ecc.)?

▶ **Answer** Dai nostri impianti transitano quotidianamente rifiuti industriali, per loro natura molto diversi dai rifiuti urbani: molto spesso si tratta di materiali declassati a scarti perché ormai fuori produzione o non più in linea con le tendenze attuali. Il nostro obiettivo è dar loro una seconda vita.

Il progetto *SCART*® nasce dalla volontà di fare comunicazione per un'azienda che svolge un'attività molto particolare, spesso demonizzata. Abbiamo scelto il linguaggio comune a tutti: l'arte. La *Trash Art* lega perfettamente il nostro settore all'arte. La nostra intenzione era ed è quella di nobilitare il nostro settore attraverso la comunicazione artistica.

Oggi il *network* degli artisti *SCART*® è in continua crescita e grazie a loro *Waste Recycling* è riuscita per esempio ad arredare un intero reparto dell'Ospedale "*Nuovo Santa Chiara*" in *Cisanello, Pisa*, a realizzare scenografie e costumi per più edizioni del Teatro del Silenzio di Lajatico, dove ogni anno il tenore Andrea Bocelli tiene uno spettacolare concerto. Nel 2016 un Barbiere di Siviglia tutto riciclato è andato in scena all'arena Fonte Mazzola di Peccioli (Pisa), in Piazza Duomo a San Gimignano, e al Comunale di Adria in provincia di Rovigo. Poi a fine anno abbiamo collaborato con i *talent show* X-Factor e ospitato artisti di fama internazionale come il cubano *José Yaque*.

Le convenzioni con le Accademie di Belle Arti di Firenze e Bologna nascono dalla volontà di sperimentare e al contempo di dare agli studenti la possibilità di lavorare su progetti di volta in volta diversi. La loro attività si inserisce ogni volta nell'ambito di un progetto concordato con i docenti responsabili dei *workshop*. Gli studenti hanno la facoltà di interpretare i materiali a loro disposizione, che essi stessi selezionano nella nostra piattaforma di stoccaggio, rimanendo però nell'ambito del tema scelto. Con loro abbiamo realizzato recentemente le scene e i costumi per il Barbiere di Siviglia, modelli di animali di dimensioni naturali, quadri che ritraggono personaggi *cult* e allestimenti natalizi per alcune delle maggiori piazze italiane: tutti rigorosamente con materiale di recupero. Gli studenti hanno sempre partecipato con trasporto e intenso interesse. La loro è stata, e continua a essere, una preziosissima collaborazione. Per noi è motivo di grande soddisfazione il far parte del processo creativo che porta alla realizzazione di opere uniche nel loro genere.

Fig. IWP 8.1 Andrea Bocelli interprets Carmen at the 'Teatro del Silenzio' (Theatre of the Silence), in Lajatico; the sopranos and the dancers were wearing hand-made recycled clothes designed by Scart Project © 12th July 2012

▶ **Question** Gli studiosi delle immagini e i filosofi del linguaggio sono concordi nel sostenere che ogni raffigurazione artistica sia in sé e per sé una "narrazione" ovvero trasmetta un messaggio che coinvolge, influenza, crea nuove idee. In una parola: comunica. Lei concorda su questa linea di pensiero?

▶ **Answer** Sì, sono d'accordo e posso dire di aver esperito in prima persona la veridicità di questa affermazione nel corso della mia attività con il progetto *SCART®*.

Ogni volta che le opere nate dai *workshop* con i giovani artisti delle Accademie di Belle Arti vengono presentate al grande pubblico il risultato lascia sempre senza parole: quando un materiale che sarebbe potuto finire a marcire letteralmente in discarica conquista il tenore dell'opera d'arte e del pezzo unico di *design*, immediatamente nasce nello spettatore la curiosità e allora quello stesso oggetto smette di essere quello che è diventato con la manipolazione artistica e si mostra come la somma delle storie di tutti i materiali di cui si costituisce.

Quando bandimmo il concorso "Diamo colore al ritmo del cuore" in collaborazione con il *Lions Club* San Miniato e l'Accademia di Belle Arti di Firenze, il numero di partecipanti superò nettamente le nostre aspettative. Le opere sottoposte al giudizio della giuria, peraltro, erano dotate di una forte valenza espressiva. Posso dire che quello è stato davvero un successo, anche perché le opere scelte, che costituiscono l'arredamento permanente del reparto di Aritmologia dell'Ospedale "*Nuovo Santa Chiara*" in <u>*Cisanello*</u>, *Pisa*, diretto dalla dott.ssa *Maria Grazia Bongiorni*, hanno proprio la funzione di creare un'interazione visiva con i pazienti e tutti coloro che per diversi motivi si trovano a transitare per quei corridoi, suscitando se vogliamo anche la curiosità per i materiali che le compongono. Ogni anno, poi, alla fiera *Ecomondo* di Rimini si ripete la magia e i rifiuti industriali rinascono ogni volta sotto spoglie diverse: lacci di scarpe assumono le forme di una volpe, tomaie mai utilizzate diventano le ali di un'aquila, perline colorate danno forma e luce al ritratto di *Marylin Monroe*. Ma c'è di più. Quest'anno l'artista cubano *José Yaque* è stato il protagonista di una mostra dedicata all'alluvione dell'Arno del 1966. L'artista, che lavora sui temi della *Sostenibilità Ambientale*, è rimasto letteralmente folgorato dall'enorme quantità di materiali di cui la società contemporanea è in grado di disfarsi quotidianamente, tant'è che per una delle sue installazioni collocate nel Centro attività espressive di *Villa Pacchiani* a Santa Croce sull'Arno, ha deciso di utilizzare migliaia di scarpe, arrivate in azienda poco prima della sua visita e destinate al macero.

▶ **Question** Può presentare alcune opere realizzate nell'ambito del progetto SCART®, focalizzandosi sul *messaggio ambientale* sotteso alla realizzazione artistica, alle ragioni della scelta dei materiali utilizzati, al suo impiego, se oggetto di *design*, e a ciò che ha ispirato la sua forma?

▶ **Answer** Ciò che ispira l'arte prodotta nei nostri laboratori è sempre il concetto di economia circolare: il desiderio tradotto in pratica di rimettere in circolo materiali scartati e per questo destinati alla discarica, questo principio sottende tutte le opere prodotte nell'ambito del progetto *SCART*®. Sono particolarmente legato agli abiti di scena, indossati da tenori di fama internazionale per gli spettacoli del *Teatro del Silenzio*, e ancora agli animali che nell'edizione 2015 hanno abitato il nostro stand nella Fiera *Ecomondo* e hanno ricevuto complimenti anche dal Ministro dell'Ambiente del Governo Italiano, *Gianluca Galletti*, e ancora ai ritratti di personaggi noti e celebri del nostro tempo, che nell'edizione 2016 hanno conquistato il pubblico di *Ecomondo* e che presto saranno oggetto di una mostra itinerante in quattro musei di altrettante città italiane.

Fig. IWP 8.2 Ecomondo 2015 Fair, Rimini, Waste Recycling Exhibition devoted to animal species: Il cane (The Dog) by Fabrizio Giorgi and La volpe (The Fox) by Monica Piazza © 2015

▶ **Question** Ritiene che i nuovi *Media* offerti dalla rete internet (inclusi i cosiddetti *Social Media*) abbiano agevolato la diffusione delle opere realizzate nell'ambito del progetto *SCART®*, rafforzandone le potenzialità comunicative? Cosa potrebbe rendere ancora più efficace il *messaggio ambientale* legato al progetto *SCART®*? Avete nuove idee per il futuro?

▶ **Answer** Siamo presenti sui maggiori *Social Network* perché sin dall'inizio abbiamo fortemente creduto nella capillarità di questo sistema di comunicazione.

Gran parte della nostra attività di comunicazione si svolge *on-line*: sulla piattaforma www.scartline.it, per esempio, raccogliamo le esperienze degli artisti *SCART*®, ma la nostra attività vuole anche lasciare un'impronta tattile oltre che visiva. *Waste Recycling*, infatti, porta il progetto *SCART*® anche nelle piazze e nei musei. Nel corso del 2017, ad esempio, una mostra itinerante esporrà quadri e animali prodotti con rifiuti industriali in quattro città italiane.

Fig. IWP 8.3 Ecomondo 2016 Fair, Rimini, Waste Recycling Exhibition for Gruppo Hera dedicated to the Circular Economy: Saint Mother Teresa of Calcutta by Gregorio Maria Mattei made of: textiles and drapes, toys, electric cables, disposable gloves and packaging material; Federico Fellini, by Ignazio Giordano made of: waste leather and plastic materials, sandpaper, wire and mother of pearl beads; Marylin Monroe by Antonella Prasse made of: beads, stoned and buttons; Luciano Pavarotti by Arianna Tosi; Lucio Dalla by Federico Niccolai; Muhammad Ali by Stefania Venuti; Nelson Mandela by Valentina Perini; Amy Winehouse by Beatrice Beneforti; David Bowie by Giulia Gigli and Frida Kahlo by Olimpia Bogazzi © 2016

Interview No. 9: Rhodri Jones

Rhodri Jones born in Gwynedd, Wales, has been based near Bologna, Italy since 2000. He has worked as a professional photographer on both personal and commissioned projects around the globe since 1989 and has had five personal photographic volumes published; *Made in China* (Logos Art, Italy 2002), *Return/Ynôl* (Seren, Wales 2006), *Hinterland* (L'Artiere Edizioni, Italy 2010), *Scambi Ferroviari* (L'Artiere Edizioni, Italy 2011) and *Cosi E'* (*L'Artiere Edizioni*, Italy 2015). Described by Magnum photographer, Philip Jones Griffiths as "a Welsh poet with a camera", Rhodri has exhibited in China, Eire, France, Greece, Italy, Netherlands, Poland, UK and USA. Distributed by *Panos Pictures* since 1992, his images have been used by leading magazines, newspapers, NGOs and publishers worldwide. His work is held in several public and private collections.

▶ **Question** In your career as documentary photographer you have been experiencing the close relationship between Nature and Humankind, two sides of the same Ecosystem, often in opposition, sometimes in harmony. We are referring, in particular, to your project "Hinterland" which shows the urban development of some areas close to big cities in northern Italy and worldwide. Environmental communication is strongly linked to visual representations which could be as effective as other *Media*. Or even stronger as *Ansel Adams* (American photographer and environmentalist) used to say: "*when Words become unclear, I shall focus with Photographs. When images become inadequate, I shall be content with silence*". Do you agree with this sentence? If yes, why? Do you think that a project like "Hinterland" would enable the public (addressees) to think about the impact of human actions on the Environment, pushing them to consider a more sustainable local development?

Rhodri Jones
www.rhodrijones.com (Web Site)

© Springer Nature Switzerland AG 2019
M. Abbati, *Communicating the Environment to Save the Planet*,
https://doi.org/10.1007/978-3-319-76017-9_16

▶ **Answer** The "Hinterland" project photographed in the lowlands around Bologna (where I have lived since 2000) was in fact a pilot project for a much larger and extensive one; E's.N.Es., Europe's New Edgelands.[1]

E's.N.Es. is basically the documentation of how the European urban/rural divide has become ever more difficult to define as our suburban boundaries have been expanding and blurring. The new landscape photographed in E's.N.Es. is the product of the neo-liberal ideology which has dominated our continent for almost three decades, an ideology which is now undoubtedly in crises and has passed its "sell by" date.

It's true; I hope the E's.N.Es. project can be a useful catalyst in the debate on peri-urban development while attempting to document how our relationship with the rural landscape, historically fundamental to our various local identities, is mutating. I also hope the project will be a useful vehicle as it documents the legacy we are leaving for future generations.

However, I consider myself to be a documentary photographer rather than a "concerned photographer"; my purpose is not to denounce something that I find disturbing but to analyze and interpret some of the most significant developments/changes that I see happening around me. Indeed, I consider myself to be a witness rather than a judge and have always attempted to photograph what I see in the most attractive or neutral available light. Ideally the images should have the necessary space to speak for themselves!

Image (and perhaps especially) photography is an intercultural language and, at its best, one of reflection. Documentary photography is obviously a subjective interpretation of reality, however the stronger (i.e. the more sincere) the image the less need for interpreters or "middle-men", indeed it can be a very powerful tool of direct communication.

Having said this, it is also true that this powerful tool has and always will be exploited for propaganda purposes. A recent example is the totally inappropriate use of agency images of Syrian war refuges taken in Slovenia as propaganda posters claiming that the UK was being flooded by economic migrants during the recent "Brexit" campaign. It is therefore necessary to carefully choose the context and manner in which it is presented; this is the photographer's moral responsibility.

You mention *Ansel Adams*, a photographic pioneer who consciously created and used his images as icons to represent the natural beauty to be found in the U.S.A. in the hope that those images would be useful tools in preserving that beauty and its perceived purity. Indeed, Adams had great success in this quest.

However, many things have changed since the days of *Ansel Adams*, not least, environmentally. The threats to our Environment are no longer simply local (e.g. the destruction caused by the commercial exploitation of natural resources) but are also global. Due to man-made global warming, we now face far worse consequences

---

[1] **Europe's New Edgelands** studies these phenomena around several important medium-sized European cities (with populations between 250,000 and 800,000 which according to OSCE statistics are where the majority of Europe's population live) and, above all, their surrounding metropolitan areas. The work presented in this project was completed between 2009 and 2013 in Croatia, France, Germany, Greece, Italy, the Netherlands, Poland, Spain and UK.

and, obviously, on a much larger scale. A contemporary photographer who has attempted to create modern-day icons in the mode of *Ansel Adams* to heighten the public's awareness to these phenomena is *Sebastiao Salgado* in his work, "*Genesis*".

▶ **Question** According to your professional career, which are the key elements suitable to catch the attention of the public opinion on sustainable issues (e.g.: climate change, overpopulation, pollution, etc.) through images, leading them to question themselves on possible solutions, including day-by-day personal actions?

▶ **Answer** Can *Salgado's* work be as successful as *Adam's* was in helping sway public opinion to positive actions; in this case, to reduce global emissions rather than protect National Parks from natural resource exploitation? Much as I respect *Salgado*'s work and motives, I'm not sure this is the best approach today. Why? Because times have changed—using the business of creating iconic images is perhaps no longer as effective. Icons involve an elitist (top down) rather than a democratic (bottom up) approach. Given the total lack of respect for our governing elites and their moral motives among, above all, the educated middle classes, I doubt this approach can continue to be effective in stimulating positive changes, globally, locally or through individual actions.

As the increasing global economic (and consequent social) inequality is undeniable proof of a failed neo-liberal ideology, so is the mass use of "post truth" reporting by mainstream traditional Media which is, in turn, proof of our governing elites' intellectual and moral hypocrisy. A banal example, within hours of his inauguration Trump's administration was using terms like "alternative facts" when claiming that the crowds present at his ceremony were bigger than those that celebrated Obama's.

Thankfully, not all changes are negative—within minutes, camera technology and *Social Media* had reduced Trump's boast to ruins! Given today's technology we are all potential witnesses.

I believe that ultimately democracy is about devolution of power, increasingly information is power and therefore this would be a truly democratic process. Perhaps the photographer's role has changed today, rather than producing icons that indirectly encourage people to react to injustice or preserve beauty, our role today is to inspire others to become direct witnesses and therefore become directly involved in the process of positive change.

▶ **Question** Can you describe some shots selected among your projects which point out their potential to communicate an environmental and/or a *sustainable message*?

▶ **Answer** During my career, I have taken literally hundreds of thousands (if not millions) of images, choosing a small representative number is a futile exercise. Not least, because I work on long-term projects that use a complex combination of images to portray a subject or theme, therefore I'd prefer to give you the link to websites where those images are presented, people can then decide for themselves if they find my work inspiring enough to themselves look at the world as a responsible witness.

.

Interview No. 10: Daniela Luise

Daniela Luise *Local Agenda 21* Coordinator and Officer for *Informambiente*, the *Environmental* institutional communication service by the Municipality of Padua including the Helpdesk *Informambiente (information on Environment)*; the management and coordination of the awareness campaign; the projects of *Environmental information* and education; the Province Lab for *Environmental education*; organization of *Agenda 21 Forum* and thematic groups; European planning (e.g.: *Life Parfum, Musec, Life Laks*); *Environmental Balance*; incentive programmes for the use of renewable energies (solar and thermal plants) ; information services on municipal contributions to promote sustainable public transport.

▶ **Question** Il processo di Agenda 21 Locale, sancito alla Conferenza di *Rio de Janeiro del* 1992 su Ambiente e Sviluppo, di cui lei ed il *team* di *Informambiente* siete portavoce, propone un modello di *governance* locale basato sulla "partecipazione", sulla "condivisione" di idee e quindi sulla "collaborazione" di tutti i settori della comunità per realizzare progetti volti a diffondere pratiche sostenibili e promuovere percorsi formativi e informativi in materia di Sostenibilità (es.: *Rete INFEA* regionale, ecc.); "saper comunicare" ai cittadini e alla comunità è dunque fondamentale. Sulla base della sua esperienza professionale, cosa non può mancare in un Piano di Comunicazione Ambientale perché sia efficace? Ritiene importante conoscere in anticipo i destinatari del *messaggio ambientale* (es.: scuole, cittadini, dipendenti dell'amministrazione comunale, ecc.)? Quale strumento/quali strumenti ritiene più adatti a veicolare il *messaggio ambientale* (es.: progetti, conferenze, forum/incontri partecipativi, corsi, ecc.)?

Daniela Luise
www.padovanet.it (Web Site)

See English version on p. 365

© Springer Nature Switzerland AG 2019
M. Abbati, *Communicating the Environment to Save the Planet*,
https://doi.org/10.1007/978-3-319-76017-9_17

▶ **Answer** Il Piano di Comunicazione è va definito di volta in volta a seconda dell'argomento da trattare, dei destinatari del messaggio, delle risorse economiche a disposizione.

Il percorso da seguire richiede una approfondita analisi iniziale di tutte le variabili e la scelta degli strumenti da utilizzare.

DESTINATARI: è fondamentali partire dalla fascia di utenza da raggiungere per riuscire a costruire un messaggio adeguato alla fascia di età e alla fascia sociale.

STRUMENTI: non esistono "strumenti migliori". A seconda dell'argomento e dei destinatari, oltre che degli obiettivi della campagna di educazione/comunicazione vanno individuati gli strumenti più adatti a raggiungere il proprio obiettivo. Spesso i risultati migliori si raggiungono utilizzando strumenti coordinati tra loro. I processi di coinvolgimento attivo (partecipativi) anche se più lunghi e onerosi sono quelli che danno i risultati migliori in quanto creano radicamento e vero cambiamento per tutte le fasce d'età ed interesse.

COSA NON PUO' MANCARE: A mio parere non possono mancare una fase informativa iniziale, una fase di coinvolgimento attivo se possibile.

Ad esempio: nel caso di una domenica ecologica è sufficiente informare sull'evento utilizzando vari strumenti integrati: dal sito internet, ai *depliant* informativi, alla comunicazione tramite i "social" (*Twitter, Facebook*, ecc). Non si devono trascurare gli strumenti comunicativi classici (il comunicato stampa, il rapporto con la carta stampata, con le riviste del settore, ecc.). A questo livello è abbastanza semplice strutturare una campagna informativa efficace.

Nel caso di una campagna di informazione alla cittadinanza sulla decisione, da parte dell'amministrazione pubblica, di introdurre la raccolta "porta a porta", naturalmente, la difficoltà è maggiore. Innanzitutto perché si deve comunicare un "cambiamento" nelle abitudini consolidate, e l'obbligatorietà del cambiamento stesso. In questo caso, più che comunicare la decisione, occorre usare tutti gli strumenti a nostra disposizione per facilitare il cambiamento che deve necessariamente essere assunto dal singolo.

La *comunicazione ambientale* non segue le stesse regole del marketing. Non mira a "vendere" qualcosa ma interviene sulla vita reale delle persone. In questo caso informare correttamente consente al cittadino destinatario del messaggio di cui assimilare il contenuto, capirlo e alla fine interiorizzarlo. Il cambiamento nel comportamento delle persone sarà reso più facile perché consapevole delle motivazioni.

Gli strumenti integrati, quindi, devono essere calibrati:

- sito *Internet, Social*;
- incontri con la cittadinanza: convegni, assemblee;
- materiali informativi;
- *infopoint* sul territorio;
- stazioni di ascolto (numero verde, e-mail, forum sul web);
- attivare percorsi partecipativi coinvolgendo le realtà sociali sensibili presenti nel territorio;
- attenzione alle categorie di cittadini più sensibili alla tecnologia, e quindi uso delle c.d. "App", per telefoni;

– utilizzare dei mediatori sociali per alcune fasce di popolazione;
– attivare percorsi educativi nelle scuole.

▶ **Question** Quanto influisce una buona *comunicazione ambientale* diretta ai cittadini, in sede decisoria (es.: attuazione di regolamenti comunali su temi ambientali "delicati", come la costruzione di un inceneritore, bonifiche di siti inquinati, chiusura di impianti industriali ad alto impatto ambientale, ecc.)?

▶ **Answer** Nel caso di azioni decisorie da parte della Pubblica Amministrazione la *comunicazione ambientale* va considerata come la prima fase di approccio di un processo partecipato e di condivisione più ampio.

Ha quindi un ruolo fondamentale per informare e aggiornare.

▶ **Question** La "comunicazione" in materia ambientale occupa il primo posto tra gli strumenti di informazione, sensibilizzazione e educazione: occorre perciò verificare l'attendibilità delle fonti su cui si basa per evitare di incorrere nel *greenwashing* o di seguire teorie ambientaliste, catastrofiche e infondate. Quali strumenti utilizzate per "certificare" l'autenticità del *messaggio ambientale*, aumentando così la fiducia dei cittadini e dei portatori di interesse, cosiddetti *stakeholder*? [es.: Certificazioni/Registrazioni ambientali, Marchi *green* di prodotto/servizio, ecc.]

▶ **Answer** Non utilizziamo certificazioni o registrazioni che sono costose e richiedono un percorso impegnativo anche a livello di risorse umane.

Utilizziamo fonti chiare e condivise citando sempre la fonte scientifica da cui prendiamo i dati.

Inoltre per aumentare la fiducia dei cittadini cerchiamo di essere sempre chiari e obiettivi nel messaggio che veicoliamo in modo da non essere attaccabili.

▶ **Question** Qual è l'approccio di *comunicazione ambientale* nei vostri percorsi formativi rivolti alle nuove generazioni? Come vi assicurate che il *messaggio/i messaggi ambientali* siano realmente compresi dai destinatari? Vi servite di *reti sociali* (es.: Facebook, Twitter)/*Internet*? Ritenete che siano strumenti efficaci che agevolano la Comunicazione Ambientale?

▶ **Answer** Nei percorsi formativi che generalmente utilizzano approcci partecipativi rendendo i giovani protagonisti del percorso stesso, la Comunicazione Ambientale ha un ruolo importante e trasversale. Viene sviluppata in tutto il percorso formativo e spesso direttamente dai protagonisti del progetto educativo.

Quindi l'uso delle "reti sociali" vengono attivate direttamente dagli studenti e specificatamente per il progetto educativo stesso, ovviamente con il supporto e la supervisione degli educatori.

Esempio:

– Green mi piace (Facebook)—Progetto Educativo Anno Scolastico 2015–2016

▶ **Question** Ritiene che le Amministrazioni Locali Italiane più vicine ai cittadini [Il Comune di Padova, in particolare] svolgano un ruolo chiave nel diffondere i *messaggi ambientali*, contribuendo così ad influenzare gli stili di vita delle comunità territoriali? Cosa andrebbe ancora fatto?

▶ **Answer** Il ruolo delle PA nel diffondere messaggi ambientali è tutt'oggi molto sottovalutato. In questa fase storica è il ruolo stesso della PA che è in crisi e con esso tutte le attività proposte.

Ritengo invece che nella *comunicazione ambientale* la PA possa e debba svolgere un ruolo fondamentale in quanto soggetto non sottoposto alle regole del mercato che non ha necessità di "lavare" con messaggi "verdi" quando veicola.

È quindi fondamentale che la PA mantenga la primogenitura della comunicazione e formazione

▶ **Question** "Comunicare green": quali sono le vostre sfide per il futuro? Su quali valori punterete? A quali soggetti vi rivolgerete?

▶ **Answer** Come pubblica amministrazione ci rivolgiamo a tutte le fasce di età con messaggi e modalità diverse a seconda dei temi da affrontare.

Priorità del prossimo futuro è il tema dei Cambiamenti Climatici che coinvolgono diversi attori sociali:

– i dipendenti pubblici che necessitano di formazione per poter agire correttamente;
– i giovani: continueremo a proporre progetti educativi alle diverse fasce d'età;
– le imprese che agiscono sul territorio, attraverso momenti formativi e condivisione di progetti da realizzare insieme anche attraverso l'accesso a fondi europei.

Elisabetta Martinelli Since 2002 she has been the Coordinator of IDEA, the Education Centre on Sustainability of the Municipality of Ferrara, where she had been traffic policewoman since 1983 acquiring competences and experience in the field of mobility. Furthermore, since 1987 she has been cooperating with the Department of Tourism of Ferrara dealing with hotel classification, trade shows and events organization and preparation of communication materials. Since 2002 she has been dealing with *Environmental Communication and Certification*, she is internal Auditor ISO 14001:2015 and ISO 9001:2008 certifications.

▶ **Question** Signora Martinelli, cosa significa per lei e per il suo *team* di lavoro "educare alla Sostenibilità" e "comunicare l'Ambiente": due azioni sempre più interconnesse come confermano alcune correnti di pensiero che parlano di "edu-comunicazione"? Condivide il ruolo chiave dell'educazione nel promuovere i valori dell'Ambiente e dello Sviluppo Sostenibile?

▶ **Answer** Il *Centro di Educazione alla Sostenibilità IDEA* è, come da lei citato un punto di riferimento per i temi della Sostenibilità dal 1998 e in questi anni è diventato anche il punto di riferimento dell'Amministrazione per l'adesione al Patto dei Sindaci, per la redazione del Bilancio Ambientale, per il Sistema di Gestione Ambientale e per tutta la comunicazione dei dati ambientali.

Elisabetta Martinelli
www.comune.fe.it/idea (Web Site)
@centroideaferrara (Facebook)

See English version on p. 367

© Springer Nature Switzerland AG 2019
M. Abbati, *Communicating the Environment to Save the Planet*,
https://doi.org/10.1007/978-3-319-76017-9_18

Nel 2012 è stato riconosciuto, ai sensi della Legge Regionale N.27 del 2009 come *"Multicentro per l'Educazione alla Sostenibilità Urbana"* inserito nel sistema regionale INFEAS, un'organizzazione a rete che coinvolge, in un modello di collaborazione attiva, soggetti pubblici e privati del territorio regionale per promuovere, diffondere e coordinare le azioni di educazione alla *Sostenibilità*.

L'"Educare alla Sostenibilità" e il *"comunicare l'Ambiente"* sono la *"mission"* del nostro Centro e uno degli obiettivi dell'amministrazione comunale per cui lavoriamo.

Il Comune di Ferrara infatti ha sempre evidenziato nelle sue politiche il tema della *Sostenibilità* dando molto risalto anche al tema della comunicazione dei "dati ambientali".

Nel 2003 ha approvato il primo Bilancio ambientale preventivo e consuntivo (anni 2000 e 2001), predisposto sulla combinazione della metodologia di *contabilità ambientale* CLEAR e da quella di *budgeting* ambientale *ecoBUDGET*. Nel corso degli anni e in seguito al confronto con altre realtà locali, il Bilancio Ambientale ha subito modifiche e aggiornamenti. Proprio in questi giorni è in fase di pubblicazione il Bilancio Ambientale Consuntivo 2014–2016.

Nel maggio del 2010 il Comune di Ferrara ha ottenuto la certificazione del proprio Sistema di Gestione Ambientale conforme alla norma *UNI EN ISO 14001:2004* ed è in corso l'aggiornamento e l'integrazione alla norma *UNI EN ISO 14001:2015*. Nel 2012 ha aderito al "Patto dei Sindaci" e nel luglio del 2013 ha approvato il suo "Piano di azione per l'Energia Sostenibile".

Condivido in pieno che l'educazione, la comunicazione e la sensibilizzazione hanno un ruolo chiave nel promuovere i valori dell'Ambiente e dello Sviluppo Sostenibile creando quella conoscenza e consapevolezza necessari ad affrontare le sfide future.

I principali obiettivi del nostro Centro:

- **promuovere nella cittadinanza lo sviluppo** di conoscenze, consapevolezze, comportamenti e capacità di azione a livello individuale e sociale, idonei a perseguire la *Sostenibilità Ambientale*, sociale, economica e istituzionale;
- **promuovere un'educazione alla Sostenibilità** nelle scuole in ambito "informale" attraverso la conoscenza e l'attività con laboratori pratici; educazione come esperienza per il cambiamento nella quale gli studenti si mettono alla prova e misurano le conseguenze delle proprie azioni. Immaginare il futuro e prepararsi a costruirlo sapendo gestire la complessità dei fenomeni ambientali, sociali ed economici confrontandosi, pensando in maniera critica e agendo in maniera responsabile;
- **promuovere la raccolta e la diffusione delle informazioni** sulla *Sostenibilità Ambientale*, sociale, economica e istituzionale del nostro territorio, favorendo la partecipazione consapevole dei cittadini ai processi decisionali.

Questi obiettivi tradotti nelle principali attività del Centro IDEA si possono così riassumere:

- **presentazione annuale di un'offerta formativa** rivolta al mondo della scuola proponendo iniziative come percorsi di promozione della mobilità sostenibile, risparmio ed uso sostenibile delle risorse, conservazione della biodiversità. Il Centro utilizza metodologie come i giochi di ruolo, i laboratori creativi, la progettazione partecipata, lezioni-conferenze;
- **sviluppo e gestione progetti educativi** locali e comunitari nel campo della Sostenibilità, in sinergia con tutta l'Amministrazione, i soggetti del territorio, la Regione Emilia Romagna e la RES;
- **coinvolgimento della cittadinanza locale e delle scuole** in manifestazioni che si svolgono annualmente e che rappresentano un filo conduttore nella comunicazione della Sostenibilità;
- **sviluppo contabilità ambientale comunale e indicatori di Sostenibilità.** Elabora e divulga il Bilancio Ambientale del Comune di Ferrara e si occupa del censimento dei dati ambientali necessari all'adesione ad indagini nazionali ed internazionali sul tema della Sostenibilità;
- **gestione della biblioteca tematica del Centro IDEA**, che contiene testi, documenti e riviste sul tema della Sostenibilità e dell'Ambiente. I libri sono consultabili on line nel Catalogo del Polo Unificato Ferrarese e disponibili al prestito presso la sede.
- **sostegno alle attività trasversali dell'Ente**, quali la Certificazione ISO 14001:2015, il Patto dei Sindaci, la Sostenibilità degli eventi culturali, gli acquisti verdi, le candidature dell'Ente ai premi sul tema della Sostenibilità.

▶ **Question** Quali sono le principali forme di comunicazione e i *Media* preferiti? Quanto sono legate le vostre azioni di comunicazione a quelle di educazione-formazione? Quali i punti di forza e le eventuali criticità che dovete affrontare? Quale risposta ricevete dalla comunità cittadina? Come verificate il grado di soddisfazione dei vostri servizi e la qualità dell'informazione ambientale veicolata, sia in forma didattica che comunicativa?

▶ **Answer** Le forme di comunicazione sono per noi strettamente legate a quelle di educazione e formazione; principalmente adottiamo forme dirette, rivolte a studenti e cittadini attraverso i laboratori proposti nell'offerta formativa del *Centro IDEA "Educare alla sostenibilità"*, nei corsi *"ActivECOlab"*—laboratori gratuiti per mettersi all'opera, per realizzare piccole pratiche sostenibili nella vita di tutti i giorni—proposti alla cittadinanza, per arrivare alle campagne di comunicazione che passano da attività diretta di diffusione del tema proposto attraverso convegni, *workshop*, *workcafè*, *info-point* ed iniziative dove vengono distribuiti materiali ed omaggi (*gadgets*), alla comunicazione via stampa locale e sito web. Esempio: *"Un albero* *per ridurre la CO_2*" in occasione della Giornata nazionale degli alberi dove vengono regalate piante ed arbusti alla cittadinanza.

Nel tempo questa attività si è integrata con l'utilizzo dei nuovi strumenti ormai di uso comune come i Social Media. Il *Centro IDEA* gestisce una pagina pubblica su *Facebook* per diffondere eventi e tematiche legate ai temi della Sostenibilità. Questo si è rivelato un canale strategico per la diffusione e per raggiungere il maggior numero di persone. Usare i *Social Media* vuol dire aumentare la trasparenza innovando modalità di relazione, di ascolto e di recepimento della voce del cittadino, rendendo i nostri servizi accessibili in una delle modalità più utilizzate dai giovani di oggi.

I punti di forza sono senz'altro quelli legati all'attività di formazione nelle scuole dove i ragazzi, dalla scuola primaria alla scuola secondaria di secondo grado, dimostrano sempre grande interesse e sensibilità per i temi e le attività proposte. Le criticità che dobbiamo a volte affrontare riguardano il coinvolgimento dei cittadini, che a seconda del tema trattato, dimostrano più o meno interesse. La risposta dalla comunità comunque è buona perché ormai la maggior parte della cittadinanza è attiva e consapevole e partecipa con interesse anche alle attività dell'amministrazione come dimostrano i progetti di partecipazione "*Ferrara mia*" promossi dall' *Urban Center* di Ferrara.

Per rispondere alla domanda di come verifichiamo il grado di soddisfazione dei nostri servizi posso dire che questi sono monitorati attraverso vari indicatori che variano dal numero di persone che hanno partecipato a una certa iniziativa, al numero di visite effettuate nel sito *Web* o nella pagina *Facebook*, al numero di adesioni alle attività proposte nell'Offerta Formativa, ai questionari di gradimento che inviamo sia alle insegnati che hanno partecipato alle nostre attività sia ai cittadini che hanno frequentato i nostri laboratori.

▶ **Question** Il vostro Centro di Educazione è parte integrante del sistema regionale INFEAS (Informazione, Educazione alla sostenibilità), un modello di collaborazione del settore pubblico-privato predisposto dalla Regione Emilia-Romagna al fine di promuovere, diffondere e coordinare le azioni di educazione alla sostenibilità, quanto è importante "fare rete" nel diffondere il *messaggio ambientale*?

▶ **Answer** Il coordinamento e l'integrazione delle diverse programmazioni ed esperienze educative sui temi della sostenibilità è fondamentale. Curare le relazioni tra diversi soggetti, mettere in sinergia gli *stakeholder* presenti sul territorio promuovendo azioni comuni, potendo confrontare il proprio lavoro, integrandolo e favorendo lo scambio con altre esperienze maturate all'interno della Rete di Educazione alla Sostenibilità dell'Emilia-Romagna (RES) è senz'altro un grosso valore aggiunto che mette a disposizione esperienze e metodologie consolidate in un'ottica di collaborazione e miglioramento continuo.

Giulia Meloncelli is an eclectic designer with a lot of curiosity. The desire to experiment led her to attend the courses of Industrial Design at I.S.I.A. (*Istituto Superiore per le Industrie Artistiche/Higher Institute for Artistic Industries*) of *Faenza* in Italy, near *Bologna*. She then improved her skills attending studies at *Kent Institute of Arts and Design of Rochester*, focusing on *Creative Model Making*, *Packaging Design* and *Graphic Design*. She realized her first art pieces in 1996 for the firm *Quattrozampe* in Milan, and successively for: the fashion firm *Opposite, Moschino, A.G.Spalding & Bros, Max & Co, Fraboso Argento, Tetra Pak* and son on, always increasing her production. In 2000 attended the Business Master *Valextra-ADI* "The leather, material and its evolutionary applications" in Milan. In this town, her experience of Industrial Designer improves and becomes refined and she works as project manager for *Selesta Ingegneria, Digicom, Swarovski* and *Lg Electronics* Italy. The ongoing search for materials and the "green" led her to plan with a method always more responsible and more aware of the *Environment*. Nowadays she lives in *Forlì (65 km far from Bologna) where she* deals in eco-sustainable *design*: her production at *Zero km* includes home furnishings and fashion accessories made with industrial waste materials post-consumer recycled materials under the *brand RICICLI.*

▶ **Question** Gentile Giulia, lei vanta una carriera pluriennale (di quasi 20 anni) nel settore del *design* cui ha dato, fin da subito, ad inizi Anni '90, un'impronta decisamente "green", precorrendo i tempi rispetto alla cosiddetta *Green Economy* esplosa nel corso degli Anni 2000. Un *know-how* che le ha permesso di creare un'impresa di successo, *Ricicli Design*. Un vero laboratorio in cui unendo creatività e talento realizza complementi per la casa ed accessori moda realizzati con materiali di recupero a kilometro "0". Che cosa l'ha spinta a fare questa scelta? Che cosa significa

Giulia Meloncelli
@RicicliDesignDiGiuliaMeloncelli (Facebook)
www.riciclidesign.it (Web Site)
recicli_design (Instagram)

See English version on p. 369

© Springer Nature Switzerland AG 2019
M. Abbati, *Communicating the Environment to Save the Planet*,
https://doi.org/10.1007/978-3-319-76017-9_19

287

per lei realizzare *Eco-design*? Quanto contano gli aspetti comunicativi nella sua attività?

▶ **Answer** La mia carriera nel mondo del *design* ha avuto inizio prima ancora di terminare i corsi accademici all'I.S.I.A. di *Faenza*, parliamo dell'anno 1995 e uno degli aspetti positivi del piano di studi era proprio quello di dare la possibilità a chi la frequentava di spaziare a 360° nel campo del *design*, partendo dalla progettazione dei prodotti fino alla loro comunicazione promozionale. Il poter sperimentare nei laboratori accademici la fattibilità dei propri progetti, "sporcandosi le mani", apre gli occhi sulle problematiche produttive e sui vantaggi e limiti dei materiali da scegliere per realizzare il pezzo. Possiamo paragonare quegli anni alla "*Preistoria*" della *Green Economy*. Le aziende che si lanciavano in produzioni ecosostenibili si contavano allora sulle dita di una mano. Gli incarichi dei primi anni avevano come *briefing* la progettazione di prodotti con materiali tradizionali. Avendo preparato una tesi che portava come titolo "Il rifiuto come risorsa" nel 1999, è chiaro che il mio intento era quello di voler convertire tutti ad una produzione rispettosa dell'Ambiente. Solo pochi clienti, tuttavia, mi avevano "concesso" di utilizzare materie prime seconde per le linee progettate. Ogni volta che conoscevo un'azienda nuova non potevo fare a meno, nel corso della visita al reparto produttivo, di sbirciare nei cassoni degli scarti pieni di "sfridi" (residui) sfavillanti di resina, tessuto, spurghi dalle forme più bizzarre in mille colori accattivanti. Rivolgevo così al tecnico di turno la solita domanda: "Ma di questi scarti quanti ne producete al giorno?" e poi continuavo: "Posso prenderlo per avere degli *input*?", mentre pescavo con le mani, in quelle che, a mio avviso, sono da sempre "miniere d'oro. Potete immaginare le occhiatacce che si lanciavano i dipendenti quando mi vedevano passare con uno scatolone pieno di ogni "*ben di Dio*" ravanato in qua e in là.

Soddisfatta da una parte perché gli "*spurghi da iniezione*" (scarti di materiale plastico), nella mia mente, nel tragitto ditta-parcheggio si trasformavano in un sottopentola in stile *Gaetano Pesce*. Allo stesso tempo, ero mortificata per le mostruose quantità di scarti prodotte giornalmente. Scarti che continuano a darmi *input* e ad alimentare il *brand* che ho fondato, RICICLI. *Eco-design* è *design* d'avanguardia, che rompe gli schemi a favore dell'Ambiente senza togliere funzione ed estetica. Esso aggiunge anzi un *quid* importantissimo, il rispetto per la Natura sia scegliendo di lavorare con dei materiali di recupero, o comunque con un basso impatto ambientale, sia utilizzando processi produttivi il meno inquinanti possibili. La progettazione per essere "*eco*" deve prevedere, infine, lo smontaggio del prodotto, al termine del suo utilizzo (*end of life*), in modo da poter essere differenziato dal cliente e poi riciclato. Per queste sue caratteristiche un prodotto di *eco-design* è portatore di una storia, un ideale, uno stile di vita che fanno parte dello stesso. Sono fermamente convinta dell'importanza di divulgare tutto ciò che succede nel "*backstage*", dietro le quinte. Al giorno d'oggi, non è sufficiente fare del pane integrale con una farina biologica "X", senza glutine, e metterlo in vendita insieme a migliaia di prodotti da forno "Y", preparati con farine derivate da semi modificati, alimentati a diserbanti, e dal costo di gran lunga inferiore.

Questo valeva alla fine dell'800 quando le farine erano tutte scure e ricche di principi attivi (e gli OGM dovevano ancora "essere progettati").

Bisogna comunicare il prodotto nella sua totalità. Nel caso di un prodotto di *eco-design*, è indispensabile l'informazione rivolta al pubblico che non sa, alle aziende che ne condividono la filosofia per creare rete e opportunità e alle aziende che ancora non hanno dato la priorità alla salute (dell'Ambiente e degli esseri viventi). Ma auspico che lo facciano presto.

Comunicare è parte del prodotto:

A. borsa in ecopelle di eco—design con tasca realizzata con manica di giacca;
B. borsa in ecopelle che noi acquistiamo come *fine pezza* dal cliente, quindi destinata a gravare sul magazzino o a essere smaltita (=inceneritore), con una tasca realizzata dalla manica di una giacca facente parte di uno *stock* di abiti invenduti. Partendo da materiale disomogeneo, ogni pezzo quindi è unico al mondo. Non solo, trattandosi di capi pregiati, per essere coerenti utilizziamo tutte le parti della giacca in modo tale da ottimizzarla e non sprecarne nemmeno una taschina nel rispetto di quelle abili mani che, taglio dopo taglio, l'hanno confezionata con maestria.

La descrizione "B" è quella che accompagna in modo indissolubile ogni mio prodotto in modo da informare il pubblico della filosofia aziendale e, allo stesso tempo, di coinvolgerlo. Quando il pubblico diventa consumatore di prodotti di *eco-design* contribuisce al benessere del Pianeta.

▶ **Question** *"Eliminare il concetto di rifiuto, non ridurlo... minimizzarlo o evitarlo...ma eliminarne il concetto stesso, attraverso il design"* questa è una citazione del celebre designer americano, *William McDonaugh*, e del chimico tedesco, *Michael Braungart*. Ne condivide i contenuti? Quali e quanti eco-messaggi trasmettono le opere che realizza? Che riscontro riceve da parte della sua clientela? Raccoglie i *feedback* dei clienti a fini statistici? In caso positivo, ci può fare un bilancio?

▶ **Answer** Condivido appieno in quanto qualsiasi materiale o oggetto viene etichettato "rifiuto" o "scarto" nel momento in cui non è più utile a chi ha deciso di sbarazzarsene ma lo stesso può essere un tesoro per un altro individuo. Chi di noi non ha mai ceduto alla tentazione di prelevare una lampada, una sedia, un comodino, una bicicletta che magari sporgeva dal cassonetto dell'immondizia? E chissà quante volte poi, con la luce del giorno, ha scoperto con sorpresa che si trattava di un pezzo storico di *design*? O semplicemente un oggetto giudicato meritevole di essere per così dire salvato dall'inceneritore? Il concetto di rifiuto è soggettivo dunque, relativo alla nostra cultura e subordinato a un determinato momento, perché lo stesso oggetto lo possiamo considerare interessante e ancora utile quando intravvediamo in esso una potenzialità, una possibile trasformazione in qualcosa d'altro. Questo è ciò che mi accade ogni volta che riesco a recuperare i materiali e gli oggetti: se non mi si accende la lampadina riguardo il loro secondo utilizzo, quel materiale od oggetto rimane ahimè scarto da smaltire.

I prodotti che realizzo sono portatori di diversi *eco-messaggi*:

1. NON SPRECO
   - il 98% dei materiali e degli oggetti che utilizziamo per creare i nostri prodotti sono rimanenze di magazzino o usato, il restante 2% comprende la ferramenta e la merceria;
   - il 98% dei materiali che utilizziamo è reperito a "km0", tant'è che a volte usiamo la bicicletta per il trasporto.
2. CONSUMO CRITICO E CONSAPEVOLE
   - incentiviamo il cliente al recupero dei propri oggetti e/o abiti trasformandoglieli con ironia.
3. PRODUZIONE A BASSO IMPATTO AMBIENTALE
   - gli scarti della produzione (pochissimi) vengono a loro volta riciclati;
   - il lavaggio dei tessuti avviene con detergenti ecologici rispettosi dell'Ambiente;
   - la promozione della nostra attività viene fatta utilizzando carta destinata al macero e carta certificata FSC.
4. LUOGO DI LAVORO SANO ED ECOLOGICO
   - abbiamo scelto la bioedilizia per contenere il nostro lavoro perché in sintonia con il nostro modo di essere e di vivere;
   - un orto ed un giardino senza veleni ci circondano di colori;
   - l'illuminazione è tutta a basso consumo e all'esterno è la minima indispensabile per non nuocere agli animali di piccola taglia e insetti notturni.

Tutti questi requisiti fanno alzare decisamente l'indice di gradimento della clientela che non manca di manifestare la propria soddisfazione attraverso continui messaggi sui *Social* e tramite *email.* Molti di questi commenti sono pubblici e si possono visionare sulla pagina *Facebook* e *Instagram* e ogni volta mi riempiono il cuore di gioia perché spero sempre di poter migliorare, attraverso la mia *produzione ecosostenibile*, la vita delle persone che scelgono di inquinare meno. Il *feedback* è stato finora sempre positivo e mi auguro che mantenga questo *trend* anche in futuro. L'indice di gradimento dei clienti è rappresentato dalle vendite di ogni prodotto. Come in tutte le collezioni, ci sono articoli che spopolano in ogni stagione e regione; altri con andamenti più ciclici. I nostri clienti apprezzano l'artigianalità e il "su misura".

Concludo citando lo scrittore *Mark Victor Hansen* che condivido appieno questo pensiero: "*La spazzatura è una grande risorsa nel posto sbagliato a cui manca l'immaginazione di qualcuno perché venga riciclata a beneficio di tutti*".

▶ **Question** Come evidenziato anche a livello comunitario (*Ecodesign Working Plan 2016-2019*) l'Eco-design, insieme all'etichettatura energetica, è una delle strategie più efficaci per promuovere l'efficienza energetica (si stima possa contribuire al 50% del risparmio energetico entro il 2020) e spingere i consumatori ed il mercato ad investire in prodotti eco-efficienti. Le sue creazioni di *Eco-design* contribuiscono a ridurre i consumi energetici e/o l'inquinamento ambientale? Utilizza indicatori come l'*impronta ecologica* (o simili) per "certificare" le *performance* ecologiche dell'oggetto o capo di abbigliamento da lei progettato?

▶ **Answer** Mi sono data come obiettivo quello di *"fare con ciò che è già stato fatto"* in particolare, lavorare e trasformare rimanenze di magazzino, "sfridi" (residui) di lavorazione, rivisitare oggetti obsoleti. Quindi tutto ciò che porta il marchio RICICLI contribuisce per forza a ridurre i consumi energetici e l'inquinamento. Ne consegue una bassissima ed empirica *Impronta Ecologica* (I.E.) per ciò che riguarda i miei prodotti. È chiaro che, trasformando materiali fabbricati da altre aziende, non è possibile certificare una determinata I.E. in quanto sono sconosciute sia le provenienze geografiche sia i processi produttivi.

▶ **Question** Quanto influisce l'aspetto "green" del suo *design* sullo stile di vita di chi lo sceglie? Secondo la sua esperienza, comunicare ai consumatori / ai clienti le caratteristiche ambientali o informarli sulla Sostenibilità della filiera di produzione, anche sulla base di certificazioni di prodotto, può costituire un valore aggiunto per un'impresa o uno *start-up*, incrementando le vendite? Perché?

▶ **Answer** RICICLI è un *brand* apprezzato da due distinte categorie di persone: la prima rimane affascinata dal *design* e ne apprezza la funzionalità e l'estetica nonché il fatto che siano prodotti ecosostenibili non influenza l'acquisto. La seconda categoria, invece apprezza in particolar modo, oltre gli aspetti pratici ed estetici, il fatto che sia stato riciclato del materiale in disuso, e per di più tramite manodopera italiana. Il *messaggio del riciclo* e dell'importanza della *seconda vita dei manufatti* è decisamente un valore aggiunto che cerco di trasmettere ad entrambe le categorie di clientela, in quanto in un mercato tanto diversificato non è sufficiente produrre oggetti accattivanti ma bisogna comunicare tutti gli aspetti della loro produzione. Sono un'appassionata del mio lavoro e, fin dalla nascita, innamorata della Natura. Parlare di come nasce l'idea di una borsa e come viene trasformata, per esempio, una giacca, incuriosisce il cliente che contribuisce a quello che si definisce "consumo critico". Quindi l'aspetto ecosostenibile dei prodotti RICICLI vuole far riflettere maggiormente sull'"altro consumo" i clienti che non hanno come priorità l'impatto ambientale. Il cliente già sensibilizzato, al contrario, si gratifica senza sensi di colpa anzi è consapevole di contribuire a ridurre l'impronta ecologica.

Infine, ritengo sia fondamentale comunicare alla clientela gli aspetti della fase produttiva, nonché dell'origine dell'idea, per testimoniare che i prodotti siano realmente frutto di uno studio e di un riciclo e non il risultato di un'operazione strettamente commerciale che, per cavalcare l'onda dell'eco-sostenibilità e del riciclo, spaccia migliaia di prodotti industriali dall'aspetto *"vintage"*, spesso prodotti all'estero, per prodotti ecologici.

Paola Poggipollini Graduated in Political Science (Economic Policy Department) at the *Università degli Studi* of *Bologna Alma Mater Studiorum*, she specialised at the *Scuola della Pubblica Amministrazione* (Public Administration School), held in the same University. Corporate professional *coach*, former head of the Sustainable Development and Participation Department, Local Agenda 21 and Technical EMAS—ISO 14001 Secretary at the Municipality of Ferrara. She gained a long professional experience at the design stage of European project management *(Life, Intelligent Energy Projects)*, peer review mechanisms, Public Administration planning and programming.

▶ **Question** *"La buona organizzazione e la modernizzazione di un ente pubblico non possono prescindere da un efficiente piano di comunicazione"*,[1] condivide questa affermazione?

▶ **Answer** Si, in quanto la buona organizzazione e la modernizzazione dell'ente pubblico richiede un impegnativo lavoro di definizione del quadro strategico comunale, contenente gli obiettivi di medio e lungo termine delle politiche territoriali e di conseguenza comporta la definizione delle nuove *vision* e *mission*

Paola Poggipollini
@ppoggipollini (Twitter)
Paola Poggipollini (LinkedIn)

See English version on p. 372

[1] Rizzo/Bordi, "La comunicazione istituzionale sul web" [The Institutional Communication on the Web], IlSole24Ore, 2009, pages 106.

© Springer Nature Switzerland AG 2019
M. Abbati, *Communicating the Environment to Save the Planet*,
https://doi.org/10.1007/978-3-319-76017-9_20

293

dell'amministrazione, che debbono essere diffuse e condivise dai dipendenti per poter diventare operative. È quindi essenziale la predisposizione di un efficace *Piano di Comunicazione* per la diffusione dei *goal setting*, fissati dall'ente, rivolti ai dipendenti, ma anche agli attori del territorio ed ai cittadini.

▶ **Question** In considerazione della sua pluriennale esperienza manageriale all'interno della Pubblica Amministrazione nel settore della Sostenibilità, quali strumenti di comunicazione ritiene siano più efficaci per comunicare l'Ambiente ai cittadini? Ci può fare qualche esempio concreto di sua applicazione che ha gestito quando ricopriva il ruolo di Dirigente del *Servizio Sviluppo Sostenibile e Partecipazione*?

▶ **Answer** Dipende molto dal tipo di comunicazione che s'intende diffondere. Se parliamo di semplici informazioni: internet, e-mail, SMS, newsletter, stampa, televisioni e radio locali rappresentano mezzi sufficientemente efficaci di diffusione delle notizie.

Se intendiamo interagire con i cittadini, come ad esempio educarli ad una corretta raccolta differenziata, dobbiamo ricorrere a mezzi più diretti come: assemblee informative, diffusione *depliant* presso centri commerciali, *focus group*, distribuzione, casa per casa, di *brochure* illustrative, messi a disposizione di personale, via per via, per fornire istruzioni corrette.

Per coinvolgere i cittadini nelle scelte dell'amministrazione è utile ricorrere ad assemblee partecipative, a *focus group*, a *workshop* di lavoro e adottare metodologie come l'*open space tecnology*, quando si è in presenza di un numero elevato di partecipanti.

Nella mia esperienza di Dirigente del Servizio Sviluppo Sostenibile e Partecipazione del Comune di Ferrara ho sperimentato tutti questi tipi di comunicazione.

Ricordo, in particolare, i numerosi workshop di *Agenda 21 Locale* che hanno coinvolto gli *stakeholder* del territorio e molti cittadini nella elaborazione di obiettivi di Sviluppo Sostenibile della città in campo ambientale, economico e sociale.

Ricordo anche l'esperienza dei Programmi partecipati di Quartiere che han rappresentato una riuscita sperimentazione di bilanci partecipativi e poi il Piano di Comunicazione legato alla diffusione del Sistema di Gestione Ambientale comunale e il Piano Energetico.

▶ **Question** Ritiene che il Processo di *Agenda 21 Locale* abbia favorito la *Comunicazione* delle tematiche ambientali, sviluppando così una maggiore sensibilità nelle comunità territoriali, in Italia, e, in particolare, a Ferrara?

▶ **Answer** Si. *L'Agenda 21* ha fatto discutere i cittadini e maturare una nuova coscienza ambientale nella città, ma come ogni buona pratica rischia di perdersi, se non viene sostenuta, nel tempo, da efficaci azioni comunicative ed educative e dalle buone pratiche messe in atto dal comune

Gli strumenti di rendicontazione come i *Bilanci Ambientali, Eco-budget* sono efficaci metodologie per dare conto dei risultati ottenuti in campo ambientale soprattutto agli *stakeholder*.

Il fatto che siano ancora strumenti facoltativi per le amministrazioni locali ne riduce l'efficacia. Molti comuni che avevano partecipato alla iniziale sperimentazione, poi, nel tempo o non hanno più messo in pratica *SGA (Emas Eco Management and Audit Scheme, Norme ISO 14001)* o ne hanno ridimensionato la portata.

▶ **Question** Ritiene che strumenti di partecipazione (es.: *Forum Climarchitettura, UNIAT Ferrara, ecc.*), di rendicontazione ambientale (*Bilancio Ambientale metodo CLEAR, Eco-budget*, ecc.), di progettazione europea (es.: *Life IDEMS, Intelligent Energy pro-ee*, ecc.) favoriscano realmente la comunicazione tra i vari portatori di interesse (*stakeholder*) e facilitino quindi il processo decisionale a livello locale, nazionale, europeo? Cosa pensa delle tecniche di "valutazione tra pari" (*peer review*), che lei ha avuto modo di sperimentare direttamente sul campo?

▶ **Answer** I *Sistemi di Gestione Ambientale* (SGA) applicati agli enti locali sono mezzi utili a favorire processi di semplificazione delle prassi amministrative, di messa a norma e di adeguamento delle strutture comunali nel campo della sicurezza, del risanamento ambientale, del risparmio energetico, degli acquisti verdi e a promuovere una maggiore trasversalità nell'analisi e nella ricerca di soluzioni ai problemi legati alla sostenibilità del territorio.

Strumenti come il Bilancio Ambientale o di Sostenibilità andrebbero estesi e collegati con i bilanci finanziari dello Stato e di quelli europei, poiché contengono significativi indicatori di risultato dei Programmi rivolti all'attuazione di politiche di eco-sostenibilità.

In Italia sarebbero quanto mai utili poiché è carente la cultura della rendicontazione delle politiche ambientali, economiche e sociali.

La valutazione tra pari è uno strumento utilissimo per perfezionare le metodologie di rendicontazione ambientale ed ha il vantaggio di non avere costi. Viene però poco utilizzata dagli enti locali italiani.

▶ **Question** La crescente attenzione dell'opinione pubblica nei confronti delle tematiche ambientali ha spinto le Pubbliche Amministrazioni a dotarsi di strumenti di gestione ambientale, i cosiddetti *Sistemi di Gestione Ambientale*—SGA (Certificazione ISO 14001/Registrazione EMAS); sono nati così nuovi documenti informativi che mirano a "comunicare l'Ambiente" all'interno e all'esterno dell'Ente; lei è stata testimone del percorso *green* che ha portato l'Amministrazione Comunale di Ferrara a certificarsi ISO 14001, nel maggio 2010, uno dei primi esempi in Italia di amministrazione pubblica locale medio grande, cosa ci può dire a riguardo? Ritiene i SGA siano strumenti utili per diffondere il *messaggio ambientale*?

▶ **Answer** L'adozione di Sistemi di Gestione Ambientale facilita le amministra-zioni locali nell'assunzione di comportamenti *green.*

Tale adozione testimonia la reale volontà di praticare e diffondere prassi virtuose in campo ambientale e rappresenta la condizione indispensabile per poi indurre i cittadini e gli *stakeholde*r ad assumere e praticare buone prassi di Sostenibilità.

L'adozione da parte di alcuni Comuni, compreso quello di Ferrara, di Sistemi di Gestione Ambientale quali *l'ISO 14001 ed Emas* ha consentito di dimostrare che tali strumenti si possono applicare con successo anche alle amministrazioni locali e che possono portare indubbi benefici alla qualità e alle buone prassi ambientali degli Enti.

Interview No. 14: Carlo Ratti

IWP 14

Carlo Ratti Architect and engineer, *Carlo Ratti* teaches at MIT of Boston, where he is also the director of *the Senseable City Lab*. He is founder of an engineering consulting firm *Carlo Ratti Associates*. He graduated at the *Politecnico of Turin* and at *École Nationale des Ponts et Chaussées* of Paris, he completed then a *Master Degree in Philosophy* and a *PhD in Architecture at the* University of *Cambridge* (England). He owns numerous patents and authored more than 250 publications. A protagonist of the international debate on design and innovation, he cooperates regularly with *Project Syndicate*; his articles and interviews have been published by mastheads like *New York Times, Washington Post, Financial Times, Scientific American, BBC, Il Sole 24 Ore, La Stampa, Corriere della Sera, Domus*. His works have been shown in exhibitions at *Biennale of Venice*, at the *Design Museum* of Barcelona, at the *Science Museum* of London, at the *Museum of Modern Art* of New York, at MAXXI of Rome. *Esquire* magazine included him among the *"Best & Brightest" and Forbes* among the *"Names You Need to Know"* and *Wired* in the list of the "50 people that will change the world". *Fast Company* mentioned him among the "50 most influential designers in America" and *Thames & Hudson* among the "60 innovators shaping our creative future". Two among his projects—*Digital Water Pavilion and Copenhagen Wheel*—have been included in the "Best inventions of the Year" by Time magazine (2007 and 2014). During Milan Expo 2015 he was in charge for the *Future Food District Pavilion*. A member of the *Italian Design Council*, he covers the co-chair of the *World Economic Forum Global Future Council on Cities and Urbanization* and he is a *special advisor* at the European Commission on Digital and *Smart Cities*.

Carlo Ratti
@SenseableCity (Facebook/Twitter)
http://senseable.mit.edu (Web Site)
Carlo Ratti (LinkedIn)

See English version on p. 373

© Springer Nature Switzerland AG 2019
M. Abbati, *Communicating the Environment to Save the Planet*,
https://doi.org/10.1007/978-3-319-76017-9_21

▶ **Question** Carlo Ratti Associati, laboratorio internazionale di *design* ed innovazione, con sede a Torino e filiali a Londra e a Boston, ove dirige il MIT *Senseable City Lab*, rappresentano un'eccellenza dal "cuore" italiano per la progettazione *high tech* nei campi dell'architettura, della riqualificazione urbana e del design. Sulla base della sua pluriennale esperienza quanto le nuove tecnologie hanno rivoluzionato il linguaggio ambientale? E quanto possono ancora fare?

▶ **Answer** Le tecnologie digitali sono entrate nelle nostre vite negli ultimi due decenni, di pari passo alla progressiva affermazione di una coscienza ecologica a livello sociale e politico. Trovo interessante questa coincidenza. E sono convinto che le tecnologie digitali possano aiutarci a sviluppare un rapporto migliore con l'Ambiente in cui viviamo—in particolare se sapremo fare un uso accorto e aperto dei dati che le stesse ci permettono di acquisire. Da parte nostra, sia al *Senseable City Lab*, sia la *Carlo Ratti Associati*, abbiamo provato a esplorare questi temi in una moltitudine di lavori. Ad esempio pochi mesi fa abbiamo lanciato *Treepedia*, una piattaforma per la mappatura delle chiome degli alberi, quella straordinaria "cortina verde" che svolge un ruolo fondamentale a supporto del benessere collettivo nelle nostre città. È un progetto che si basa sui dati in arrivo da *Google Street View*, e ci permette di avviare uno sguardo comparato su metropoli come Boston, Singapore o Torino in modi che sarebbero stati quasi impossibili, fino a pochi anni fa.

▶ **Question** Se dovesse stilare una lista di priorità per rendere la *comunicazione ambientale* più efficace ed adeguata ai "codici" di linguaggio moderno, che cosa metterebbe ai primi 5 posti…e perché?

▶ **Answer** Direi di adottare un linguaggio propositivo e aperto al futuro, non un linguaggio di minacce e prospettive catastrofiche—per quanto amara la realtà spesso possa essere. Sono un grande ammiratore di *Edward O. Wilson*, il biologo americano che per primo, negli anni Ottanta del Novecento, ha formulato a livello scientifico, la cosiddetta "ipotesi della biofilia", ovvero l'idea che gli uomini siano 'programmati' per provare felicità quando si trovano immersi tra gli elementi della natura. Ecco: credo che questo potrebbe essere un buon punto di partenza per aiutarci a fronteggiare sfide enormi come il Cambiamento Climatico.

▶ **Question** Quale messaggio o quali messaggi ambientali ha voluto trasmettere al pubblico di EXPO Milano 2015 attraverso il progetto del *Supermercato del Futuro*, creato nel *Future Food District*, realizzato da lei e dal suo *team*? A cosa si è ispirato nella sua realizzazione? Ritiene che "il supermercato del futuro" possa diventare presto "il supermercato del presente"?

▶ **Answer** Per il *Future Food District* ci siamo ispirati all'immagine del signor *Palomar* di Italo Calvino che, immerso in una *fromagerie* parigina, ha l'impressione di trovarsi in un museo o in un'enciclopedia: "Dietro ogni formaggio c'è un pascolo d'un diverso verde sotto un diverso cielo (…)". Questo negozio è un museo: il signor *Palomar*, visitandolo, si sente come al *Louvre* e percepisce, dietro ogni oggetto esposto, la presenza della civiltà che gli ha dato forma e che da esso prende forma."

Ecco, è stato questo uno dei punti di partenza del nostro progetto: cercare di usare nuovi strumenti per permettere ai prodotti di raccontare le loro storie—e in ultima analisi per stimolare un consumo più informato e consapevole. Gli articoli, esposti non su scaffali bensì su grandi tavoli come in un antico mercato, ci raccontano la loro storia in modo immediato. Una maggior tracciabilità dei prodotti permette anche l'instaurarsi di nuove relazioni tra le persone. Grazie alle maggiori possibilità di condivisione offerte dalle reti, perché non pensare al supermercato come un luogo di scambio aperto a tutti?

▶ **Question**   Lo sviluppo di tecnologie di *Internet of Things* ha permesso a lei ed al suo *team* di lavoro di creare ambienti ed oggetti in grado di interagire, e spesso "capire" le esigenze di una comunità più o meno grande di persone. Pensiamo all'"Ufficio 3.0" (progetto Fondazione Agnelli 2016) o al progetto HubCab o a MONiTOUR solo per citarne alcuni (presentati in occasione degli Stati Generali all'edizione 2016 di *Ecomondo*). La *comunicazione ambientale* è, per sua natura, un processo interattivo, circolare che implica una risposta da parte di chi riceve il messaggio, come verificate che questo sia stato compreso correttamente? Quanto conta garantire un'adeguata formazione-informazione del pubblico in fase progettuale?

▶ **Answer**   Conta moltissimo. Per gran parte del Ventesimo secolo, nel clima culturale Occidentale, ha dominato il paradigma dell'architetto-eroe, che lottava per imporre al mondo la sua verità—con risultati spesso opinabili per la vita dei cittadini. Oggi sta emergendo un paradigma nuovo, quello dell'*"open source"*, titolo di un libro di cui sono autore, pubblicato dalla Casa Editrice Einaudi [Ratti Carlo, Claudel Matthew, *Architettura open source—Verso una progettazione aperta*, Torino, Einaudi, 2014, pp. 142]. Ci piace allora pensare a progetti in codice aperto, da portare avanti con il contributo attivo di un team multidisciplinare e degli utenti finali, magari abbandonando l'idea dell'*"archi-star"* e aprendo invece la porta a un architetto "corale", capace di armonizzare le diverse voci in un accordo consonante. Per questo fondamentale comunicare con i cittadini—un filo diretto che va in entrambe le direzioni.

Maestro Niccolò Ronchi is an Italian young pianist and composer who graduated at the famous *Accademia di Santa Cecilia* (Academy of Saint Cecilia) in Rome under the guidance of renowned pianist *Maestro Benedetto Lupo*. He has started playing the piano since he was a child and he was soon noticed by the critic for his great talent able to give virtuosity, expressiveness and great communicative interpretation to every piece performed. Under the guidance of prestigious Maestri like: Isabella Lo Porto, Franco Scala, Director of the International Academy of Imola (near Bologna), Vincenzo Balzani, Leonid Margarius, student of Regina Horowitz, Vladimir Horowitz's sister, Russian famous pianist of Ukrainian descent. In 2006 he won the "Roma International Piano Competition" dedicated to Chopin, in Rome, to which more than sixty Maestri coming from all over the world participated. He starts then a brilliant career at the age of twenty, that sees him perform, as a soloist and in orchestra, in more than two hundred concerts in the most prestigious theatres and halls in Italy ("*Teatro La Fenice*" in Venice; "*Auditorium Parco della Musica*" in Rome, "*Sala Verdi*" in Milan), in Germany, in France, in Spain, Great Britain, United States, Japan; China, Russia and Australia.

Niccolò Ronchi
@NiccoloRonchi (Twitter)
niccolo.ronchi.pianist (Instagram)
Niccolò Ronchi (LinkedIn)

See English version on p. 375

© Springer Nature Switzerland AG 2019
M. Abbati, *Communicating the Environment to Save the Planet*,
https://doi.org/10.1007/978-3-319-76017-9_22

Premise to the Questions: The "green market pressure" is involving the music world. Sustainable music concerts and festivals are spreading at European and International level, confirming this trend. Being "green" means not only the eco-planning (e.g.: energy efficiency, waste eco-management, etc.) and a suitable eco-location (e.g.: urban gardens or a protected natural area) but also a "green" music event, such as the Festival Øya at Tøyenparken (Oslo, Norway).

▶ **Question**   Maestro, quali pezzi musicali del passato, del presente e del futuro (compresi brani da lei composti) sceglierebbe per trasmettere efficacemente un *messaggio ambientale* come la protezione della biodiversità, la lotta ai Cambiamenti Climatici e la tutela della risorsa vitale "Acqua"?

▶ **Answer**   Innanzitutto ritengo che la tutela della biodiversità e delle risorse vitali come l'Acqua, nonché la lotta ai cambiamenti climatici, siano tematiche estremamente importanti che affondano le loro radici nel macro-concetto di "Rispetto". Rispetto verso l'altro, rispetto verso il diverso, rispetto verso ciò che non ci appartiene ed infine rispetto verso sé stessi.

Con il trascorrere del tempo, avendo da poco superato i trent'anni ma essendo ormai venticinque anni che dedico anima e corpo allo studio della musica, mi rendo conto in prima analisi che lo studio della musica sia per l'essere umano una incredibile scuola di vita e un veicolo attraverso il quale apprendere e fare proprio il concetto di "Rispetto". La cosa incredibile è che tutto ciò avviene in modo implicito, tanto più se lo studio della musica e di uno strumento musicale viene proposto fin dall'infanzia.

Volendo pensare al pianoforte, strumento al quale sono particolarmente legato, imparare la polifonia è fondamentale, fin dagli esordi. Dall'etimologia della parola stessa (dal greco classico: "πολύς—φωνή", "polùs—fonè", "molti—suoni") si capisce l'importanza di gestire, fin dall'inizio, le diverse "voci" che compongono un brano musicale. Capire cioè come talvolta sia non solo opportuno ma necessario condurre una linea musicale che ad un certo punto dovrà inevitabilmente "farsi da parte", affievolendosi o tacendo del tutto. Per poi lasciare spazio ad un'altra linea musicale, ad un'altra voce che potrà ribadire la stessa idea, lo stesso concetto, lo stesso tema o proporne uno completamente diverso. All'interno dello stesso accordo, suonato con la stessa mano al pianoforte, è assai raro che i singoli suoni abbiano la stessa intensità. E, anche se succedesse, bisogna rispettare le peculiarità del singolo suono all'interno dell'accordo stesso, per trovare quell'equilibrio che va proprio a creare l'armonia, metaforicamente ed acusticamente parlando.

Gli esempi musicali sono innumerevoli. Basti pensare ad una forma musicale come "la fuga", in cui conta la polifonia, il saper dare risalto o affievolire le single voci per concorrere al raggiungimento di un risultato chiaro e armonioso. Emblema della forma musicale stessa. Pensiamo a cosa avviene in un'orchestra, dove a rispettarsi vicendevolmente non sono le voci di una fuga ma intere sezioni di orchestra con caratteristiche e peculiarità molto diverse tra loro, e spesso diametralmente opposte. Ci si renderà facilmente conto di come lo studio della musica possa diventare la migliore scuola in cui apprendere le regole del rispetto reciproco e di ciò che concorre al bene finale comune.

Volendo essere forse un po' meno filosofici e un po' più "pratici", sono molti i brani che si potrebbero proporre ad un vasto pubblico in un tentativo di sensibilizzazione alle tematiche ambientali. Con riferimento all'Acqua, per esempio, alcuni dei più grandi compositori del passato hanno composto musica cosiddetta "a programma" o descrittiva, nel tentativo di creare suggestioni riconducibili proprio a questo prezioso elemento liquido. Pensiamo a *"Jeux d'eau"* di **Maurice Ravel**, a *"Jeux d'eau à la villa d'Este"* tratto dagli *"Anni di Pellegrinaggio"* di **Franz Liszt**, a *"Reflets dans l'eau"* di **Claude Debussy** o al celeberrimo preludio *"La goccia d'acqua"* tratto dai *"Preludi op. 28"* di **Fryderych Chopin**. Opere in cui è il titolo stesso a richiamare l'Acqua. Per non parlare poi dell'opera orchestrale *"La Mer"* di *Claude Debussy*. O ancora il poema sinfonico *"La Moldava"* di **Antonin Leopold Dvořák**, composto per celebrare la bellezza del fiume Moldava e la sua epopea, dalla nascita nei boschi della selva boema, al suo confluire nel fiume Elba fino alla foce, nel Mare del Nord.

Altrettanto numerosi sono i brani ispirati alla "Natura". Il grande **Ludwig van Beethoven** ne era positivamente ossessionato perché vedeva nella Natura un elemento di "bontà", metafora di quel Dio capace di ispirare l'Uomo e l'umanità nella ricerca interiore per affermare la forza dell'animo umano. Molte delle lettere ai suoi amici raccontano di questo "amore", lo descrivono come appassionato del mondo agreste, delle passeggiate nei campi, in contemplazione di alberi, colline, ruscelli. Da qui nasce la sua Sinfonia n. 6 op. 67, con la dicitura "Canto pastorale. Sentimenti di benevolenza e ringraziamento alla Divinità dopo la tempesta". La tempesta vera e propria, la tempesta metaforica che il "Beethoven uomo" visse dentro di sé a causa delle avversità continue che ebbe nella sua vita prima tra tutte la sordità. Ma anche la "tempesta" che il nostro Pianeta sta attraversando in conseguenza dei Cambiamenti Climatici. Fenomeni naturali violenti ed imprevisti quasi per sopperire ad un disequilibrio che il genere umano sta creando nell'Ecosistema Terra, la nostra sola ed unica casa.

Proseguendo nel viaggio musicale ispirato al mondo naturale non possiamo non citare altri brani più o meno conosciuti: le celebri *"Quattro stagioni"* di **Antonio Vivaldi**; *"Pierino e il lupo"* del compositore russo **Sergej Prokofiev** in cui il suono emesso da diversi animali viene "abbinato" ad alcuni strumenti dell'orchestra ed al loro timbro; la suite *"All'aria aperta"* di **Béla Viktor János Bartók**, la *"Sacre du Printemps"* di **Igor Strawinskij**, *"Il cantico del Sole"* della compositrice **Sofija Gubajdulina**, *"Le stagioni"* op. 37a del compositore russo **Pyotr Ilyich Tchaikovsky**. E poi tutta la tradizione compositiva dei compositori del Nord Europa che vede in **Edvard Grieg** il rappresentante più illustre e che si ispira profondamente alle suggestioni della natura, a metà tra il magico e il reale. Come non citare *"Il mattino di Grieg"* o alcuni brani tratti dai suoi meravigliosi *"Pezzi Lirici"*. Infine, come inno vero e proprio alla *diversità biologica* il *"Catalogue des oiseaux"* (Catalogo degli Uccelli) di **Olivier Messiaen**. Già, gli amati uccelli che *Messiaen* considerava i più grandi musicisti al mondo. Tale la sua passione che il compositore si considerava più un ornitologo che un compositore.Musica e Natura. Natura e Musica. Del resto l'uomo stesso è Natura ed era difficile immaginare che la creatività artistica umana non aspirasse, più o meno coscientemente, ad un ricongiungimento con la propria "Divinità", a cui appartiene.

Premises to Questions: Music has been more than shown to be inspired by the natural world, translating the natural sounds into the "music language" or even introducing some original natural tunes within the piece of music, thanks to the new technologies. Some pop-stars became an icon of the *eco-message* (e.g.: the Icelandic songwriter *Björk* or the American Group, *Marron 5*). Nevertheless, the "green" trend in music is rooted in the classical music Let us think of the baroque "Four Seasons" by Antonio Vivaldi, you have already mentioned, or by *Ludwig Van Beethoven* (e.g.: Simphony No. 6), by *Claude Debussy* (e.g.: "Deux Arabesques" and "Moonlight"), by *Frederic Francois Chopin* (e.g.: "Prelude in D flat", known as the *Raindrop Prelude*) or more recently by *Edvard Greig* (e.g.: "Morning Mood" composed for the musical work Peer Gynt) and by *Nikolai Rimsky-Korsakov* (e.g.: "Flight of the Bumble Bee").

▶ **Question**   Condivide questa riflessione? Ritiene che queste opere possano fungere da *medium* di comunicazione e che possano essere impiegate in moderne campagne di sensibilizzazione su tematiche ambientali e sostenibili? Come le contestualizzerebbe o "modernizzerebbe" affinché siano efficaci anche per le nuove generazioni?

▶ **Answer**   Come coinvolgere l'umanità intera per far capire quanto sia importante l'impegno attivo da parte di ogni individuo in merito alle tematiche ambientali e sostenibili? È vero, al giorno d'oggi fortunatamente anche alcuni musicisti di musica leggera e quindi, forse, più vicini alla sensibilità di un pubblico più vasto, sono divenute icone del *messaggio eco-sostenibile*. Questo penso che sia un bene. Eppure ci tengo a fare una precisazione. Gira voce che la musica classica sia anacronistica, se non addirittura "vecchia". Credo che le persone confondano il concetto di "vecchio" con quello di "immortale". La musica cosiddetta classica, o meglio "colta", è immortale. Tutta la Musica, da sempre espressione diretta dell'Uomo, cerca di esprimere le infinite sfumature emozionali dell'animo umano e le sue altrettanto infinite coniugazioni ontologiche.

L'Uomo, nella sua essenza, non è cambiato. Dai tempi in cui non esisteva la notazione; alle prime parti cantate nelle tragedie della Grecia classica; al canto gregoriano; alla musica rinascimentale e barocca; al periodo classico, e a quello dei due secoli successivi, fino ad arrivare ai giorni nostri.La musica, da sempre, accompagna l'Uomo. Ognuno di noi, appena nasce, viene in contatto con una *cellula musicale*: il battito cardiaco. Il cuore. Quell'elemento ritmico che rappresenta la Vita. Vita che è strettamente collegata al Pianeta in cui viviamo.

Ecco perché, se penso ad una campagna di sensibilizzazione musicale, immagino ad un qualcosa che sia un inno alla vita. Immagino *Beethoven* accostato ai "*Giardini di Marzo*" di *Lucio Battisti* senza la necessità di dover snaturare l'uno in favore dell'altro. Immagino coinvolgere altre arti: come la fotografia, la pittura, la regia, la scrittura. Viviamo in un mondo in cui il *caos* e il bombardamento sonoro e visivo a cui siamo costantemente sottoposti hanno abbassato la nostra capacità di percepire attraverso i sensi. Forse non siamo più in grado di cogliere la Bellezza della musica, la Bellezza di un gioco di luci, la Bellezza di un'immagine, la Bellezza di un testo. Si potrebbe quindi creare un qualcosa che unisca e sommi tutte queste Bellezze. Creando un'unione di forze comunicative in grado di non lasciare nessuno indifferente e di fermare l'insensibilità di questo mondo caotico.

▶ **Question**   Se le chiedessero di realizzare un brano pianistico (assolo strumentale o musica e voce) per promuovere l'importanza del rispetto per l'Ecosistema, di cui l'*Uomo* fa parte, ispirato ai principi della *Sostenibilità* così come definita dal Rapporto della Commissione Mondiale per l'Ambiente e lo Sviluppo intitolato "*Il Nostro Futuro Comune*" [Rapporto *Brundtland* 1987] quali scelte stilistiche (es.: genere musicale, suoni gravi/acuti, ecc.) farebbe e perché?

▶ **Answer**   La composizione, che altro non è che il prodotto della creatività e sensibilità umana, appartiene a quella ristrettissima cerchia di espressioni umane che possono ancora influenzare l'uomo. Se dovessi comporre un brano per promuovere l'importanza del rispetto per l'Ecosistema, di cui l'Uomo fa parte, certamente abbinerei la voce umana all'utilizzo di altri strumenti. Questo perché la voce umana è lo strumento a cui ogni altro si è sempre ispirato e di cui ha cercato di riprodurne le infinite sfumature.

Non sceglierei nessuna lingua ma la semplice riproduzione di suoni. Sceglierei, infine, una forma musicale che non sia una vera "forma" musicale ma piuttosto un contenitore emozionale. E per questa ragione caratterizzerei il brano con forti contrasti: suoni acuti e suoni gravi; dinamiche dal pianissimo al fortissimo. Questo perché credo sia importante far capire a tutti quanto la diversità sia un valore aggiunto e non un problema. Infine concluderei con pochi suoni e un qualcosa che possa mettere in pace con sé stessi e far vibrare l'animo umano fino a far concentrare tutto sulla vibrazione e nient'altro. Dimenticarsi di sé stessi per sentirsi parte di un tutto. Tornare all'inizio. C'è un brano che mi suscita un'emozione simile. "*Spiegel Im Spiegel*" di **Arvo Pärt**. Commovente Bellezza, nient'altro. Quello che dovrebbe provare ognuno di noi di fronte alle meraviglie del nostro mondo.

Antonio Salinari Graduated in Engineering at *Politecnico* of Turin, he gained extensive experience as a Manager in Multinational leader within computer science and technology. In 2006, he met the sculptor *Franco Alessandria*, who became his mentor, so in tune with *Richard Sennett*'s theories, he started to grow old passions: drawing, sculpture, furnishing design and photography. The artist then combined his experience, his skills, his design skills to the typical industry passion and the crafts "know how": since 2009 he has been directing the School of Sculpture in *Alta Valle Susa*, Susa Valley, oriented to the recovery of ancient crafts. <u>Awards</u>: he took part to Group Exhibitions in Italy, in France and Austria with more than one hundred and fifty sculptures, some of them are monumental, in the framework of the *Public Art. Mostra*, a personal public art exhibition in Turin. He was selected for the Biennial of *Montecarlo* 2016 and 2018 and for Florence Biennial in 2017.

▶ **Question** "C'è bisogno di bellezza formale e di coinvolgimento emotivo se ci si vuole mettere davvero in comunicazione con il cittadino consumatore di cultura, spettacolo, narrazione" (Erik Balzaretti, Benedetta Gargiulo, "La comunicazione ambientale: sistemi, scenari e prospettive. Buone pratiche per una comunicazione efficace", Franco Angeli Editore, 2009). Quando un artista affronta il tema Ambiente non può dunque limitarsi al suo "sentire personale" ma deve necessariamente suscitare un'emozione positiva che sia compresa a livello globale. Si identifica in questa affermazione?

Antonio Salinari
www.antoniosalinari.it (Web Site)

See English version on p. 378

© Springer Nature Switzerland AG 2019
M. Abbati, *Communicating the Environment to Save the Planet*,
https://doi.org/10.1007/978-3-319-76017-9_23

▶ **Answer**   Condivido in generale la necessità che l'arte susciti un'emozione positiva e che sia compresa a livello globale.

Ritengo che per emozionare in modo positivo, l'arte dovrebbe essere narrazione e avere una forza di generazione del senso, coinvolgendo il pubblico nell'identificazione e nell'interazione con un valore, qualunque esso sia.

Penso anche che, per essere compresa a livello globale, l'arte dovrebbe contare su una figurazione: tutte le narrazioni hanno un significato e una forma, se questa è figurativa diventa più intelligibile. Ciononostante il messaggio artistico, proprio perché deve indurre alla riflessione, difficilmente potrà essere immediato come quello pubblicitario.

▶ **Question**   Cosa la ispira nel realizzare le sue opere?

▶ **Answer**   In accordo con il pensiero di *André Derain*, "*bisognerebbe aver penetrato intimamente la vita delle cose che si dipingono*", per me le principali fonti d'ispirazione sono il mio vissuto, il *background* tecnologico e la passione per la *Natura*.

Più in particolare sono ispirato da:

- la personale esperienza sulle idee mitizzate del nostro tempo, che ormai diamo per scontate;
- la necessità di ricostruire dei valori di riferimento, bisogno che ciascuno di noi sente per poter vivere;
- la bellezza in generale e, in special modo, quella della natura. Qui intendo il concetto di bellezza greca: *kalogakathia*, letteralmente il "bello e buono", in quanto valore certo, che crea emozioni.

▶ **Question**   Utilizza elementi simbolici o altre forme di comunicazione?

▶ **Answer**   Quando creo le mie opere provo a raccontare una storia attraverso quegli straordinari differenziatori che sono i simboli, la metafora e la materia.

Inoltre ritengo che anche il processo di lavorazione, basato su tecniche manuali e su significativi tempi di realizzazione dell'opera, possa rappresentare un gesto espressivo nell'ambito della comunicazione. Un atto che, a dispetto delle apparenze, non è nostalgico nei riguardi del passato, ma vuole tendere a coniugare gli aspetti positivi del mondo in cui viviamo con la necessità di cambiare approccio per un futuro sostenibile.

▶ **Question**   Quali emozioni prova nel realizzarle?

▶ **Answer**   Dare forma a un pensiero, soprattutto perché nasce da un'urgenza interiore di esprimersi, è per me un'esperienza estremamente gratificante.

Altrettanto gratificante è utilizzare le mani per costruire qualcosa che non esisteva, sentire il profumo del legno e la sensazione tattile che si ricava nell'accarezzare le superfici dell'opera.

▶ **Question** Quali emozioni vorrebbe suscitare nel pubblico?

▶ **Answer** Mi piacerebbe far riflettere sul valore che ha per l'uomo ritrovare la sua collocazione nella natura, proprio in quanto parte di essa.

Ben venga se le opere riescono a suscitare emozioni, bisogni e desideri: dalla gioia e l'amore per l'incanto del creato alla fiducia nell'essere capaci di sviluppare un atteggiamento responsabile verso l'Ambiente.

▶ **Question** Perché la scelta di utilizzare principalmente il legno (ma non solo) come materiale per le sue opere?

▶ **Answer** Il legno è materia calda, viva, rinnovabile e lo trovo altamente simbolico nel rappresentare la natura; per me esprime la forza di costruzione della bellezza. In particolare nel progetto *"ricerca della stabilità"* è l'armonia dei legni pregiati, provenienti dai differenti continenti, assemblati tra loro e scolpiti, a predominare sulla forma, confermando il pensiero di *Marshall McLuhan*: *"medium is message"*.

▶ **Question** Come influisce l'artista sulla materia?

▶ **Answer** Direi che nel caso del legno si tratta di un'influenza reciproca, di un'interazione: se da una parte lo scultore dà forma alla materia, dall'altro questa influisce sull'opera, con i disegni creati sulle superfici dalle sue venature, fondamentali per la bellezza dell'opera stessa, venature che, a volte, spingono l'artista a "rivedere" la forma per assecondarle.

▶ **Question** Lei ha creato parallelamente un laboratorio dedicato alle nuove generazioni che consente a molti giovani della Val di Susa di sperimentare l'arte di modellare il legno e creare opere di interesse artistico, quale *messaggio/quali messaggi ambientali* si prefigge di trasmettere attraverso questa attività artistico-didattica? Quali "risposte" riceve dai giovani?

▶ **Answer** Lo sviluppo della creatività, l'utilizzo di tecniche manuali di lavorazione, la ricerca di materiali naturali, la necessaria osservazione della natura promuovono un *modus vivendi* sano, che porta a trascorrere più tempo a contatto con l'Ambiente (peraltro, per un caso fortunato, il laboratorio si trova al confine con il *bosco di Oulx*). Tutto ciò vuole generare nei giovani un'empatia con la natura stessa e un ruolo attivo, che non implica la rinuncia alla cultura moderna e alla tecnologia, ma valorizza uno dei nostri bisogni primari che abbiamo in parte dimenticato.

Grazie a queste attività e al contesto, apprezzo nei giovani un aumento del risp-etto per l'Ambiente e l'attitudine a valorizzare le risorse disponibili, a riconoscere la bellezza, a criticare gli interventi umani di maggior impatto ambientale. Passare più tempo tra laboratorio e natura, consente anche di distinguere i fabbisogni essen-ziali da quelli indotti dal consumismo, focalizzandosi sulla qualità della vita.

▶ **Question** Come si può trasmettere un *messaggio ambientale* attraverso un'opera d'arte, frutto della creatività dell'artista? Ci può fornire qualche esempio, descriv-endo alcune sue opere e/o quelle prodotte dai suoi allievi?

▶ **Answer** Più che rappresentare aspetti negativi, anche se certamente reali e di impatto immediato, come lo spreco delle risorse e l'inquinamento, preferisco trasmettere messaggi sulla necessità di sentirsi parte della natura.

Ad esempio l'opera "*1938—Metamorphosis*" raffigura l'incanto e la meraviglia della natura, attraverso il racconto di un ipotetico sogno di Escher, che poi sfocer-ebbe nel suo capolavoro "*Cielo e Acqua*" del 1938. Essa rappresenta il concetto che tutto ciò non in armonia con la Natura, per quanto bello e tecnologico, è destinato ad esserne sopraffatto, umanità inclusa.

In "*Awareness*" viene messa alla prova la capacità dell'Uomo di identificarsi con gli altri esseri viventi, atteggiamento che diventa un aspetto fondamentale per favorire il cambiamento di mentalità verso l'Ambiente.

Il significato del gesto, viene allora amplificato e diventa così l'emblema della scelta e dell'impegno; la libertà vera di scegliere, in contrapposizione alla mentalità nichilista.

L'ultima opera fa riferimento al filosofo greco *Anassimandro*; egli riteneva che, in origine, tutte le cose fossero armoniosamente unite nell' "*Apeiron*", che per lui raffigurava l'universo, fino a quando, per colpa dell'Uomo, questa coesione si ruppe.

Nella scultura i legni pregiati dei diversi continenti, che per me simboleggiano anche le culture e i popoli, si fondono in una forma classica a rappresentare la pos-sibilità per gli uomini di scontare questa colpa, ritornando a essere parte dell'*Apeiron* e, quindi, della Natura.

Fig. IWP 16.1 © Antonio Salinari, 1938—METAMORPHOSIS 2016—Swiss pine wood, silver leaf, fabric belts; cm 130 × 80 × 7

Fig. IWP 16.2 © Antonio Salinari, **AWARENESS** 2014—Swiss pine wood, steel; cm 100 × 135 × 7

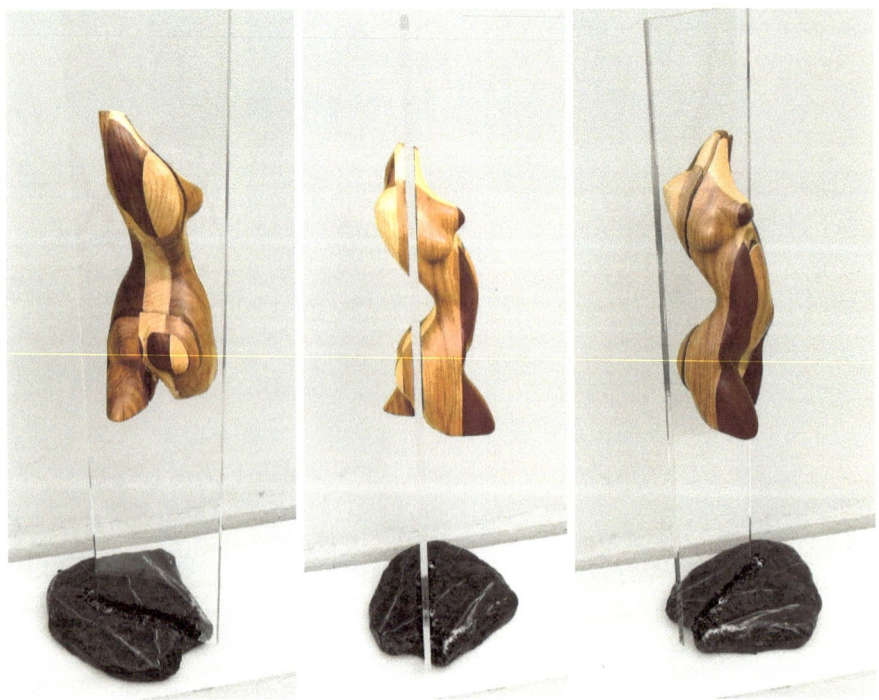

Fig. IWP 16.3 © Antonio Salinari, **APEIRON** 2016—various precious woods, glass, stone river; cm 73 × 24 × 20

Omero Soliman Architect, town planner and *designer*, he has been dealing with Holistic Bio-architecture for ten years. The Holistic Firm he founded takes its inspiration from the "Green Building" or "green House" a well-Kown *Leed certification* of the *Green Building Certification Institute*, set up to certify the "performances" related to health and wellness for those living in closed spaces. Expert in *Feng Shui* at the University of Ferrara (Italy), Soliman specialized with architect *Claudio Melloni*, director of the Uruguayan Institute of *Feng Shui* (among the leading experts in South America) and *Howard Choi*, *Feng Shui* Master at the College of Sydney, graduating at the *Feng Shui Research Center*, famous Canadian Centre of Research directed by *Joseph Yu* Master. He cooperates as an expert on healthy living to the new project *Biometric Point*®, a series of specialized centres whose goal is to spread an innovative philosophy of well-being of man thanks to the improvement of the Environment around.

▶ **Question** Quale importanza ricopre la Comunicazione Ambientale nella sua professione? A chi si rivolge principalmente?

▶ **Answer** Oggi più che mai, la Comunicazione Ambientale nella mia professione (architettura), è diventata di estrema importanza se non addirittura una priorità assoluta. Questo perché la visione olistica di tutto ciò che mi circonda ha influenzato in modo radicale il mio "essere architetto", attraverso un percorso consapevole che affonda le sue radici in un profondo legame di rispetto e di amore con la Madre Terra che ci ospita e ci permette di esistere. Per questo motivo la mia Comunicazione

---

Omero Soliman
www.studio-olistico.com (Web Site)

See English version on p. 380

© Springer Nature Switzerland AG 2019
M. Abbati, *Communicating the Environment to Save the Planet*,
https://doi.org/10.1007/978-3-319-76017-9_24

Ambientale è rivolta alla collettività. Ognuno di noi appartiene a questo pianeta e ognuno di noi può fare la sua parte per preservarlo e custodirlo. Purtroppo però, trattare argomenti scientifici legati all'Ambiente non è sempre facile, quindi i *target* sperati a volte tardano ad arrivare. Questo perché, nonostante la consapevolezza di fronte a queste tematiche da parte del soggetto finale, la comunicazione non è sempre semplice da attuare e richiede impegno, costanza e tempo.

▶ **Question** Cosa significa per lei Comunicare l'Ambiente? Che cosa non può mancare per realizzare un efficace Piano di Comunicazione? Che cosa va evitato?

▶ **Answer** È un impegno costante nell'informare il cittadino attraverso una ricerca metodica di un linguaggio comunicativo il più semplice possibile, chiaro e comprensibile ma nel contempo efficace per stimolare quella giusta dose di sensibilità interiore che ogni individuo possiede nei confronti dell'Ambiente. Contrariamente però, il più delle volte, questa sensibilità viene a mancare perché condizionata dalla quotidianità di eventi esterni che ci distraggono da quell'attenzione che dovremmo porre con più forza di volontà. È fondamentale quindi che vi sia la presenza di linee guida adeguate in grado di tracciare un percorso specifico. Semplicemente anche come quelle adottate dallo Studio Olistico, ispirate alle "*Green Building*" o "*Green House*", nota certificazione *Leed del Green Building Certification Institute*, studiata appositamente per certificare le "*performance*" relative alla salute e al benessere di chi vive in spazi chiusi.

Per questo motivo credo che un efficace Piano di Comunicazione debba essere prima di tutto una chiara identificazione degli obiettivi, accompagnata da una mirata pianificazione strategica ma soprattutto la costruzione di una fluente rete di relazioni tra gli attori protagonisti nel Piano di Comunicazione, in ambo i sensi. Quello che cercherei di evitare è di sottovalutare o sottostimare il processo organizzativo post concepimento, in particolare modo quello della valutazione finale, ovvero la verifica dei risultati ottenuti o dei lori impatti generati sul soggetto destinatario.

▶ **Question** Ritiene che i soggetti destinatari del suo Piano di Comunicazione siano adeguatamente preparati a riceverlo? Che cosa andrebbe ancora fatto, secondo lei, per migliorare la comprensione dei messaggi ambientali? Che cosa può potenzialmente "disturbare" la corretta interpretazione di tali messaggi?

▶ **Answer** Stiamo vivendo un momento particolare dell'esistenza umana, difficile e caotico per certi aspetti. Fortunatamente però, alcuni di questi disagi sociali, stanno in qualche modo "risvegliando" l'attenzione per le tematiche ambientali. Pertanto, da parte del soggetto destinatario, c'è la volontà di voler ricevere più informazioni perché dettate dalla necessità di voler comprendere maggiormente i risvolti ambientali per poter agire con maggior consapevolezza ed efficacia. Ritengo che il soggetto destinatario al quale mi rivolgo stia acquisendo la giusta maturità e abbia ormai un'adeguata preparazione per ricevere qualsiasi tipo di messaggio.

Sicuramente la ricettività dei *messaggi ambientali* può essere e deve essere migliorata. Per quanto mi riguarda mi sono assunto una maggior responsabilità come libero professionista, cercando di rendere il soggetto destinatario più partecipe nel mio impegno ambientale, informandolo e rassicurandolo costantemente, concedendogli anche la possibilità di esprimere un giudizio in merito. Ad esempio inviandogli un questionario nel quale poter manifestare il suo grado di soddisfazione o suggerire alcuni consigli per migliorare la qualità dei servizi offerti. Oppure aggiornandolo con la nostra *newsletter* su nuove normative entrate in vigore oppure su iniziative e programmi che lo Studio Olistico intende promuovere.

Cercando, inoltre, di allontanare quella sorta di "disinformazione ingannevole" che spesso con estrema facilità si insinua in certi contesti depistando il soggetto destinatario dal giusto percorso, fino a condurlo all'abbandono di uno stile di vita adeguato.

▶ **Question** Ritiene importante il ruolo dei portatori di interesse (*stakeholder*) nella diffusione del *messaggio ambientale*?

▶ **Answer** All'interno di una rete di relazioni ben ramificata i "portatori di interesse" diventano strategicamente importanti se non fondamentali. Sono il collante o se vogliamo il ponte di collegamento della capillarità della rete stessa. Sono quel "*feedback*" necessario per comprendere la giusta politica da adottare per un efficace raggiungimento dell'obiettivo. Nel mio lavoro è di estrema importanza il coinvolgimento degli *stakeholder*, da quelli più "tradizionali" come fornitori e clienti, a quelli più "attuali" come investitori, partner, la comunità stessa (intendendo anche quella dei *Social*), la stampa, i *Media*. Senza di loro non saprei come fare.

▶ **Question** *Command and control* (strumenti legislativi/sanzionatori) o Strumenti Volontari (es: *eco-design*, *eco-label*, *energy-label*, *sistemi di gestione ambientale*, ecc.) qual è, secondo lei, l'approccio più efficace per comunicare tematiche *green*?

▶ **Answer** A mio modesto parere ci dovrebbe essere un giusto equilibrio da entrambe le parti. L'uno non può esistere senza l'altro. Sono due approcci che per quanto diversi si compensano reciprocamente bene. Mi trovo spesso a dovermi confrontare con entrambi. Sicuramente sono molto più legato all'eco-design, che prediligo e preferisco. Ma non posso prediligere l'uno piuttosto che l'altro. È un processo inevitabile.

▶ **Question** Quali sono le sue prospettive future? Come intende sviluppare il suo Piano di Comunicazione?

▶ **Answer** In primo luogo dando l'esempio con un impegno costante sul campo. Queste sono le mie prospettive future. Un'azione programmata e determinata per portare dei risultati concreti e tangibili a tutti. Come architetto mi sono sempre adoperato per adottare le migliori strategie, atte ad ottenere soluzioni eco-sostenibili in tutti i miei progetti. Conseguentemente, il mio studio di architettura olistica ha reso noto la sua "*mission*" comunicando pubblicamente il suo impegno ambientale attraverso un Piano di Comunicazione suddiviso in nove semplici punti, consultabile sul nostro sito internet. Per citarne qualcuno: riducendo ad esempio i viaggi d'affari e prediligendo le video conferenze per ridurre le emissioni di carbonio nell'Ambiente. Oppure scegliendo (quando possibile) quei ristoranti che propongono il "piatto del clima", cibo preparato con ingredienti regionali e di stagione, con provenienza da fattorie agricole biologiche a km zero. Nove obiettivi da raggiungere per uno stile di vita più *"green"* a favore dell'Ambiente che intendiamo perseguire nei prossimi anni iniziando fin da subito.

Il mio impegno professionale prosegue nel veicolare il messaggio ambientale curando i dettagli dei servizi offerti dallo studio di architettura che ho fondato. Infatti dalla sua "rinascita" avvenuta nel 2015, la visione olistica ancora più evoluta, ha incentivato e stimolato lo Studio che dirigo a compiere un altro salto di qualità, implementando la ricerca sostenibile nei confronti dell'Ambiente con nuove metodologie. Il suo obiettivo primario è diventato il benessere abitativo, ovvero cercare di raggiungere un livello di qualità in grado di soddisfare e assicurare la tutela della "salute" dei suoi fruitori, attraverso l'individuazione di tutte quelle condizioni di tipo igienico-sanitarie e/o di sicurezza all'interno dei cosiddetti "luoghi di vita" (casa, lavoro, ambienti ricreativi ecc.). Non solo nel rispetto della normativa vigente, ovviamente, ma con una visione più olistica, cioè partendo dal presupposto che ogni individuo è unico nel suo genere e merita quell'attenzione strettamente legata alle sue specifiche esigenze fisiche. Prendendosi cura della sua salute attraverso un protocollo olistico che accompagna nel percorso il soggetto finale, dall'habitat ad esempio, inteso questa volta come luogo in cui si vive, il luogo in cui si dorme. I suoi materiali, il modo in cui l'abitazione è costruita, il luogo su cui sorge, il modo in cui è orientata. Ma anche l'alimentazione, motore del metabolismo, benzina per il corpo. L'acqua che si beve con il difficile e meraviglioso compito di regolare l'equilibrio elettrolitico, trasportare tutte le sostanze ed eliminare le scorie del nostro organismo.

Fino ad arrivare a considerare anche aspetti intangibili della realtà come quelli legati alla fisica delle energie sottili. Quelle che ancora oggi risultano difficili da misurare con uno strumento tecnologico ma che esistono e sono percepibili dall'uomo e possono fungere da *Media* in grado di comunicare l'Ambiente.

▶ **Question**   Ritiene che le Reti Sociali, i cosiddetti *Social Network* (es.: *Facebook*, *Twitter*, *Instagram*, ecc.) possano essere strumenti affidabili di Comunicazione Ambientale?

▶ **Answer**   Ritengo che i *Social Network* siano strumenti comunicativi assai potenti ma al contempo estremamente delicati e fragili e che non devono essere sottovalutati. L'avvento delle "reti sociali digitali" ha introdotto un nuovo modo di comunicare grazie allo strumento della condivisione. La loro grande forza è di "parlare" con le persone e non alle persone. Questo aspetto implica la scelta di un linguaggio comunicativo estremamente ponderato, misurato, pensato, mirato, efficace. Ma quello che ha maggior valore, a mio avviso, è la sua capacità di poter interagire direttamente con il soggetto destinatario. Ti permette nel tempo di monitorare la reale opinione del singolo cittadino e di capire veramente il suo pensiero e le sue necessità. D'altro canto però il rovescio della medaglia è che basta commettere un solo errore di comunicazione per ottenere il risultato completamente opposto a quello sperato creando un danno a volte irreversibile.

È quindi doveroso utilizzarli saggiamente e con la giusta preparazione.

▶ **Question** Come giudica le campagne pubblicitarie che veicolano *messaggi ambientali*? Ritiene che siano sufficientemente efficaci? Che cosa ritiene essenziale nel messaggio pubblicitario per poter dare fiducia ad un prodotto/servizio *green* e allontanare ogni dubbio di *greenwashing*?

▶ **Answer** Non è mai semplice realizzare una campagna pubblicitaria sulle tematiche ambientali, soprattutto quando i soggetti destinatari sono molto eterogenei. Far leva sui bisogni e sugli interessi della collettività o del singolo individuo è sempre molto arduo se l'intento che sta dietro è veramente genuino, e purtroppo non è sempre così. Partendo da questo presupposto vi sono ovviamente campagne più efficaci e altre un po' meno. Credo che un buon messaggio pubblicitario debba essenzialmente toccare interessi a lungo termine a livello globale, legati ai temi della "Sostenibilità". Intesi come uno stile di vita che porti a un progresso senza danneggiare o abusare delle preziose risorse del territorio, volto alla sopravvivenza delle future generazioni nel breve termine. Ma anche come qualità di servizi in grado di far risparmiare tempo, offrire convenienza e soddisfare le esigenze del cliente. Normalmente questo approccio comunicativo infonde nel soggetto destinatario tranquillità e fiducia. Perché in una previsione di Sostenibilità Ambientale nei confronti delle future generazioni vi è una sorta di speranza di lasciare una buona eredità; mentre nella soddisfazione e convenienza di un servizio immediato il consumatore si sente appagato in quel momento. Credo che questo connubio di visioni renda la campagna pubblicitaria quasi sempre efficace.

Jeremy Tamanini Graduated in Foreign Service at Georgetown's School of Foreign Service with a Master's degree in Foreign Policy, an Honour's certificate in International Business Diplomacy and a Bachelor's degree in Urban Studies and Economics at the Columbia University, Jeremy Tamanini founds "*Dual Citizen Inc.*" in 2009, a qualified Consulting Firm based in Washington DC (USA). The Company provides strategic communication and data analytic tools to government agencies, international organizations and private firms, worldwide, in order to promote advanced economic and policy programmes. Dual Citizen publishes annually the *Global Green Economy Index (GGEI)* which has become a global reference to measure national Green and Circular Economy (through a comparative analysis among eighty Countries around the World). Jeremy Tamanini has a long professional career as Business Advisor in private sector and start-up. From 2002 until 2006 was Director of Marketing at *Armani Group*. Being awarded a Fulbright fellowship in the United Arab Emirates, he explored the interaction of place branding and economic development in Dubai, connecting with local government officials.

▶ **Question** Mr. Tamanini—On the basis of *Dual Citizen* expertise dealing with *strategic data communications* aimed at promoting "green growth", at international level, by addressing to governments, ministries, international organizations and

---

Jeremy Tamanini
@DualCitizenInc (Twitter)
www.dualcitizeninc.com (Web Site)
Jeremy Tamanini (LinkedIn)

© Springer Nature Switzerland AG 2019   319
M. Abbati, *Communicating the Environment to Save the Planet*,
https://doi.org/10.1007/978-3-319-76017-9_25

private firms, etcetera, which is the best approach to communicate corporate eco-performances? What would be the best way to prevent *green* communication "noise" (what influences the interpretation of each message, individually) and test the audience perception? What are the possible solutions?

▶ **Answer**   In the realm of companies, building the robustness and adoption of *Corporate Social Responsibility* (CSR) is one way to communicate corporate eco-performance. It is surely the case that some instances of CSR are in fact not very authentic. An example of this is a company that publishes a superficial accounting of CSR activities only to be recognized as "green" by its customers and shareholders. But by scrutinizing CSR activities and demanding greater data reporting—particularly for activities related to the Environment—consumers and the public can pressure companies to disclose more, and reform their business practices in a genuinely sustainable manner. For example, many companies now report carbon emissions from their primary operations. Yet many global companies have the largest emissions "footprint" in other parts of their supply chain. By pressuring companies to disclose these emissions too, consumers and the public not only understand the company's overall environmental footprint better, but also pressure the entire supply chain, an activity that could introduce greater Sustainability practices to other markets.

▶ **Question**   Since 2010, Dual Citizen has been publishing the *Global Green Economy Index*™ (GGEI) which is measuring national performance in the *Green and Circular Economy*, worldwide. According to the latest edition of GGEI (September 2016) what really matters in order to have a right perception of *Circular Economy* driving forces with reference to the main strategic issues? What about the situation in Italy in relation to the rest of the World?

▶ **Answer**   Looking into the *Global Green Economy Index*™ (GGEI) drawn up, every year, by *Dual Citizen* based upon worldwide economic data coming from eighty Countries around the World, it seems clear that what makes *Green Economy*, or better *Circular Economy*, a success is a combination of many driving forces. Environmental quality, energy efficiency, sustainable action plans help with improving "green performances and assets" but they are not enough. What makes the difference at national level is the "green reputation" which represents a Country business card to the World. A keen communication strategy involving the Media is needed. It becomes essential, then, to convey proper messages enhancing the attractiveness of investing in eco-innovation and/or renewable sources and/or green start-ups, just in that geographical area. This may result in establishing proper government policies aimed at improving market conditions and foreign economic investments. Moreover, a sustainable future development depends on it.

The proof is the *Green Economy Italian Report*, on request of the *Italian Foundation on Sustainable Development* (Fondazione per lo Sviluppo Sostenibile). According to the analysis of four main components: Leadership; Climate Change;

Market and Investments and Environment, Italy is ranked 15th in the Green Economy performance list and 29th in the Green Economy perception classification (2016 Report referring back to 2015 data). The final result, consisting of the weighted average rate of all components, is based on a comparison of eighty Countries, around the World.

In spite of any margin of uncertainty, the resulting difference highlights a fundamental problem: internationally speaking, too little attention is given to the Italian green reputation. An outcome strongly linked to communication which must be seriously considered in order to give the right credibility to the excellent results achieved by most of the Italian green companies (corporate performance is ranked tenth out of eighty). Consequently, a more productive cooperation would be highly recommended with special regard to multi-level institutions, Media, research centres, green companies and any other organisation. A new well-coordinated communication approach targeted to find the true face of the environmentally friendly Italian branding. Italy, as a matter of fact, needs to focus on a suitable public communication strategy besides improving its environmental performances.

As stated in my overview to *Global Green Economy Index™ Report*, Italian policymakers should boost the idea of "Green and Circular Economy" informing the national and international audience about the Italian opportunities for investment. Media and national leaders should, then, play a key role in storytelling the advantages of investing in green projects pushing a stronger value of Italian Green and Circular Economy both on the international economic market and in the framework of international forums like UN Climate Change Conferences [e.g.: COP21 (Paris, 2015), COP22 (Marrakech, 2016)].

▶ **Question** What would you suggest in order to avoid *greenwashing*, from a consumer and producer point of view? How much communication matters?

▶ **Answer** Interestingly, communication is sometimes the reason for "greenwashing," rather than a solution to avoid it. Many companies now have whole departments—often constituted from the public relations or communications units—who are responsible for CSR and Sustainability issues. As already mentioned, this work can be real and impactful, and particularly after COP21, many companies are doing real work in reducing emissions and natural capital use, as well as purchasing greater amounts of renewable energy. However, these activities are still too often merely communications, where the company points to the fact that a CSR department exists to satisfy shareholders, when often these efforts are not too substantive.

▶ **Question** Which are the new challenges in terms of *green* communication? What should be done in order to improve an effective communication? Which role can *Social Networks* (e.g.: *Facebook*, *Twitter*, *Google+*, etc.) play in disseminating *environmental messages* and/or data? Have you ever experienced *Social Media* communication in disseminating environmental data? Would you trust them? Under what conditions?

▶ **Answer**   *Social Media* are becoming more mainstreamed and many organizations that are very authoritative disseminate new research and data through these networks. So, on the one hand, these networks are really just another communications channel like email or the *Web* etc. That said, I think that *Social Media* offer new opportunities for the future in two main areas. The first is activism, and leveraging the "viral" aspect of *Social Media* to spread a message related to *Green Economy* or the *Environment*. One example from another realm is the ALS "*Ice Bucket Challenge*" where people made videos of themselves to raise money and awareness for ALS (Web link: www.alsa.org that describes the program more). The second is data collection, and using *Social Networks* to gather real-time data and other information. For example, air quality in cities has become increasingly concerning, particularly in Asia. If citizens could post air quality readings and photos of their daily activities being impacted by poor air on *Social Media* platforms, it could have the effect of building awareness but also "*crowd-sourcing*" data on where in geographic areas air quality is worst, and during what times of the day.

Joaquim Tarrasó Climent has been fulfilling the role of Senior Lecturer in *Urban Design, Architecture* at the *Chalmers University of Technology, Göteborg* (Sweden), since 2015. Furthermore, he has been acting as International Correspondent of the *Catalan Association of Architects in Sweden*, since 2013. *Tarrasó Arkitektur*, own practice *Barcelona* (Cataluña) since 2003, *Göteborg* (Sweden) since 2011 *Architect at Espinàs i Tarrasó, Arquitectura Disseny I Paisatge*, Barcelona (Cataluña) since 1999. He studied Architecture at the Escola Tècnica Superior d'Arquitectura de Barcelona (Cataluña), *Universitat Politècnica de Catalunya* and *Sint-Lucas School of Architecture* in *Ghent, University of Leuven* (Belgium). He used to lecture as Design and Construction Teacher in *Building Engineering Degree*, ELISAVA, *School of Design and Engineering of Barcelona*, *Universitat Pompeu Fabra, Barcelona* (Cataluña), 2007–2008; and as Course Tutor and Master Thesis Supervisor in the *Master Program Urban and Architecture Design Laboratory*, *Chalmers University of Technology, Göteborg* (Sweden), 2010–2014.

Premise to Questions: "The world will not evolve past its current state of crisis by using the same thinking that created the situation" stated *Albert Einstein* highlighting a key aspect of communicating and informing on *Environmental issues*: the continuing need to develop in order to adapt to new "linguistic codes", those belonging to the *Circular Economy*, the *Social Media*, the *Internet of Things*, etcetera. In this perspective, *Environmental communication* is a multimedia process that aims at presenting new environmentally-friendly values.

Joaquim Tarrasó Climent
www.tarraso.se, http://espinasitarraso.com/index.html (Web Site)
Joaquim Tarrasó (LinkedIn)

See English version on p. 382

© Springer Nature Switzerland AG 2019
M. Abbati, *Communicating the Environment to Save the Planet*,
https://doi.org/10.1007/978-3-319-76017-9_26

An informative, educational and creative "strategy" which allows to face most of *Environmental matters* consciously and responsibly. It is also an ever-changing "tool" able to express both orally and through non-verbal *Media*: shapes, symbols, materials and any other natural element which affects human beings can "revolutionise" the citizens or countryside lifestyle changing the public opinion's point of view on "ecosystems".

▶ **Question** ¿Estimado Arquitecto *Joaquim Tarrasó Climent*, según su experiencia plurianual realizando proyectos a nivel internacional, cuales son las fuerzas impulsoras de la arquitectura atenta y coherente con los principios de la Sostenibilidad, es decir la *bio*-arquitectura? ¿Como comunicar eficazmente a las partes interesadas los beneficios a favor del "Ecosistema urbano" en la fase de diseño, a pesar de un enfoque "*business oriented*"? ¿Puede el diseño de un edificio de acceso público, así como de una plaza o de un jardín urbano, enviar mensajes "*green*" (o mejor durables) a los usuarios o a quienes admiran su estructura? ¿En su opinión, qué importancia tiene la comunicación escrita y oral en transmitir los rendimientos ambientales de una obra arquitectónica con respecto a la parte visual?

▶ **Answer** Al tratarse de *comunicación medioambiental* y en relación a las preguntas que formulas he pensado de dividir mi respuesta o agrupar mis respuestas en tres bloques diferentes: (1) Concepción del proyecto; (2) Comunicación Gráfica; (3) Proyecto Construido.

1. **La concepción del proyecto y su desarrollo**: puedo de algún modo vincularlos a temas de relación con clientes y con todos aquellos actores que estén envueltos en el proceso. Hay diferentes maneras de comunicar en esta fase. La experiencia a través del proyecto realizado y de la experiencia en diferentes ámbitos es que, en general, es muy conveniente y muy interesante poder involucrar a los que se llaman los diferentes *stakeholder* desde las primeras fases, es decir también desde la primera concepción del proyecto. Eso tiene diferentes perspectivas. Una primera podría ser un proceso participativo. Puede ser una excelente oportunidad de análisis, o sea una manera muy buena de extraer información, de entender las condiciones de un proyecto y también los parámetros que pueden, de algún modo, ayudarnos a desarrollar las soluciones que dan respuesta a los requisitos, a las preguntas y a los otros aspectos. Un proceso participativo también puede ser una herramienta de eco diseño. No solo come extracción de información sino también como elaboración propia del proyecto. A nivel del diseño tenemos que especificar como es una palabra bastante amplia. No se refiere únicamente a la materialización de la forma, pero si como a la definición del concepto, igualmente. Con lo que este proceso de eco diseño puede ser muy interesante en todos los aspectos. Es evidente que eso también es una herramienta muy importante para involucrar y, de alguna forma, para compartir ese proceso. Os tiene que evitar de considerar el papel del arquitecto como un elemento externo que aporta soluciones de una manera individual sino promover un acercamiento más compartido. Desde ese punto de vista y resumiendo, diría que siempre involucrar a las diferentes partes,

non solo clientes sino personas que formen parte de ese ámbito, de ese contexto, tiene muchos beneficios. Por tanto, sería un elemento que destacaría en esta fase de concepto y desarrollo del proyecto. Otro elemento que tiene que ver con esta fase es el contexto. Es evidente que todos los proyectos deben dar respuestas. Es algo condicionante. Es una necesidad que siempre está vinculada a unos aspectos concretos y específicos al lugar, a todos los niveles. A nivel de usuario, por ejemplo: cuales son aquellos actores que van a utilizar o que van a ser usuarios de esta construcción y de ese proyecto. Pero también, si hablamos de Sostenibilidad: cuales son los elementos que refinen ciertos Ecosistemas. Y hablando de Ecosistemas podemos ampliar al máximo el sentido de esa palabra. Además, podríamos hablar de ámbitos sociales, obviamente. Por tanto, cuesta a priori pensar en proyectos que sean verdaderamente sostenibles sin vincularlos a partir de unos condicionantes sujetos al contexto. Puedo extender este concepto más tarde. Hay un elemento también en cuanto a concepto y comunicación entre los más importantes a la hora de tratar el Medio Ambiente y de su implementación en proyectos arquitectónicos y de paisaje: el valor añadido que los elementos medioambientales pueden aportar a todos niveles. Desde hace ya bastante años se trabaja, se analiza y se estructuran estas ideas a partir de los que se llaman *ecosystem services*, en cuatro categorías diferentes. Es decir, se intenta relacionar, de algún modo, ese vínculo dentro los Ecosistemas y su repercusión a nivel social. Yo creo que sea una buena manera de comprender y también de comunicar. En cuanto a Medio Ambiente es evidente que hablamos de sistemas, de estructuras que consisten en hacer comprender esas capas, esos sistemas que convergen en un proyecto o en un entorno y, de cualquier forma, compatibilizarlo. Eso representa un reto fundamental por un arquitecto.

2. **Comunicación gráfica**: En esta fase de realización del proyecto a nivel técnico y grafico hay que tener en cuenta, con mayor razón, las implicaciones que tiene a nivel comunicativo. Como arquitectos tenemos varias herramientas. Y no solo la comunicación verbal sino también la comunicación gráfica. Nuestro medio natural, nuestro "lenguaje" habitual es el grafico. Es un arma poderosa. En vez de los idiomas empleamos los dibujos, los diagramas, los gráficos. Y todos tienen la capacidad de comunicar. A nivel de desarrollo técnico, y de desarrollo gráfico, es siempre importante utilizar esta documentación como un arma muy poderosa. Cuando nosotros hacemos un proyecto no solo organizamos esos dibujos para cumplir, de algún modo, con los requisitos técnicos, legales, relativos a los permisos municipales sino también como trasmisores de conocimiento para que ese proyecto se materialice a nivel constructivo. Pero más allá ad eso está también la posibilidad de trabajar sobre esos documentos gráficos, dibujos, esquemas como una herramienta divulgativa. Cuando antes hablábamos de sistemas que se sobreponen en el contexto es evidente que explicar esa condición mediante unos diagramas que hablen de ciclos medioambientales, pero también de su capacidad de interactuar con los usuarios, se tiene la posibilidad de hacer comprender las lógicas del nuestro entorno.

Voy a poner como ejemplo unos casos muy claros en cuanto a tipología de proyecto. Podemos, por ejemplo, hablar de proyectos de gestión del agua. ¿Como podemos utilizar la gestión del agua en un entorno urbano como un elemento de diseño? Si tratamos, de algún modo, el agua en superficie, digamos la de la lluvia, de manera que sea un elemento visible y que configure un paisaje urbano, automáticamente estamos interactuando a diferentes niveles sistémicos. El sistema del agua y lo del usuario o del sistema social. Como el proyecto "*Théâtre Évolutif*" concebido por **OOZE Arquitectos** en *Bordeaux* (Francia), que desarrolla a nivel grafico de manera muy buena a lo que me refiero. Non solo explica el proceso, sino que lo divulga. Eso fomenta el entendimiento del nuestro entorno y por tanto puede ser la implicación de algún modelo usuario. O por ejemplo en cuanto en materia de biodiversidad el proyecto de la **Escuela Primaria para Ciencias y Biodiversidad** en **Paris**, diseñado por *Chartier Dalix Arquitectos* intenta explicar a los usuarios de esta escuela primaria los diferentes hábitats de las especies animales, a través de los dibujos. Así que esas creaciones artísticas pueden comunicar de una manera muy efectiva el Medio Ambiente.

3. **Proyecto construido**. A este respecto puedo mencionar, como ejemplo, el proyecto del Despacho de Arquitectura *Urbanisten* en los **Países Bajos** donde la cuestión del agua es imprescindible para que esa sociedad pueda desarrollarse. Por tanto, es una sociedad muy avacada che favorece el encuentro entre usuario y agua. En una plaza que se llama *Waterplein* en *Rotterdam* queda, de algún modo, tanto a nivel grafico que a nivel de proyecto construido. Ese entorno gestiona el agua a la vez como soporte y escenario. Un caso en el que el uso se realiza mediante diferentes actividades. Como elemento-entorno construido, por tanto, es importante siempre pensar en la relación directa, considerando el usuario como elemento activo de su entorno. No como espectador únicamente pero también como usuario. Yo creo verdaderamente que es un elemento interesante, como demuestran los proyectos a los que me refería.

Premise to Questions: Sustainability is a concept which influences many aspects, including the everyday object design, although of decorative nature. Thus, the eco-design goes further than being "eco-friendly" since it affects the whole production chain such as: the manufacturing process, the future scenarios of the "product life cycle" (more and more circular), the innovation and the creativity. All that with the view to act more responsibly.

▶ **Question** ¿De qué depende interpretar de forma correcta los mensajes "ecológicos" asociados a los proyectos arquitectónicos y a los objetos de *eco diseño*? ¿Cómo se pueden reducir los gastos que afectan al precio de los eco productos?

▶ **Answer** Por supuesto depende de muchos factores, depende por ejemplo de lo contaminado que está en el entorno y que puede ser antiguos o zonas industriales. En este último caso, las concentraciones de metales en el entorno son altísima y por tanto requieren un tipo de soluciones específicas y otras en cambio son nuestros contextos que tienen otros elementos: por ejemplo, la polución acústica. Es difícil establecer una relación entre eco diseño y precio. Por tanto, no puedo darle ninguna respuesta al respecto. Más allá, es fundamental hacer un análisis exhaustivo de cuáles son las condiciones, el contexto; de cuáles son las necesidades; de cuales son aquellas prioridades y aquellos elementos que nos pueden ayudar a traer soluciones que sean válidas y sostenibles. Cuando trabajo en los proyectos con mis colegas la comunicación se necesita como un entendimiento especifico de cuáles son las condiciones a nivel de Sostenibilidad, así que tratamos con consultores medioambientales. Nosotros, sin ser expertos de la materia, lidiamos con estos elementos y, evidentemente, como arquitectos, colaboramos con expertos en la materia de *comunicación medioambiental*, de la misma manera que con ingenieros hidráulicos, expertos en tráfico o con biólogos o antropólogos y sociólogos cuando trabajamos con temas más culturales.

Paolo Taticchi Graduated in Mechanical Engineering, he has attained a Master in Business Administration—MBA (focused on Innovation) and he is a researcher in Industrial Engineering (with specialisation in Operational Management). He is Professor and Researcher in performance measurement and management, business networks and Sustainability. He is Editor and Co-Author of various books and scientific articles, including: "*Business Performance Measurement and Management: new contexts, themes and challenges*" and "*Corporate Sustainability*", both published by *Springer Internationally*. He is responsible of corporate management addressed to academic students and researchers. He holds seminars as part of the MBA and EMBA Master programme at international level: Europe, Africa, Asia, North and South America. He has been acting as a technical adviser on the industrial corporate field. He is appointed Visiting Teacher at the New York University (New York, USA) and Honorary Visiting Teaching Fellow at the *Bradford University School of Management* (UK).

▶ **Question** Professore Taticchi, lei ha maturato un'esperienza pluriennale, a livello internazionale, in tema di ricerca e formazione nel settore *business economics*, con particolare riferimento alle strategie sostenibili nella filiera economica e societaria nonché nella catena di fornitura di beni e servizi, quale ruolo ricopre la *comunicazione ambientale* nella *Green Economy* e quanto può influire sulle scelte dei consumatori-cittadini-utenti?

Paolo Taticchi
@taticchipaolo (Twitter)
Paolo Taticchi (LinkedIn)

See English version on p. 385

© Springer Nature Switzerland AG 2019
M. Abbati, *Communicating the Environment to Save the Planet*,
https://doi.org/10.1007/978-3-319-76017-9_27

▶ **Answer** Sicuramente la *comunicazione ambientale* gioca un ruolo importante tanto nella *Green Economy* quanto nelle strategie di *Sostenibilità* e *comunicazione ambientale* delle aziende. Il quesito apre sostanzialmente due discussioni: da una parte qual è la rilevanza dell'*e-corporate* (tutto ciò che riguarda il *brand* ed il *brand identity* di una realtà aziendale che utilizza la Rete); dall'altra qual è la sua percezione da parte dei cittadini (consumatori/utenti). Dal punto di vista dell'*e-corporate* la comunicazione è fondamentale in quanto la filiera comunicativa fa sostanzialmente parte del *marketing*, derivante da quelle azioni strategico-operative promosse dalle aziende per posizionare i propri prodotti nei mercati e creare i propri *brand* al fine di aumentare le vendite di certi prodotti e servizi. Per quanto concerne, nello specifico, l'aspetto ambientale dei prodotti vi sono svariate pubblicazioni e ricerche che dimostrano che la prima ragione per cui le aziende vanno a sviluppare strategie di *Sostenibilità* ed azioni in *engagement*[1] con le tematiche della *Sostenibilità* è strettamente legata al *brand*. Ciò è conseguenza del fatto che essere un'azienda sostenibile significa andare a sviluppare un marchio che necessita di essere comunicato adeguatamente per riflettere l'impegno ambientale e sociale delle aziende, ovvero la loro trasformazione ecosostenibile. Una strategia che fa leva sulla comunicazione è dunque l'unica strada da percorrere per garantire il successo del prodotto/servizio.

Ulteriori studi evidenziano come i cittadini/consumatori siano oggi sempre più informati e sempre più attenti ai criteri di scelta negli acquisti. Il preferire un prodotto/servizio rispetto ad un altro dipende da una molteplicità di fattori che girano intorno all'idea stessa di *brand*, elemento identificativo per eccellenza. Avere un marchio che ha una componente di *Sostenibilità Ambientale* è dimostrato che porta aumenti di *market share* (quota del mercato: la percentuale di mercato che è controllata da una impresa) più importanti. È dunque una leva di competizione aziendale da non sottovalutare. D'altro canto, però, il *branding* non è da solo sufficiente a disseminare le qualità ambientali del prodotto/servizio. Per informare i consumatori/cittadini è necessario predisporre un adeguato Piano di Comunicazione capace di veicolare dati di natura tecnica.

▶ **Question** Quanto conta il "saper comunicare l'Ambiente" nel *marketing*? Come rendere fruibili al largo pubblico informazioni e dati ambientali di natura tecnica per sottolineare l'impegno sostenibile nella "filiera" di produzione e/o di erogazione di servizi?

▶ **Answer** Comunicare in maniera efficace e credibile in materia di Ambiente è qualcosa di estremamente complesso. Qui si tratta di saper educare i consumatori

---

[1] Literally "active and passionate involvement" which implies, in an economic & corporate environment: *inclusiveness*, namely an issue to encourage the participation of the *stakeholders* in order to respond responsibly and strategically to Sustainability; *materiality*, namely the importance of a specific environmental issue in respect of the organization and the *stakeholders*; *compliance*, namely the ability of the organization to comply with the environmental issues proposed by the *stakeholders* which the performances, the decisions, the action plan, the results and the internal and external communication depend on (Reference: *Marco Minghetti*, Professor at the *Università di Pavia*, journalist and blogger for *IlSole24Ore*).

sulle tematiche di Sostenibilità Ambientale. Ciò presuppone anche che ci sia un certo livello di collaborazione nell'industria. Se tante aziende all'interno della stessa industria iniziano a sviluppare tante etichette e gestiscono la loro comunicazione in maniera indipendente, si rischia di bombardare il consumatore con centinaia di messaggi, simili ma diversi, che sostanzialmente creano più confusione che altro. Per questo motivo vengono chiamate alcune associazioni, organizzazioni di servizio che si fanno carico di creare degli *standard* per quanto riguarda la *comunicazione ambientale* e ciò rappresenta un passaggio particolarmente importante. Mi riferisco, per esempio, a tutto quanto concerne il "mondo" dell'*organic food* (il biologico) piuttosto che le etichette energetiche (es.: *Energy Star*) o l'*Eco-label*, a livello europeo. In questi casi si tratta di un piano di azione a livello industriale, ma anche a livello governativo (Governi nazionali), che sta cercando di creare una standardizzazione del *comunicare l'Ambiente* affinché la comunicazione di natura tecnico-ambientale sia più semplice e comprensibile per i suoi stessi destinatari (clienti, utenti, consumatori).

▶ **Question** Come riconoscere un autentico prodotto o servizio *eco-sostenibile*, basandosi sul *messaggio ambientale* veicolato attraverso il piano di comunicazione? Ci sono delle "parole chiave" che allontanano il rischio di essere vittima del *greenwashing*?

▶ **Answer** Credo che i consumatori siano ancora soggetti al rischio di diverse forme di *greenwashing* e quindi potenziali vittime di una comunicazione che veicola messaggi che millantano un'immagine di prodotti e servizi aventi un impatto positivo sull'Ambiente. In realtà, non esistono parole chiave codificate per identificare la veridicità delle informazioni comunicate. Molto dipende dall'educazione e dalla formazione ecologica dei singoli consumatori. Ciò che può facilitare il compito è basarsi su forme di certificazione o eco-etichette, ormai riconosciute ufficialmente dal mercato mondiale ed autorevoli, dal punto di vista comunicativo. Al di là di questo, regna una generale confusione in quanto è ancora difficile comprendere e valutare l'autenticità del *messaggio ambientale* se non supportato da dati e fonti certi e tracciabili.

▶ **Question** Gli strumenti volontari di certificazione ambientale di processo (es.: *sistemi di gestione ambientale: Norme ISO, EMAS, ecc.*) e di prodotto (es.: *Eco-label, Energy label, Energy-star, Certificazione di Prodotto Sostenibile*, ecc.), che implicano azioni di Comunicazione Ambientale, garantiscono realmente sulla qualità dell'informazione ambientale?

▶ **Answer** In realtà, purtroppo, non tutelano totalmente il consumatore rispetto alle prestazioni ambientali. Il mercato di prodotti e dei servizi "green" è divenuto un grande *business*; quindi sostanzialmente molte aziende che vanno a sviluppare delle certificazioni ambientali legate ai processi, poi, in realtà, operano spesso una sorta di esercizio più teorico-sperimentale che pratico. Spesso può capitare di visitare aziende il cui manuale qualità suggerisce determinate *best practice* nei processi che poi non corrispondono alla reale struttura interna dell'organizzazione. Personalmente, dunque, non sono particolarmente sostenitore delle certificazioni anche se, in alcuni casi, le aziende utilizzano questo strumento per riflettere sui temi ambientali e riorganizzare i processi aziendali creando un reale miglioramento della struttura organizzativa. Di certo, l'elemento positivo di tali certificazioni è quello di sensibilizzare i governi nazionali e il settore industriale ad adottare azioni a tutela dell'Ambiente e nel rispetto dei valori dello Sviluppo Sostenibile. Questo ha giocato (e sta giocando) un ruolo di catalizzatore dell'attenzione dell'opinione pubblica sulle tematiche ambientali, a livello internazionale. E, come effetto domino, innesca un processo virtuoso a livello industriale.

▶ **Question** La predisposizione di un corretto Piano di Comunicazione Ambientale può avere ripercussioni positive in termini di prezzi di vendita e di fatturato di un'impresa?

▶ **Answer** Assolutamente si. In molti casi, più che di *comunicazione ambientale* si tratta di comunicazione legata al *brand* quindi oggi sostanzialmente tutte le strategie di impresa sono collegate alla creazione di *brand*. E crearlo "sostenibile" può sostanzialmente portare ad un incremento da *market share*, ad una competitività maggiore o ad una marginalità maggiore.

▶ **Question** Conoscere a priori gli *stakeholder* (es.: fornitori, distributori, trasportatori, consumatori, operatori in *outsourcing*, aziende concorrenti, ecc.) può influire sulla buona riuscita di un Piano di Comunicazione Ambientale?

▶ **Answer** Sicuramente si. Quella che viene chiamata in gergo industriale la *stakeholder analysis* è una conoscenza importante per lo sviluppo di qualunque strategia aziendale anche perché la comunicazione è efficace quando si conosce chi ascolta e si comprendono le sue esigenze. Di conseguenza, le azioni di comunicazione che interessano alla categoria dei fornitori possono, per esempio, non suscitare alcun interesse tra i consumatori. È necessario, dunque, creare piani di comunicazione mirata per ogni gruppo di destinatari.

▶ **Question** Quali sono le nuove frontiere nella strategia di *Corporate Communication*?

▶ **Answer** Parlare di prodotti o servizi "*green*" e di "Ambiente", a livello aziendale, non è pienamente corretto. Nel settore *corporate*, si parla sostanzialmente di *sostenibilità di impresa* ovvero si parla di modelli di *business* che mirano a svilupparlo tenendo presente la dimensione ambientale, sociale ed economica (*Triple Bottom Line*), con l'obiettivo di sviluppare il *business* con il fatturato piuttosto che con l'utile. Minimizzando così quelli che sono gli impatti ambientali e sociali della produzione. Il comparto "*green*" non è in realtà una sfera indipendente ma lavora in simbiosi con le altre due sfere (economica e sociale). La nuova tendenza è ciò che si definisce *corporate Sustainability* o *Sustainability business* che si traduce nella contemporanea gestione dei *trade off*[2] economici, ambientali e sociali, sviluppando così modelli di *business* più sostenibili. Questo comporta un approccio totalmente diverso rispetto al *decision making* classico. Fino a dieci anni fa il processo decisionale nel mondo degli affari era totalmente improntato su variabili economiche. L'obiettivo ultimo era dunque quello di aumentare il fatturato, l'utile, il ritorno sugli investimenti o il dividendo per gli investitori. Nel momento in cui alle variabili economiche si sono aggiunte quelle ambientali e sociali, si sono raggiunti dei *trade off* diversi da gestire, comportando altresì una modifica dello scenario che, da breve-medio, diviene "a lungo termine". Ma ecco che gli stessi processi decisionali risultano oggi diversi: una rivoluzione, che possiamo definire positiva, e cha ha portato molte aziende ad intraprendere dei percorsi di riorganizzazione aziendale legati all'essere più sostenibili anche grazie al supporto di parte dei ricercatori. Di sicuro, dunque, possiamo affermare che oggigiorno l'essere più sostenibili, dal punto di vista economico-ambientale-sociale, porta le imprese ad essere più competitive sul mercato. La nuova frontiera è sicuramente l'implementazione della cosiddetta *Business Triple Bottom Line* su cui costruire nuovi modelli industriali che riflettano i valori della Sostenibilità.

---

[2] In Economy, functional relationship between two variables so that the growth of one is incompatible with the growth of the other, casing a downturn (Reference: Enciclopedia Treccani).

Paula Cristina Cayolla Morais Trindade Graduated in Technological Chemistry at the Science Faculty at the *Universidade Nova* of Lisbon (Portugal), she attended a Master in Health and *Environmental Engineering* at the Faculty of Sciences and Technology at the *Universidade Nova* of Lisbon (Portugal). Since 1989, she has been acting as a researcher in health and *Environmental sector*, focusing on *Environmental Auditing*, Eco-efficiency, Eco-design, Integrated Product Policies (IPP) and Green Public Procurement (GPP), coping with prestigious institutions such as: the *Núcleo de Estudos de Impacte Industrial* (The Industrial Impact Assessment Unit)—and the *Environmental Technologies Department* at the *Instituto Nacional de Engenharia, Tecnologia e Inovação* [National Institute of Engineering, Technology and Innovation]—INETI. Since 2009, She has been officer at the Investigation Unit of Sustainable Consumer Goods at the Portuguese National Laboratory on Energy and Geology Laboratory, intervening as an expert on the drawing up of the *National Plan on Green Public Procurement* 2008–2010. Furthermore, she has been fulfilling the role of *Project Manager* in eco-sustainable projects promoted by the European Commission (e.g.: SMART-SPP, *Life+*, *Horizon*—former I*ntelligent Energy, SPP Capacity Building*, etc.).

▶ **Question** Mrs. Trindade, you have been developing, for many years, Environmental projects at European level, within the framework of LNEG, the *Portuguese National Laboratory of Energy and Geology* which carries out research, testing and technological development, mainly in the areas of Energy and Geology. According to your professional multi-annual experience to which extent does "green" communication affect the successful outcome of a European project? Which are the key elements and main activities to include? Can you give us some practical examples, briefly?

Paula Cristina Cayolla Morais Trindade
http://www.lneg.pt (Web Site)

▶ **Answer** *Green Communication* is very important in the outcome of a European project. From my point of view the most important aspects are:

- To develop communication materials in English and also in national languages that can reach both stakeholders at European and national level. This was the case in pro-EE project.
- To identify key targets for the project and adequate the materials to different types of targets.
- To consider the involvement of a specialised organisation in communication right from the beginning of the project, in order to succeed in taking the message to different types of stakeholders.

▶ **Question** In a project, usually, an efficient and effective communication is an essential requirement for the functioning of its management, with special regard to the continuous improvement, who do you address to in the dissemination of a project? What is the difference in communicating the environmental outcomes to different stakeholders (e.g.: executives, employees, clients, suppliers, citizens, project team, etc.)?

▶ **Answer** Indeed, it is important to define in the beginning of the project different types of targets and to adapt the message to each of them. I think that we should be careful and try to clearly define a small number of targets, otherwise it is difficult to achieve good results.

▶ **Question** According to your professional experience, which are the best tools to be used in order to guarantee a continuous and systemic flux of information, ensuring feedbacks, transparency and accessibility of data (ref.: Aarhus Convention, European Directive 2003/04/EC)? Can you give us some few practical examples?

▶ **Answer** At LNEG we had good experiences in establishing a network with different stakeholders, to keep during the entire project duration. Regular meetings were organised and this ensures that stakeholders can actually keep up with project results and give their feedback. This was done with Building SPP project, a LIFE+ financed project and results were good.

▶ **Question** Would you consider the new Media (*Social Networks*) effective in conveying a *green message*? What about the traditional Media (e.g.: newspapers, television, radio, press offices, etc.)?

▶ **Answer** My experience with *Social Networks* is that it only works for general public and not to special targets. For instance, public officers have difficulties in accessing internet during work timetable in some countries, so in some cases this is not the right approach.

▶ **Question** How do you usually check that the *environmental message*/information has been fully understood by the audience?

▶ **Answer** We usually have questionnaires to access the level of interest and the quality of the message transmitted during the network meetings.

▶ **Question** Generally speaking, do you think that the knowledge of environmental issues in Portugal are satisfactory enough? What LNEG is going to plan in order to implement the "green background" of the public opinion on Energy, in the coming years?

▶ **Answer** My personal opinion is that knowledge on Environmental and Energy aspects has progressed a lot in the last 15 years in Portugal, however we have to keep working and to improve it. LNEG is deeply involved in this by working in new European projects with practical implementation in Portugal.

Eleonora Vallone Founder and Artistic Director of Aqua Film Festival (first edition from 6th to 9th October 2016, Rome; second edition—Elba Island from 22nd to 25th June 2017), an International Festival devoted to the theme of WATER, Eleonora has always cultivated an interest in many fields being a cinema, television and theatre actress, a journalist, an author, a painter and a fashion designer. In the meantime, she plays many sports with passion and particularly swimming. She is the creator of a new discipline: GymSwimming and also founder of the Patent used both in Italy and abroad [Certified by C.M.A.S. *Confédération Mondiale des Activités Subaquatiques*, founded by Jacques Cousteau and acknowledged by C.I.O. and U.N.E.S.C.O]. She can practice: ski, water-skiing, horse riding, tennis, classical dance, modern and South-American dance, belly dance, skating and ice-skating, trapeze, jogging, yoga, cycling, football, rowing, rally race, kick boxing *Savate* (feet boxing) and golf. She has obtained the boating license beyond 20 nautical miles; she is a diving instructor and lifeguard, confirming her strong connection with the Water and the Marine Ecosystem.

Premise to Questions: Mrs. *Vallone*, you have been committed for a long time in projects that enhance the importance of the sustainable use of Water, as a natural resource, but also as an element linked to the whole eco-system and Man's survival. To reinforce and make this *environmental message* more effective you have conceived and directed the first *Aqua Film Festival*, a review completely devoted to the short films and "very very short films" (of max 3 m) whose main theme is Water. An event that had a big success in terms of participation at international level, also thanks to the word of mouth conveyed through the Social Media.

Eleonora Vallone
www.aquafilmfestival.org (Web Site)
@eleonoravallone (Twitter)
@www.aquafilmfestival.org (Facebook)
aquafilfestival (Instagram)

See English version on p. 387

© Springer Nature Switzerland AG 2019
M. Abbati, *Communicating the Environment to Save the Planet*,
https://doi.org/10.1007/978-3-319-76017-9_29

▶ **Question** Cosa l'ha ispirata nell'ideare questa originale *kermesse* cinematografica? Che cosa può comunicare l'elemento Acqua? Quale il suo legame con il mondo del Cinema?

▶ **Answer** Il progetto internazionale dell' *"Aqua Film Festival"* è il risultato della mia vita! Sono arrivata a realizzarlo a seguito di molto impegno e combattendo in prima linea. Per me l'elemento Acqua è praticamente una seconda "Mamma". Un'acqua fisica e mentale. Provenendo da una famiglia che ha vissuto il cinema in prima persona, mi riferisco sia a mio padre, il noto attore italiano *Raf Vallone*, che a mia madre, l'attrice italiana *Elena Varzi*, ho nelle cellule questa nobile arte e il suo insegnamento di disciplina e di vita. Ho voluto così **l'Acqua** ed il **Cinema** perché questa preziosa risorsa naturale mi ha aiutata in un momento particolarmente delicato della mia vita. È stata infatti l'elemento essenziale per recuperare le mie abilità motorie, a seguito di un grave incidente stradale (nel 1984). Dalla mia rinascita fisica e morale sono nati più di tremila esercizi ginnici in acqua, frutto della mia creatività unita a studi fisioterapici che ho codificato tramite brevetto e che sono stati oggetto di cinque libri da me pubblicati. L'occasione per esprimermi con gioia attraverso la comunicazione del corpo in acqua come elemento fisico.

Ma l'Acqua è molto di più! Ha da sempre un valore simbolico profondo. E la sequenza delle immagini di un film ci possono ben raccontare questo. Partendo proprio da queste premesse, il Festival si ispira a quattro aree tematiche: (1) *l'Acqua del mare* che costituisce il più grande bacino esistente; (2) *l'Acqua dolce* di laghi, fiumi e sorgenti che rappresenta il nostro alimento principale; (3) *l'Acqua termale* che da sempre fornisce cura e sollievo per il nostro organismo; e *l'Acqua del nostro Ambiente* e di tutto quello che lo circonda. E si struttura in due categorie per me significative: i "Corti", film di venticinque minuti (cortometraggi) e i "Cortini", filmati brevissimi di massimo 3 minuti. L'obiettivo principale, ovvero la *mission*, è che tutti possano concentrare l'attenzione attraverso i propri occhi, o meglio puntare l'obiettivo del proprio *Smartphone* ad un elemento che è straordinariamente fotogenico; anzi, forse quello che più si adatta a rendere uno scatto fotografico unico. Se uno ci pensa, infatti, l'Acqua ha tanti aspetti, così come i suoi tanti riflessi, ed ognuno di essi "comunica" uno o più messaggi allo spettatore. L'Acqua è anche metafora della nostra vita che scorre come un fiume o un mare. Ce lo ricordano scrittori, filosofi e pensatori come *Buddha, Eraclito, Lev Tolstoj, Giuseppe Ungaretti, Herman Hesse, Jorge Louis Borges, Fabrizio Caramagna, Alessandro Baricco*. Ce lo ricorda anche la **Dichiarazione Universale dei Diritti dell'Uomo** che lo considera strumento per assicurare *"un tenore di vita sufficiente a garantire la salute ed il benessere personale e della propria famiglia"* (art.25).

Questa risorsa comunica di per sé un divenire continuo. È anche il nostro primo elemento e la prima fonte di sostentamento, il corpo umano è formato da un'alta percentuale di acqua che arriva fino al 75% del peso corporeo nei neonati. Non a caso, anche nelle spedizioni spaziali uno dei primi elementi che gli scienziati e gli astronauti ricercano è la presenza di acqua sulla superficie di pianeti, satelliti e altri piccoli corpi celesti. L'Acqua è dunque Vita!

Per me l'Acqua comunica anche il ricordo della mia infanzia, di quando andavo con mio padre a tuffarmi in mare. Un mare cristallino troppo spesso deturpato da una sorta di "rete" di sporcizia. Che ancora oggi, purtroppo, si ripropone ciclicamente e si manifesta sulle nostre spiagge durante il periodo estivo. Le cronache ci riportano spesso notizie di lamentele di turisti che denunciano la presenza di strane chiazze lattiginose galleggianti sulla superficie del mare, anche in note località balneari del Bel Paese. E questo danneggia il turismo ed ogni altra attività legata al territorio. Tutto per il fatto che anche in quelle aree si tende ad accumulare rifiuti anziché gestirli in maniera sostenibile, e questo vale sia per le popolazioni costiere che per gli equipaggi a bordo delle imbarcazioni.

L'*Aqua Film Festival* non nasce, quindi, da un'idea esclusivamente ecologica ma da una grande passione personale. Nutro davvero un immenso amore per il mare che rappresenta tutto per me.

Premise to Questions: In order to make the *Environmental Communication* really effective it is necessary that the contents of the message are listened to, understood and remembered by the recipients. This is what most of the *Environmental scholars* think. Internet Network offers curious attempts to make people be aware of the *Environmental* and *social issues* but often they tend to lose their original communicative function and become the target of the "current fashion". Last but not least the viral effect of "The Ice Bucket Challenge", the campaign launched by ALS, American association for SLA (Amyotrophic lateral sclerosis) patients and their families. A charity message—challenge spread out so much as to produce 24 million of videos and as many buckets of cold water, coming from all over the world, shot both by amateurs and VIPs.

▶ **Question**   Secondo la sua esperienza, potrebbe accadere qualcosa del genere per una causa ambientale? Quale gesto, o gesti, potrebbero essere più efficaci per sensibilizzare l'opinione pubblica mondiale sulla necessità di agire per tutelare un Ecosistema vitale come l'Ambiente marino?

▶ **Answer**   Al di là dell'esempio che ha citato e che è stato largamente inflazionato sulla *Rete* a tal punto da snaturare, in alcuni casi, il suo stesso obiettivo, credo che un'adeguata campagna di comunicazione, da diffondersi anche attraverso la Rete, dovrebbe sensibilizzare l'opinione pubblica sugli sprechi della risorsa Acqua. A partire dai gesti quotidiani più semplici. Lavarsi i denti o fare la doccia e lasciare aperto il rubinetto ad oltranza determina un consumo insostenibile [si è calcolato che lasciare il rubinetto aperto mentre ci si lava i denti consuma circa sei litri di acqua al minuto mentre si arrivano a consumare fino a circa 15–16 litri di acqua al minuto lasciando il getto della doccia aperto senza inserire alcun frangi—getto]. Per non parlare poi di quando si fa il bagno nella vasca, pratica assolutamente salutista e benefica per il corpo umano ma che potrebbe essere fatta una volta a settimana, alternandola con la doccia [per riempire una vasca da bagno occorrono circa 150 litri di acqua]. Lo stesso vale per il lavaggio a mano o in lavatrice dei panni sporchi o ancora di piatti e posate in lavastoviglie. Gli elettrodomestici moderni ci facilitano il compito consentendo lavaggi rapidi, o addirittura *"eco"*, con risparmio di consumi energetici ed idrici. E lo stesso vale per gli scarichi dei WC a flusso differenziato, a seconda della necessità, al fine di razionalizzarne i consumi di acqua. Credo che si dovrebbero incrementare forme di comunicazione sempre più efficaci su queste tematiche che sono già oggetto di studio di esperti comunicatori per società private, aziende pubbliche, catene di supermercati, e tante altre realtà.

A tale proposito **Legambiente**, ha già introdotto una speciale classifica, gli *"Oscar dell'Ecoturismo"*, che premia annualmente le strutture turistiche, le aree marine protette e i parchi nazionali e regionali più virtuosi dal punto di vista ambientale e sostenibile. Ed il risparmio di acqua è tra i principali parametri presi in considerazione per stilare la lista dei finalisti. Tra l'altro l'*Isola d'Elba* (dove si è svolta la seconda edizione dell'*Aqua Film Festival*) è tra le prime realtà insulari ad aver valorizzato il turismo sostenibile, garantendo una ricettività alberghiera certificata dal punto di vista ambientale. Un ritorno alla natura nel rispetto dell'economia locale ancora più significativa perché coincide con un'area particolarmente quotata dal punto di vista turistico. E che si avvale anche delle certificazioni ambientali, come *Ecolabel*, per comunicare la sua immagine eco-compatibile agli abitanti dell'isola e ai suoi visitatori.

▶ **Question** Quanto può influire un *Festival Cinematografico*, come *Aqua Film Festival*, sul pubblico per aumentare la consapevolezza generale sull'Ambiente, spingendolo ad evolvere verso uno stile di vita più sostenibile?

▶ **Answer** Può avere una forte influenza! Il Cinema è uno strumento estremamente efficace per esprimere *messaggi ambientali* e, in particolare, per trasmettere l'amore per il mare ed il suo Ecosistema. Perché mettere insieme l'elemento Acqua e improvvisarsi registi? Perché il Cinema cattura e fa "vivere" qualcosa a cui noi non daremmo peso nella vita frenetica che conduciamo. Ecco, dunque, che attraverso l'immagine, che può essere anche quella immortalata da un *selfie*, secondo la moda del momento, ognuno di noi può prendere consapevolezza del fatto di non essere l'unico, o meglio il principale essere vivente ma di far parte dell'Ecosistema Terra. Da qui l'idea di far partecipare registi ed amatori da tutto il mondo. Questo si è rivelato vincente fin dalla prima edizione con la partecipazione di più di mille cinquecento contributi filmati, arrivati da ogni angolo del nostro Pianeta grazie alla diffusione della *Rete Internet*. La promozione attraverso gli sponsor infatti è stata minima, in partenza.

Credo che il pubblico abbia recepito in modo corretto il nostro messaggio che è anche quello di vivere l'Ambiente attraverso l'arte del Cinema. Nella prima, così come nella seconda edizione del festival, si è data la possibilità ai giovani di sperimentare la regia, attraverso laboratori guidati da esperti, studiando le migliori tecniche di ripresa per rappresentare il mondo naturale che ci circonda in cui l'acqua è un elemento fondante. Ma anche soggetto difficile da filmare o immortalare in fotografia per la sua capacità di riflettere la luce solare in modo dinamico.

La realizzazione di video *curricula* (video racconti di sé stessi) a sfondo naturale ed acquatico ha sicuramente accresciuto la loro creatività artistica, la consapevolezza di sé nel rapporto stesso con il loro apparire in video ed il grado di conoscenza e consapevolezza ecologica dello stretto legame che unisce la vita di ciascuno di noi con questa risorsa liquida, in continuo movimento. Un'esperienza di grande valore sociale che ha consentito di utilizzare lo *Smartphone* in mondo non superficiale.

Anche la seconda edizione, che si è da conclusa da pochissimi giorni, è stata un successo. Centocinquanta cortometraggi provenienti da trenta Paesi in rappresentanza dei cinque Continenti. Si è svolta in un luogo simbolo del Mediterraneo, l'Isola d'Elba (la più grande delle isole dell'arcipelago toscano). Qui hanno concorso tre giurie: una tecnica; la seconda composta da un'università di ragazzi ed una composta dal grande pubblico.

In questa seconda edizione, inoltre, abbiamo previsto una nuova sezione fuori concorso, chiamata "**Fratello Mare**", organizzata a scopo benefico dai volontari di *UNIVERSI AQUA* in *partnership* con *Legambiente*, responsabile della fase operativa. Questa iniziativa ha permesso a chiunque di cimentarsi nella "regia", pur non essendo esperto. L'obiettivo era quello di filmare attraverso il proprio *Smartphone* dei *video documenti* finalizzati a *denunciare* situazioni di degrado come l'inquinamento che mette a rischio la biodiversità marina, in violazione di leggi o regolamenti. I video, sottoposti alla nostra valutazione, hanno così consentito di localizzare le aree incriminate, indagare sulle cause dell'inquinamento e identificarne i potenziali autori al fine di avviare azioni di bonifica e risanamento ma anche di promuovere adeguati piani di comunicazione a forte impronta informativo-didattica, per prevenire in futuro nuove situazioni di degrado. Il progetto ha voluto così consentire al singolo cittadino di trasformarsi in *reporter ambientale* innescando un processo comunicativo che, attraverso i documenti filmati, ha contribuito ad aumentare la consapevolezza sull'importanza di tutelare l'Ecosistema marino. Rappresentando dunque un *medium* aperto a tutti per ragionare sulle criticità ambientali e sulle loro possibili soluzioni.

L'idea è proprio quella di sensibilizzare l'opinione pubblica ed "educarla" alla protezione del mare che è essenziale per la nostra esistenza. Una missione educativa sostenuta da tutti gli ospiti d'onore intervenuti nelle due edizioni del festival, tra i quali ricordiamo: *Antonietta De Lillo* (regista), *Ludovico Fremont* (attore), *Simonetta Grechi* (Legambiente), *Paola Gassman* (attrice), *Enrico Magrelli* (critico cinematografico), *Sara Serraiocco* (attrice), *Filippo Scicchitano* (attore), *Sebastiano Somma* (attore), *Cinzia TH Torrini* (regista).

Inoltre, a suggellare il nostro impegno in tal senso, a partire dalla seconda edizione abbiamo raggiunto un accordo con *Legambiente*, una delle associazioni ambientaliste italiane più attive, presente su tutto il territorio, proprio per poter intervenire nelle situazioni di degrado, in tutta Italia e, in prospettiva, anche fuori dai suoi confini nazionali, dato che il concorso è internazionale.

Un impegno alla sensibilizzazione dell'opinione pubblica sulle tematiche legate all'Ambiente particolarmente caro anche al *Governo del Principato di Monaco* e, in particolare, alla *Fondazione Principe Alberto II di Monaco*, nostro partner ufficiale, impegnata, fin dalla sua nascita (nel 2006), nella promozione di progetti internazionali di alto valore volti a salvaguardare l'Ecosistema Terra, grazie al costante impegno di S.A.S. Principe Alberto II di Monaco. Non a caso la terza edizione del *Aqua Film Festival* (nel 2018) si svolgerà proprio nel Principato, uno dei territori più densamente popolati al mondo (più di diciassette mila abitanti per kilometro quadrato) nonostante la sua esigua estensione (2 km^2). Ma che è da sempre particolarmente attento alla tutela ambientale dell'Ecosistema urbano costiero, compreso tra la riserva marina di *Larvotto*, ad est, e l'area protetta del fondale coralligeno delle *Spélugues*, a sud-ovest.

▶ **Question** Se dovesse girare Lei stessa un breve video per sensibilizzare l'opinione pubblica e valorizzare l'Ambiente marino, e la sua biodiversità, quali parole, suoni ed immagini userebbe?

▶ **Answer** Mi concentrerei nel scegliere il *medium* migliore per far capire a tutti la gioia che provo quando entro in acqua. È un sentimento sconfinato che mi fa sentire al centro del mondo, se non dell'intero universo. Tutto diventa perfetto. Ritengo davvero che l'Acqua sia un elemento magico che fa sentire il tuo corpo leggero, che allieva i tuoi dolori fisici e intimi. Con l'acqua, soprattutto quella di mare, è come se ritornassi veramente alle origini. Certo sarebbe molto difficile poter esprimere tutte queste emozioni. Di sicuro richiederebbe del tempo ed un lavoro in stretta collaborazione con il mio aiuto regista.

E poi, sicuramente, troverei il modo per far sperimentare al pubblico quali bellezze nascondono i fondali marini per farli riflettere sull'importanza degli *eco-gesti* quotidiani finalizzati a preservarli. La mia esperienza di istruttrice di *GymNuoto* e di *sport* acquatici ha dimostrato quanto sia efficace comunicare l'amore per il mare condividendo con gli altri le proprie emozioni. Molti allievi mi hanno confidato, a fine corso, di aver preso reale coscienza di quanto sia essenziale preservare la biodiversità marina e di aver cambiato il loro stile di vita per evitare di deturparla.

Fig. IWP 22.1 Official Logos of *Aqua Film Festival* held at Isola d'Elba and Principality of Monaco Third Edition © 2018

Fig. IWP 22.2 *Aqua Film Festival*—Second Edition—at Isola d'Elba—The film-maker *Cinzia Th Torrini* and *Eleonora Vallone*, artistic director of #aquafilmfestival, are announcing the special prize awarded by the documentary film *Bacio Azzurro* (Blue Kiss), directed by *Pino Tordiglione*— they are accompanied by the artistic cast: the Italian actors *Sebastiano Somma* and *Morgana Forcella*—© 2017

APPENDIX – INTERVIEW TRANSLATIONS (IT)

IT 1   Translation of Interview No. 1: Mounir Bouchenaki

IT 1.1 Premise to Questions

Mr. *Bouchenaki*, the importance of the World Human Heritage is at the heart of UNESCO at an international level, and it includes cultural and natural targets to be achieved that underline a sound link among archaeology, art and ecosystem. It is not by chance that the first forms of human artistic expressions, in the Pre-history, represent drawings whose depicted subjects are taken from Nature (e.g.: *Cueva de Altamira*, Spain; *Grotte dei Balzi Rossi*, Italy; *Lascaux et Chauvet*, France; *Wadi In Djeran*, Algeria; *Tadrart Acacus*, Sahara—North Africa; *Serra de Capivara*, NW Brazil; *Bhimbetka*, India etc.). Therefore, at the end of the years Nineties, we started to talk about *archaeology of tourism* (in the modern sense) a topic which implies many aspects among them the Sustainable Development: the necessity to preserve the eco-cultural heritage for the future generations.

IT 1.2 Questions

▶ According to your long-standing experience in the field of preservation of the World Human Heritage, how can we communicate effectively the importance of a Sustainable Management of the archaeological sites or other favourite places towards tourism and eco-systems that host them?

▶ "The image uses languages that the reason of words doesn't know…" *used to say Frédéric Lambert*, professor of Semiotics of the image and of *Media* at *Institut Français de Presse (IFP)—Université Paris 2—Panthéon Assas*. What do you share of this viewpoint? In your experience at UNESCO, do you think that an awareness campaign addressed to tourists underlying the great link between the cultural and natural heritage could use a visual language more than a written message to clarify the natural and *Environmental nuance* (subtle difference)?

© Springer Nature Switzerland AG 2019
M. Abbati, *Communicating the Environment to Save the Planet*,
https://doi.org/10.1007/978-3-319-76017-9_30

▶   For a long time, the role of the *Media* in comparison to communication has been the study subject by anthropologists, sociologists, semiologists, etcetera, divided into those who consider them a precious *medium* to enlarge knowledge and those who have a tendency to "disturb" the quality of information. What is your idea about the new *Media* and in particular of *Social Media* in the spread of key-values as the importance of culture, interdisciplinary education, preservation of heritage, symbol of the cultural diversity to be respected?

▶   *"Tourism is an immensely popular global social phenomenon [...] part of 'human exploratory behaviour' that serves as a diversion from the ordinary and helps to make life more interesting and 'worth living'"* (source : Cameron Walker and Neil Carr: Tourism and Archaeology: Sustainable Meeting Grounds, Routledge– Taylor & Francis, New York, 2016), how can we push such a huge world audience to take respectful actions of the human heritage in order to avoid the recurrence of actions that, in a few seconds, can destroy the socio-cultural and *Environmental context* (I am referring to the *Buddhas' of Bâmiyân* destruction in Afghanistan, in March 2001, or the Palmyra archaeological site destruction in Syria in 2015)? Which actions are to be carried out at institutional, cultural and social level?

IT 1.3   Answers in a Nutshell

First of all, the question of the management of archaeological sites starts from their enrolment in the *List of the World Heritage* to be considered as a first step in view of a rational and integrated management of the registered sites. The *Orientations* to enter this *Agreement* are anything but easy and they refer to two specific items whose contents can be summed up like this: each site should have an adapted Management Plan or a documented Management System with the specification of its universal value. The purpose of the Management System will be to ensure the effective protection of the proposed property. Recently, *Kishore Rao*, Director of the World Heritage, in his foreword to a UNESCO publishing book whose title is "To manage the World Natural Heritage", has reminded that with more than a thousand of registered sites the challenge of the *Agreement* is to ensure that the values with which the sites were classified are maintained in the context of a world increasingly globalized and rapidly developing. All this in cooperation with all the participating States that must guarantee the uttermost levels of management. The acceptance of the classification "dossiers" is then linked to the need of supplying a detailed *Management Plan* for the proposed sites. At this point, several programmes of *Environmental awareness* have been drawn up. We remember among others: "*the Heritage Days*", "*the Open-Door Days*", *the* public administration *laboratories,* the *competitions* for *the best practices,* and so on, for the award of prizes in their respective areas. We cannot help thinking of the "*Cultural Heritage Award*" in Asia, or the video on *YouTube* in the U.S.A. about why the American sites are so important for the world public.

As far the *Communication Plan* is concerned, some agreements have been concluded with the broadcasting chain such as: *Arte.tv, BBC, NHK, TV5, History Channel*, etcetera.

The management of the World Heritage and Sustainable Development have taken on a relevant importance, emphasized by the role of Culture Economy specific studies that have shown the social and economic improvement of the people living near the sites. Consequently, a lot of meetings of experts have developed *indicators* to evaluate how the Preservation and Management of the Heritage can contribute to the Sustainable Development. On the matter, one of the most meaningful example is the *Touristic Development* that we can see in the brochures published by the Industry of Tourism.

Italy is one of the big tourist destination and it has seen the development of Institutions like *Romualdo Del Bianco* Foundation, in Florence, promoting a cultural tourism as a base for the renaissance of a new humanism. Another reality that has been working in this sector for more than 20 years, "The Mediterranean Exchange for the Archaeological Tourism" where each year, at the end of October, archaeologists and representatives of the Industry of Tourism from all over the world gather together. Also, other international organizations such as the ***World Tourism Organization*** (*UNWTO*), based in Madrid and **UNESCO,** based in Paris, have been interested in tourism development of the *World Heritage* sites for the last 10 years. Finally, we cannot forget that the *Saudi Organizations* held the first **World Conference in Cambodia**, in 2015, about "Culture and Tourism" in the small city of *Siem Reap*, near the prestigious site of *Angkor*.

The press release of OMT (*Organisation Mondiale du Tourisme*) pointed out that more than 900 participants, 45 ministers and vice-ministers of tourism and culture, lecturers and experts from more than 100 countries came to the event.

Talking about the importance of the visual message in comparison to the written, professor *Frédéric Lambert* stressed the importance of *Media*, and in particular of *TV, Smartphone* and *Tablet* where the image is always present and has completely changed our perception of reality. "*The image is worth as far as it is able to change our thought, that is to renew its language and our knowledge of the world*" answered Georges Didi-Huberman to professor Frédéric Lambert in his book "*The condition of ima*ges" (*Mediamorphoses*, 2008).

In this regard, I would like to tell about an experience lived at the time in which the UNESCO launched the International Campaign for the safeguard of the old city of *Sanaa* in Yémen with *Marco Livadiotti*, responsible of an Italian travel agency set up in that country. The safety situation was normal at that time and *Marco Livadiotti* had rented an old traditional mansion and, after a suitable redecoration, used it as hotel and organizing tourist visits from Milan, preceded by special information lessons with photos and films included *Pasolini's, Decameron*. A very original experience indeed of this documentary called "*Le mura di Sanaa*" ("Sanaa walls"), pointing out the need of a UNESCO intervention to safeguard the small city still unknown in the West. Unfortunately, since March 2015, the role of the image that had been so relevant to raise awareness and appreciation for a unique heritage has given way to images of destruction and tragedies of the cultural sites.

Of course, we are wondering whether the modern technology can help us to preserve these hidden world treasures offering the same scientific guarantees of the traditional tools. It is clear that the new technologies can open a lot of possibilities in the field of *"virtual restitution"* but in the same time they make questions about the integrity and authenticity.

The recent 3D reconstruction of Palmyra destroyed sites (Syria) made by the architect Yves Ubelman (*Société ICONEM*) has shown how the remains could be set up again.

On the other hand, the heritage, once become cultural, is subjected to the development of derived products that represent it and that can be used for teaching and informative purposes. The destruction however leads to a *dematerialization of the cultural heritage traces* and one moves from the real to the virtual world. An example the virtual visit proposed by the Olympia Museum in Greece that helps the understanding of the site very much, depriving, nevertheless, the visitor of using his senses to listen to, to see how the time has affected all that remains of the past. Certainly, the computer resources attract especially the new generations for their interactive and exciting possibilities. Talking about Socio-Economic Development, the General Secretary of the World Organization for Tourism, *Taleb Rifai*, reminds that *""tourism" means that each year more than one billion people cross the international borders offering the tourism an immense socio-economic development"*. *Tourism* has thus demonstrated its capacity of increasing competitiveness, of creating new jobs, of reducing the rural exodus and of generating incomes to be reinvested. The UNESCO General Director, *Irina Bokova*, has declared that culture forges our identity and facilitates the respect and tolerance among peoples. Furthermore, it improves the people's lives and reinforce the mutual comprehension. Hence the safeguard of our cultural heritage will have to go ahead with the Sustainable Tourism. Unfortunately, each year there are increasingly more conflicts that keep away from tourist destinations and often the symbols of memory and identity of a people are destroyed.

UNESCO has been monitoring for more than 30 years all the wars and, as far as I am concerned, I must say that one of the most frustrating experiences has been undoubtedly the destruction of the *Buddhas of Bamyan* in March 2001. This destruction has made the international community react with a unanimous condemnation gasp. All the voluntary destructions have been classified both by the UN General Secretary *Ban Ki Moon*, and by the UNESCO General Director *Irina Bokova,* as "**war crimes**" and "**crimes against Humanity**".

IT 2   Translation of Interview No. 2: Mariaelena Camerini

IT 2.1   Premise to Questions

The *Environmental matter* is, for its nature, very complex since it includes different subjects thus overcoming the traditional subdivision of the knowledge. For this reason, everybody who communicates the *Environment* should do it responsibly whatever the means of communication. Words, images, sounds, films, symbols, graphic representations, designs, etc. As many communication scholars remind us each action is able to influence, inform and educate a specific "public" on condition to involve it emotionally pushing it to change its lifestyle in order to help the *Environmental issue*. "Architecture is an art fact, a phenomenon arousing emotions. Architecture is meant to "move" used to say *Le Corbusier*, a real revolutionist together with *Frank Lloyd Wright* for including Nature into his buildings.

IT 2.2   Questions

▶ Do you share this assumption? According to your experience, what does creating a "green" architecture sustainable project mean? Which elements can't be missed in designing a sustainable project?

▶ Nature and its forms are often source of inspiration for Architects unless there were some philosophies more focused on human than natural world (e.g.: *Semper* inspired by *Goethe* e and German philosophers). With the birth of modern environmentalism (traditionally linked to the publication of "*Silent Spring*" by the American biologist *Rachel Carson*), Bio-architecture totally revived. The energetic crisis, the official definition of Sustainable Development (*Bruntland's Report*, 1987) and some International Conferences devoted to the *Environment* (starting from the UN *Rio Summit* in 1992) boosted, then, an irreversible process aimed at raising awareness worldwide on *Environmental issues* both at urban and industrial design stage.

▶ At world level, different eco-Sustainable projects have been designed such as: Bosco *Verticale/Vertical Wood, by Studio Boeri Architetti;* Many Milan EXPO, 2015 Pavilions etc. or others that got their inspiration from Nature e.g.: *Aqua Tower by Jeanne Gang, Chicago.* According to your experience, an architectural design can communicate one or more non-verbal *Environmental messages* through what it represents or the materials used? Or through its shapes?

▶ If you were given the task to design a building for housing, according to the rules of bio-architecture. Which element could be integrated with the surrounding nature (a forest, a Mediterranean coastline or an Alpine, insular *Environment)* how would you imagine it? Which the priorities? Which *Environmental message* would you like it to communicate?

IT 2.3   Answers in a Nutshell

A black and white photo dating 1924 brings us back to a window, 11 m in length. This is the site from which *Le Corbusier* used to watch the *Lake Léman* facing the great "*Petit Maison*" estate. A perfect liaison between Nature and Landscape and the Human *Environment*. The architecture is meant to create emotions making us be able to follow a path for creative thinking aimed at increasing the eco-awareness and the respect of the *Environment*, that is the Sustainability. Both in a constructive or renovating phase, acting in a sustainable way must not be misinterpreted. Being "sustainable" is equally an intention, an approach and an action addressed to a method. Bio-architecture is like "designing on the mirror": on one side a pencil, on the other the Planet Earth, nowadays and in the future. The mission is to find solutions taking into account the scarcity of natural resources. Consequently, an architect should consider some fundamental aspects: (1) holistic approach; (2) basic minimalism; (3) bioclimatic approach; (4) building approach devoted to living beings and reuse of components; (5) efficient consumption strategy; (6) use of renewable sources of energy.

Since Nineties, many architects have been seeking for a new formal "linguistic code" in *Environmental architecture*. If we think, for example, to the *Tjibau* cultural centre, designed by *Renzo Piano* in New Caledonia or to *Heliotrope*, a spinning cylindrical residential building in the shape of a sun-flower, designed by *Rolph Disch*, we realise that architectural forms are means of communication. Generally speaking, nevertheless, the architectural language is still based on traditional shapes, despite *Environmental certifications* or technologies. On the contrary, if you consider architecture as a "sculpture" that includes Man (*Bruno Zevi*), its symbolic and metaphoric contribution is pure communication and a perfect "building material" able to forge ecological emotions, reflections and behaviours. Let us think of the "Cardboard Cathedral" or of the "Hualin School" designed by *Shigeru Ban* (awarded with Pritzker Prize); or even of "*Earthship*" building by *Michael Reynolds*. It is not by chance that *Le Corbusier* used to consider "the sun, the space and the trees" as the fundamental materials for the town planning. We need an architectonic vocabulary which includes a natural element as the cornerstone of the respect for the *Environment*. The "Vertical Forest" designed by *Studio Boeri Architetti,* as you mention, communicates a clear message: we must promote, at any level, a proper relationship among Man, Building and Nature.

I would think of small units hiding in the natural *Environment*: an organic, soundless, minimalist architecture that fosters and promotes a simple and healthy life style. I would use local materials and techniques. I would focus on local elements such as sand, vegetation or rocks by creating a car-free area. I would keep the parking area at distance in order to remind us that Life means walking and hiking that is time to think, healthy rituals and lower *Environmental impacts*. Last but not least, I would reuse existing buildings: I can't tell you how big the abandoned architectural heritage is.

IT 3   Translation of Interview No. 3: Francesca Carminati

IT 3.1   Premise to Questions

"To communicate" implies, in its very etymology, an incredible intuitive value. Communication is considered by many a perfect example of social expression that finds its completion when the message reaches its recipients and is understood by them, becoming thus common heritage for building a discussion, a knowledge or a culture.

IT 3.2   Questions

▶   Do you share this "social function" of communicating?

▶   How much does the Communication influence your job? Do you use *Environmental "good practices"* to sensitize the readers?

▶   Which tools do you use to verify the efficiency of the Communication, be it visual, verbal, by signs, semiotic, iconographic, etcetera?

▶   The *"Environmental communicator"* is a new professional figure who deals with the "green themes, referred to the *Environmental Sustainability*. What kind of contribution can the publishing houses give in the dissemination of *Environmental messages*? The structural organization of a book, graphic included, can really influence the readers toward a "green" world?

▶   Which elements contribute to give you confidence in the authority of the message, conveyed by a publicity campaign or spot, or an ecological label of a product or process?

▶   Which are the challenges for the future of the publishing house?

IT 3.3   Answers in a Nutshell

Nowadays, to communicate is an essential skill, more than ever. Nevertheless, "communicate well" is what matters most. These days, the way to interact has significantly changed, thanks to new technologies. Anyone is able to catch everybody's attention on what he would like to transmit. Maybe we are not completely aware of the fact that any single gesture, word or action always "communicate" something to the audience, for better or worse.

Nowadays, thanks to new communication systems, we risk to be overwhelmed by a large quantity of information that, whether misrepresented, can create great confusion. Whoever communicates has to adapt his communication to the *communicative context*; similarly, the receiver needs to make a difference between good and bad communication.

There are two aspects to consider from a professional point of view. The first affects the inner Communication, part of the social system involving a working group. A proper communication may encourage a good organisation of all activities, providing then the achievement of the fixed targets.

The second is linked to the external Communication that has to be targeted, effective and widespread, thanks to the various *Media* at our disposal which facilitate to inform.

As far as editing is concerned, I always adopt a specific kind of paper coming from responsibly managed woods, which means they respect particular *Environmental standards*, in accordance with the FSC (*Forest Stewardship Council*) certification, whose logo is usually stamped on books. Furthermore, for aesthetic and *"Environmental"* reasons, I avoid designing plastic-coated covers. Last but not least, on the occasion of events I am responsible for, I undertake to raise the eco-awareness through new forms of communication by sending digital invitations only, without the need to print them.

There are, then, best practices involving the internal organisation I would like to communicate to readers by describing them in a dedicated *Internet* page through a specific section devoted to suggestions and proposals.

Creating a "green" distribution via eco-vehicles will be my target to be achieved in the next few months.

I rely upon different tools aimed at monitoring and evaluating the efficacy of our communication. In respect of the book launch, an efficient index is the people turnout plus the number of books sold during those events. *Social Networks* are a further marker to be considered, through Media such as "likes" or the number of "visualisations", "retweets" or else. Those tools help us to understand which communication strategy is more appealing.

The Communication Editorial Campaign for the product is fundamental. The publishing field, sharing with the audience *Environmental best practices* and a "greener" way of production, is able to encourage the dissemination of the "green" message.

Marketing experts' research, and not only, confirm that specific graphic images, or chromatic effects, may stimulate our emotions.

Let us think of colours, for example. Green colour is relaxing for eyes and comforting. It is no coincidence that it matches with anything is referred to the *Environment* and the natural world. Trees, then, are always associated with "green" issues.

Visual elements are perfectly capable to induce the costumer to purchase a specific "green" product more than another.

Certifications are certainly important to verify the seriousness and truthfulness of a message. Referring to commercials and to advertising campaign, they undoubtedly influence the receiver and any institution, public or private (e.g.: *Environmental organisation*, ESCO *Energy Service Company*, etc.), which transmit the message.

The publishing world has been suffering financial difficulties for many years due to several factors. Consequently, a communication strategy is crucial to attract wider audiences and capture their attention.

Nowadays, Internet is playing undoubtedly a leading role in communicating. Social Networks, in particular, are efficient *Media* to present new products, stimulate curiosity and to attract an ever-increasing audience.

Dealing with any kind of event, involving readers in many ways, could be the right way to increase the attention and share emotions.

Thinking of the great publishing and culture market development, we must consider each book as a single product to be disseminated through a suitable communication plan according to the audience it is addressed to: to the whole market or to a large public, to a specific market segment or public, to several market segments or kind of public.

IT 4   Translation of Interview No. 4: Cristina Carretero González

IT 4.1   Premise to Questions

The legislative communication is generally associated to the institutional communication considering the level of knowledge of the juridical act; the structural and linguistic nuances are then of vital relevance if referred to *Environmental legal issues* including also different technical sectors.

IT 4.2   Questions

▶   How can the contents of a legislation be effectively communicated helping the information on key-principles on the *Environment*? How can we verify the term correctness and the clarity and quality of the *Environmental messages*?
▶   To communicate properly the "green" message of a legislation, can explicative notes, schemes, synthesis be useful? Which value is attributed to this documentation? How can we verify its authenticity (not influenced by other particular interests)?
▶   Can the legislative contents affect significantly the lifestyle of citizens toward a more sustainable behaviour? What do they need in order to achieve this target? Can you give some concrete examples at a local, national or Community level?
▶   Can the use of Internet in communication increase the public opinion understanding and sensitization on the *Environmental legislation*? What role can the *Social Network* play (e.g. Facebook, Twitter, Google+, Instagram, etc.) in the explanation process of the *Environmental regulations*? What about the visual images and all the possible multimedia applications?
▶   Next to the legislative system called "command & control" there are other regulations that follow the same objectives by the use of voluntary but compelling instruments [e.g. *Environmental Management System* (EMS), Eco labels, Energy Efficiency Certification etc.], is there the possibility to integrate them into the *Environmental legislation* to establish an ecological communication?
▶   What would you improve or change in the present juridical system to implement the communication sector about the *Environmental issues*?

IT 4.3   Answers in a Nutshell

It is essential that the legislative technique is correct and adequate. Inside the good technique, a fundamental factor is the clarity of the norm. To verify this element and the quality should be necessary to include a protocol of legislative drafting wherein the clarity parameter was regulated and contemplate the essential issues with clear language, quick and easy to read.

As far as the explicitness of the "green" message I think that your question hit the mark. If we want a good message to reach a large number of people it will be more effective as explanatory as possible. But the explanation for some people differs from that of others. I mean that there are people that understand a chart better; others prefer a summary and others long and detailed explanations. Nevertheless, there are people that consider more complete a message in which there are images, photos, or icons. In my opinion, the value of its authenticity will be associated to the institution or the person who launched it.

Hypothetically, I mean that each regulatory content that appears mandatory will affect the acquired habits. Those which are not mandatory but only simple recommendations, can have a good welcome provided that a good, motivated, clear advertising campaign is made, basing it on convincing data. I am sure that every day we gain more awareness and eco-gestures are made more naturally.

I think that the *Media Network* can increase the understanding and awareness on the importance of looking after our *Environment*. They are means close to us and of quick access to all phone users and if it is done with clear and short messages the better. Nowadays I feel that there are more readers of short and clear messages. The role of the Net is then very important.

The images and the multimedia applications, as I already said, reinforce the message and they make it accessible to more people than a piece of information conveyed by letter can do.

I think that the legislative system "*command & control*" and other systems you are referring to could be integrated into the *Environmental* legislation and, in case they are not yet, could complement the *Environmental policies*. To consolidate the "ecologic communication" the best form should be studied to spread it through the rules.

I think that the basic foundation, as already stated before, is the best regulatory system so that the message to be conveyed is simple and easy to read and its clarity derives from the reliability of the completion, almost natural by the citizens.

IT 5   Translation of Interview No. 5: Alice Comble

IT 5.1   Questions

▶ Ms. *Comble,* your project reinvents the collection of waste, creating a communication network at different levels, can you explain the targets to communicate effectively the *Environmental message* (focused on the importance of recycling small waste like the cigarette butts) and in the meantime raising awareness into the public opinion?

▶ Which are, according to your experience, the cornerstones of the communication from the point of view of the message contents relating to the addressees? Which roles are going to play the, as known as, new *Media* (e.g.: *Internet, Social Media*, etc.)? Are they going to make the *Environmental Communication* more effective?

▶ "You can have the greatest idea in the world, but if you can't communicate your ideas, it doesn't matter" used to say *Steve Jobs*, Co-founder, general director and Chairman of the Management Board of *Apple Inc.* Do you agree with this statement?

▶ Communication implies also an exchange of knowledge, what knowledge does *GreenMinded* transmit?

IT 5.2   Answers in a Nutshell

The idea to communicate the recycling while giving concrete and playful solutions is in my opinion necessary nowadays. It has been proven that repressive and moralizing solutions (e.g.: fines and bans) do not help any change in the long term. If we want to change the citizens' lifestyle, making not recycling "abnormal", we need to create an automatism and a fashion. In the modern society, the "common sense" has disappeared and so the mentality based on "I consume, I throw away" must change into "I consume, I valorise".

According to our *GreenMinded* project the cornerstones of the communication reside on a sweet mixture of seriousness and lightness. We want to give practical but shocking information without being moralizers. The new *Media* (*Internet*, *Social Net*, etc.) make the transmission of information in the world more effective.

I perfectly agree with *Steve Jobs* when he says that if we have ideas we need to communicate them because it is very difficult to make people familiar with a project or an idea if there is no communication and cooperation. So, thanks to Internet somebody is able to implement his project by himself, on condition that the message is kept alive afterwards, otherwise it will be forgotten.

The information conveyed by *GreenMinded* is about the real ecological impact of small waste, among which the cigarette ends and the actions that can be activated to overcome this type of pollution. Consequently, we intervene, in the universities, in the companies in order to raise awareness on the cigarette end impact. We stick posters and notices in strategic places, where there is waste disposal, using the as known as *"Green Nudges"* (sweet push) to achieve a sustainable lifestyle. We try also to give statistical data easily understandable such as: "a cigarette butt pollutes as much as 500 L of water, equal to three baths" or "each minute, in France, cigarette butts are thrown away as many as to reach the *Mont Blanc* top".

Finally, our terminals are able to supervise the percentage of answers and people's profiles who answered to our questionnaire understanding if they are smokers or not. Last but not least, we often give free small portable ash-trays and we organize collection of waste.

IT 6   Translation of Interview No. 6: Edoardo Croci

IT 6.1   Premise to Questions

Some recent studies on the state of art of the *Green Economy*, or *Circular Economy*, show a good position of Italy as far as the economic performance but a simple result in terms of value perception of the Italian "green trademark" still underestimated at an international level [see the report of Global Green Economy Index™ (GGEI), 2016].

IT 6.2   Questions

▶   These results, apparently contradictory, could they be an outcome of an insufficient apparatus of information and communication and national and European level? Which actions or *driving forces* would be necessary to improve the quality of the *Environmental Communication* and defend ourselves from *Greenwashing*?
▶   Which possible actions or solutions to improve the awareness of the *Environmental Communication* could be activated in Italy but also at EU level?
▶   Do you think that the tools like the *Environmental Social Balance* or the documents under the *Environmental Management System* (EMS) are still valid?
▶   Do you think that a tool as *Local Agenda 21* is still effective?
▶   Do you think that *Media* of *Environmental Reporting* should become biding for the Public Administration or for the companies in order to guarantee more transparency and a better access to the *Environmental information*? Do you foresee any new legislations or proposals on the matter?
▶   The Web Net is increasingly used as a tool of information and *Environmental Communication* to which is added the "boom" of Social Media use. According to you, could they improve the quality of the *Environmental Communication* or on the contrary be a problem, considering the lack of a suitable global regulation of the Web?
▶   How do you see the Italian and the international stage referring to the *Environmental Communication*? Which actions should be done to raise the awareness of public opinion on *Environmental issues*?

IT 6.3   Answers in a Nutshell

Italian position in the field of Green Economy is relevant for the Management of Waste Disposal and Renewable Resources (10 billion euro each of value). An important result achieved thanks to a system of incentives for renewable energies or *Environmental taxation* for the waste disposal. Italy has now a *leadership* in the *Green Economy* towards other Countries both European or international.

It is not by chance that Italy, at international level, is a historic leader for its *resource efficiency* and the capacity to use the raw materials in a very efficient way (thanks also to new *Circular Economy* strategies).

From the point of view of the *Environmental awareness*, we know that Italians (*data* from an annual survey by *Eurobarometro*, a European opinion poll firm born in 1973) do not have environmental sensitivity but only generic. As far as education is concerned starting from schools, *Environmental information and communication* are far behind. This is partly a public responsibility and partly of the companies.

Referring to the Public Sector, the most suitable means of communication is the *Environmental Reporting* linked both to the territories and different-level public administrations. In Italy, they have started talking about the State of the *Environment Report* since the Eighties (in Italy the Ministry of Environment was established in 1986). Different forms of *Environmental reporting* were thus born and so different *Environmental agencies.* Among them: *ISPRA,* Italian *Institute for Protection and Environmental Research*, at national level and *ARPA, Regional Agencies for Environmental Protection,* at regional level. Nowadays, there are new trends of eco-communication different from the *State of the Environment Report*, we mean the availability of Web tools and their continuous flux of monitored *Environmental data* and updating, outclassing the *una tantum* old data collection system.

Hence the need to think of new instruments communicating easily and quickly as "online estate registries" accessible to everybody.

On the other hand, we cannot help considering that most of the information services and *Environmental Communication* is managed by private agencies since the matter is still considered too technical and only for experts of the sector. It is therefore important, on the contrary, to reinforce its "public communicator" role. For these reasons, *Environmental Agencies* of the protection of *Environment* with a National Net System are being realized in Italy since 2016 (Legislation No. 132, 28th June 2016).

Another aspect to deal with is the *Environmental Communication* made by the *business companies* in the logic of *social responsibility* but also of the *marketing* linked to the market production. From the competitive point of view, the Centre for Economics, Energy and *Environmental Policy* (IEFE) at *Bocconi* University Milan (Italy) has made a specific study on *Environmental Communication* in Italy updated 2010. It showed that most of the cases based its main "claim" *on boasting* the qualities of the product or of the service. Not to talk about the cost-effectiveness of the product or service: *Environmental performances* mean savings.

It is always more evident the value of *social communication. Environment* seen as an element whose responsibility is to be taken by each Country.

Furthermore, the "claims" are not usually supported by a qualified certification at the national and international level to stress the objectivity. So, it is part of the Competitor Authority to be responsible of specific supervisions against any form of misleading advertising. In the last few years, newspapers have shown cases of false *Environmental performances.* Generally speaking, we can talk then of the real risk of *green washing*. It is therefore very important to invest on the *quality of the Environmental Information and Communication* to underline the importance of *credibility towards* the increasing effectiveness of the whole system and towards the citizens. Last but not least, the use of well-defined mechanism such as *EMAS Internal Environmental Statement* or other *EMAS* aspects linked to *ISO 14001* regulations or declarations related to the product statements or *Environmental trademarks* at a European level. It is then necessary to make use of codified instruments and verified information.

On this issue in Italy there has been an evolution in time, both for the available instruments of the Public Administrations and private business.

Nowadays we increasingly speak about *Environmental Balances* that is Communication Non-Financial Balances integrated with economic relationships on the part of any entrepreneur. These integrations have the forms of codified *reporting* and guidelines of the GRI—*Global Reporting Initiative.*

In Italy, we have tried to codify the *Environmental Accountancy* for a long time. The nature of the voluntary commitments made them to be perceived as rewarding factors both by the private and public sector.

On condition to accede to the international protocols such as *ISO*, *EMAS* and *Ecolabel*.

Nevertheless, the matter is in continuous evolution. In 2015, for instance, in U.S.A were approved by the UN the *Sustainable Development Goals* (SDG). Many companies then conformed to the same American sustainable goals. On the other hand, to speak of *Environmental accountancy* is already outdated by what happens at an international level (see Paris Agreements) and this need new forms of reporting and *performance indicators* not neglecting the Web use. There is a continuous innovation process.

Global and Local Agenda 21 (born in 1982 at the Rio Conference) has had a big success above all in Europe and in Italy it is still active a National Coordination Agendas 21.

Even though the path of Agendas 21 is constantly evolving.

There is no doubt that the *Environmental Communication*, but generally any form of communication, **invests more on the Web and Social Media** at the expense of the printed paper and even Television. We should reflect on the relationships between *peer* to *peer* established on the Net among *non-identifiable people* and so not reliable. Nowadays there is no world authority who can really guarantee the internet-users about the quality of what runs on the Net or *Social Media*.

To state the source of the information and any form of communication should be binding. The risk for the *Social Media* to become source-editor is therefore a threat.

The editor's responsibility of the press system to verify always the quality of information is then vanished. The "communicator" should be able to give the addressee the correct tools to judge the quality of the received message.

The *Environmental Communication* "scenario" is being changed and we can face what is as known as (aka) "*rubbish*" or "*trash*" with viral phenomenon like the "*cold water bucket*" (born with a good purpose but then distorted in its contents). It is essential to verify the context, the source and how the information is build up.

The paradox is that the political party should be motivated more than the Italian citizens as, for example, the *Eurobarometro* survey has proved with its survey at European level who are more sensitive to the *Environmental issues*. Some sectors, as the biological, are going to demonstrate that the "eco-sustainable spirit" is well in mind the consumers. The key point towards citizens is the *reliability* and *truthfulness* and *credibility* of the *Environmental information*. A further aspect is the one of the Italian *Environmental Policies*. Those who promote *Environmental actions* meant to change the citizens' lifestyle will become "*uncool*". These biases must be debunked because Italians are very open to changes on conditions that to their behavioural change corresponds an informative *feedback* showing the advantages of such eco-actions.

IT 7   Translation of Interview No. 7: Barbara Frateschi Moreno

IT 7.1   Premise to Questions

As Umberto Eco stated in the *Opera Aperta* (1962) "the literary and artistic work never stops being modified by his reader and spectator" providing different interpretations as a result of a dialectic between the form and the movement of the interpretation.

IT 7.2   Questions

▶ Mrs Barbara, do you agree with this statement? Your art production is strictly linked to the natural elements being inspired by the Mediterranean *Environment* which communicates through colour and light sensations. In your opinion, is pictorial art capable of influencing the audience eco-sensitivity, contributing to generate a personal "growth" towards a more sustainable life style?

▶ "The appearance of an image, independently to his "power" and effectiveness, 'affects' us, then undresses us" states *Georges Didi-Huberman*, art historian, philosopher and professor at the *Ecole des Hautes Etudes en Sciences Sociales* in Paris. This statement clearly describes one of the key elements of the *Environmental Communication* process, concerning the awareness or the new knowledge of the addressee by the communicator via the *Environmental message*. If you would like to transmit an eco-message through the pictorial art, what would you represent?

▶ Can you describe some of your paintings inspired by *Environmental* and sustainable values, offering us an "interpretative guide" on their ability to influence the audience's perspective on the natural world?

IT 7.3   Answers in a Nutshell

I agree with Umberto Eco's point of view, considering that a piece of art is the expression of a thought, a state of mind or a sensibility on a given matter that you would like to present and share with the public. The latter could then come to an interpretation different from the author's viewpoint, through his emotions, knowledge and maturity. A piece of art is able to raise the audience awareness on the greatness of the natural heritage. A good *Environment* has a positive effect on the personal development. The relationship with a healthy and harmonious *Environment* helps human beings to act properly in everyday life; a ravaged *Environment*, on the contrary, causes deterioration of the human quality. If an art work succeeds in stimulating emotional senses, both visual and olfactory, in its members of the audience, maybe it is always capable of encouraging their desire to preserve the wonderful treasure of Nature.

As already done in some of my paintings, I would probably represent the sea and the air, essential elements for life that should be more respected among the natural key resources. The pictorial language lends itself well to represent the sea which shows up through a rich colour palette, both boosting the artist to highlight its strength and helping him to enhance its beauty, while the air creates lighting effects reflecting on the water surface, through light overlaid and transparent nuances which reveal its purity.

The sea bequeaths us a huge heritage and it is our responsibility to preserve and pass on it to the next generation as a treasure of rare beauty.

The contemporary art has different ways to raise the awareness on *Environmental issues*.

The Land Art, popular in the Seventies (of the last century), put art and *Environment* together, pushing the Nature to be involved directly in the piece of art. The masterpiece made of natural elements, usually of huge dimensions, using the landscape as the art framework, was often located in a remote place difficult to achieve. *Robert Smithson*, American artist, was one of the greatest representative. "Spiral Jetty" is one of his finest piece of art, located at the "Great Salt Lake" in Utah (United States of America).

Smithson ordered to place tons of basalt volcanic stones, black in colour, creating a circular dock, approximately half-mile long, designing a spiral. Nature, then, as time goes by, took over and became the main artist. The masterpiece, submerged by the lake highly saline water, resurfaced occasionally 30 years later due to the water level. But salt and algae carried on the art work: the black dock converted into white, after its crystallisation, and the algae, the only living organism in such a lake, coloured the water pink and then green for some natural and sublime reasons.

Smithson succeeded in drawing the attention of the worldwide public opinion on this inhospitable and isolated location. Consequently, Nature can play the role of an improvised artist showing off its extraordinary adaptability. Thus, if we reverse and destroy a million-year-old ecosystem we will jeopardize our safety and the next generation survival.

IT 8   Translation of Interview No. 8: Maurizio Giani

IT 8.1   Premise to Questions

Besides managing one of the Tuscany leader companies in the treatment and disposal of Industrial Waste, you have been taking charge of the Communication Project *SCART*®, for 18 years, which was born from a very creative intuition. The idea that from anonymous industrial waste, exclusive art crafts could be born, an expression of art and design (e.g.: components of furniture, clothing musical instruments, sculptures, mosaics, paintings, etc.). To do it at its best you have conceived artistic projects able to involve Italian and foreigner creators and some Academies of Fine Arts teachers and students from Florence and Bologna.

IT 8.2   Questions

▶ What inspired this project with strong *Environmental value*? What about the young artists' response with giving new life to a material considered just a "waste"? What were their reactions from the communicative point of view (e.g.: they were inspired by forms from Nature; *Environmental issues* like pollution, Climate Change, proper waste management were pointed out; an educational-informative message meant to valorise the recycling was taken into account etc.)?

▶ The image scholars and the philosophers of the language agree on stating that each artistic representation in and of itself a "narration" that is it conveys a message involving, influencing creating new ideas. In one word: it communicates. Do you agree with this line of thought?

▶ Can you introduce some of the artworks made in the framework of the project SCART®, focussing on the *Environmental message* behind the artistic realization, the choices of the materials used, its use, if object of design, and what has inspired its shape?

▶ Do you think that the new Media offered by Internet (*Social Media* included) could help the spread of the works made in the framework of the project SCART® emphasizing the communicative potentialities? What could make the *Environmental message* linked to the project SCART® more effective? Do you have any new ideas for the future?

IT 8.3   Answers in a Nutshell

In our plants, everyday industrial waste of different types is smuggling, for its nature very different from urban waste and downgraded because out of production and not in line with the current trend anymore, our target is to give it a second life.

SCART® project originates from the willingness to communicate for a company that carries on a peculiar activity, sometimes demonised. We have chosen a language common to everybody: art. The *Trash Art* links perfectly our sector to art. Our intention was and is to raise to the nobility our sector through artistic communication.

Nowadays the *SCART®* artists network is constantly growing and thanks to them *Waste Recycling* is able for example to furnish a whole department in the Hospital *"Nuova Santa Chiara"* in *Cisanello, Pisa*, to prepare stage sets and costumes for more editions at the *Lajatico Theatre of Silence*, where every year the tenor *Andrea Bocelli* holds a spectacular concert.

The agreements with the Academies of Fine Arts of Florence and Bologna are an attempt to experiment and in the same time to give the students the possibility of working on projects from time to time different. The students have the faculty to interpret the materials at their disposal, selected by themselves in our platform of storage, always being sensitive to the theme chosen. Recently we have realized the scenery and the costumes for the *opera "The Barber from Sevilla"*; models of animals of natural size; paintings portraying *cult* characters and Christmas decorations for some of the Italian largest squares: everything made of recycled waste material.

Each time that the works of art created in the *workshop* by our young artists from the Academies of Fine Arts are introduced to the big audience the result is always amazing: when a material that could have rotten in landfill gains the content of the artwork and one-piece design, immediately in the audience the curiosity arouse and then that same object stops being what it has become by the artistic manipulation and it shows itself as a summary of the stories of all the materials it is composed of.

When the competition "*Let's give colour to the heart rhythm*" in cooperation with the *Lions Club San Miniato* and the *Academy of Fine Arts of Florence* was open, the number of participants surpassed sharply our expectations. I can say that was really a great success, also because the selected works, being permanently the furnishing of the department of *Arhythmology* of the Hospital "*New Santa Chiara*" *at Cisanello, Pisa*, are specifically intended for creating a visual interaction with the patients and all those that for different reasons are passing through the corridors, triggering the curiosity for the materials they are composed of. Each year, at *Ecomondo Fair in Rimini* (Italy) the magic is repeated and the industrial waste reborn in new shapes: shoe strings become a fox; shoe uppers, never used, become the eagle's wings; small coloured pearls give shape and light to a *Marylin Monroe*'s portrait. The Cuban artist, *José Yaque*, who works on the themes of the *Environmental Sustainability*, has been shocked by the big amount of materials abandoned by the present society every day and so for one of his installations located in the Centre of Expressive Activities of *Villa Pacchiani* in *Santa Croce on Arno*, he has decided to use thousands of shoes, arrived at the factory just before his first visit and destined to the shredder.

What inspires the piece of art produced in our labs is always the *Circular Economy*: the wish to put in practice the recirculation of discarded materials and for this reason destined for landfill sites, this principle is implied in all the works produced in the framework of the project *SCART*®. I am particularly linked to the scene costumes worn by the internationally well-known tenors for the performances at the *Theatre of Silence*, and also to the animals set in our stand at *Ecomondo* Fair, 2015 receiving congratulations even by the Italian Minister of *Environment*, *Gianluca Galletti*, and even more to the portraits of famous characters of our time, that were very appreciated by the public in the *Ecomondo Fair* edition of 2016 and they soon will be the subjects of a *travelling exhibition* in the museums of four Italian towns.

We have been present on the most important *Social Network* for many years because since the beginning we have strongly believed in the extensiveness of this communication system.

The majority of our activity of communication is developed on-line: on the platform www.scartline.it, for example, we collect SCART® artists' experiences, but our activity wants to leave a tactile as well as visual imprint.

IT 9   Translation of Interview No. 10: Daniela Luise

IT 9.1   Premise to Questions

The Local Agenda 21 process, ratified at the Conference of *Rio de Janeiro in 1992* on *Environment* and *Development*, which *Informambiente* team is mouthpiece of, proposes a model of local *governance* based on "participation", on "sharing" ideas and so in the "collaboration" of all sectors of the community to draw up projects devoted to disseminating sustainable practices and promote educational and informational paths on Sustainability (e.g. the regional *Rete INFEA*, etc.); "to know how to communicate" to the citizens and to the community is then fundamental.

IT 9.2   Questions

▶ On the basis of your professional experience, what cannot be missed in a Communication Plan to make it effective? Do you think that it is important to know in advance the *Environmental addressees* (e.g.: schools, citizens, administration officers, city administration employees etc.)? Which tools are more suitable to convey the *Environmental message* (e.g.: projects, conferences, *forums*, participatory meetings, courses, etc.)?

▶ The *Environmental "Communication"* is number one among the tools of information, awareness raising project and education: we need to verify the reliability of the sources on which it is based to avoid the *greenwashing* or to follow environmental theories, catastrophic and unfounded. Which tools do you use to "certify" the authenticity of the environmental message, thereby increasing the citizens' and stakeholders' trust? [e.g.: Environmental Certification/Registration, green Trademark products or services, etc.]

▶ How much does a good *Environmental Communication* addressed to citizens influence in decision-making (e.g.: implementation of municipality rules on "delicate" *Environmental issues*, as the building of an incinerator, rehabilitation of polluted sites, closing of industrial plants at high *Environmental risk*, etc.)?

▶ Which is the *Environmental Communication* approach in your education paths addressed to the new generations? How do you ensure the *Environmental messages* are really understood by the addressees? Do you use social networks (e.g.: Facebook, Twitter, Internet) Do you think that they are effective tools that help *Environmental Communication*?

▶ Do you think that the Italian Local Administrations nearer to citizens [The municipality of Padua mainly] have a key role in disseminating the Environmental messages, thus contributing to influence the lifestyles of the territorial communities? What should still be done?

▶ To communicate "Green": which are your challenges for the future? What values are you aiming at? Which subjects are you going to address to?

IT 9.3   Answers in a Nutshell

The Communication Plan must be defined from time to time according to the topic to deal with, the message, the addressees and the economic resources available.

The route to follow requires an initial thorough analysis of all the variables and the choices of the tools to be used.

ADDRESSEES: it is fundamental to start from the users group to reach in order to create a suitable message for age and social stratum.

TOOLS: there are not "better tools". According to the topic and addressees, beside the targets of education-communication campaign, the most suitable tools to achieve the goals must be identified. The best results are often achieved using *coordinated instruments*. The active involvement process (participated), though longer and more expensive, are those that give the best results because they create roots and a real change of all the age groups and interest.

WHAT CANNOT BE MISSED: In my opinion, an initial information phase and possibly an active involvement phase cannot be missed.

For example: in the case of *an ecologic Sunday* it is sufficient to inform about the event through integrated tools: Internet Website, brochures, *Social Media (Twitter, Facebook,* etc.). Nevertheless, traditional communication tools (press release, written reports, specialized magazines, etc.) are to be used as well. At this level, it is simple enough to plan an effective information campaign.

In the case of an information campaign by the public administration, addressed to the citizenship to decide for the "door to door" collection, it will be more difficult, of course. First of all, because we must communicate a compulsory *change of established habits*.

The *Environmental Communication* does not follow the same rules of the marketing. The purpose is not meant "*to sell*" something but to intervene in the people's life. A correct information allows the citizen, addressee of the message to assimilate the contents, understand it and finally internalize it. The change in the people's behaviour will be easier because they are aware of the motivations.

In decision-making by the Public Administration, the *Environmental Communication* **is to be considered** as the first phase of approach of a participated and shared process as wide as possible.

It has therefore a fundamental role to inform and update.

We do not use certifications or registrations which are expensive and they need a demanding route also at human resource level. We use clear and shared scientific sources for our data. Furthermore, to increase the citizens' trust we try to be clear and objective in the message we convey so as to be unassailable.

In the educational routes, generally we use participated approaches making the students feel protagonists of the route itself, the *Environmental Communication* has a main and transversal role. It is developed throughout the educational route and often directly by the characters of the educational project. So, the use of "Social Networks" are activated directly by the students and specifically for the educational project itself, obviously with the support of the educators.

The role of the Public Administrations (PA) in spreading messages has been so far underestimated. In this historical phase, the role of PA is in crisis and with it, all the other proposed activities. I think that in the *Environmental Communication* the PA can play a fundamental role not being a subject to the rules of the market without the need of "washing" with "green" messages what it is conveying. It is then essential that the PA keeps the birth right of communication and education.

As a PA, we address to all groups of age with messages and different manners according to the different themes to deal with.

A future priority is the theme of the Climate Change that involves different social factors:

- the public officers that need education to act correctly;
- young people: we will keep on proposing educational projects to the different age groups;
- the companies that are doing business in the territory, through educational and shared moments for projects to be carried on together also by the access to European funds.

IT 10   Translation of Interview No. 11: Elisabetta Martinelli

IT 10.1 Questions

▶ Mrs. Martinelli, what does "teaching Sustainability" and "communicating the *Environment*" mean to you and your team? Do you support the key role of Education in promoting *Environmental* and *Sustainable Development* values?

▶ Which are the main types of communication and favourite *Media*? How much are your actions linked to communication and to education-training? Which are the strengths and criticalities you have to face? What response did you get from the town community? How can you verify the degree of satisfaction of your services and the quality of the *Environmental information* conveyed both in teaching and communicating?

▶ Your Education Centre is part of the INFEAS (Information, Education to Sustainability) regional system, a model of cooperation between the public and private sector drawn up by the *Regione Emilia-Romagna* in order to promote, disseminate and coordinate the education to Sustainability actions, to achieve all this, how is the capacity of "Networking" important in spreading out the *Environmental message*?

IT 10.2   Answers in a Nutshell

The *IDEA* Education to Sustainability Centre in Ferrara, Italy, has been a benchmark point since 1998 and nowadays it is also a reference point for the Public Administration and for the support to the *Covenant of Mayors*, for the drawing up of the *Environmental Balance*, for the *Environmental Management System* (EMS) and for the Environmental Data communication.

"Teaching Sustainability" and "communicating the *Environment*" are the *missions* of the Centre.

In 2003, the first *Environmental Balance*, budget and final (years 2000–2001) was approved. It was based on the methodology of *CLEAR* (*City and Local Environmental and Reporting*) *Environment accountancy* and on *ecoBUDGET*, *Environmental budgeting*.

In May 2010, the Municipality of *Ferrara* (Italy) got the Certification of its *Environmental Management System* (EMS) consistent with the standard *UNI EN ISO 14001:2015,* successively updated in 2015. In 2012, Ferrara adhered to the "*Covenant of Mayors*" and in 2013 approved its "Sustainable Action Plan".

Therefore, I personally support the key role of Education in communicating the values of *Environment* and the *Sustainable development* as well as raising awareness in the community in order to create knowledge to face the future challenges.

The main goals of our Centre are:

- Promote into the community the development of knowledge, awareness and behaviours;
- Promote an education to Sustainability in the schools also with specific laboratories;
- Promote the collection and dissemination of information about the *Environmental*, social, economic and institutional Sustainability;

For our *IDEA Centre*, these objectives can be put into practice as follows:

- Submission of a yearly provisional training addressed to the schools;
- Development and management of local education and European Commission projects in the field of Sustainability in synergy with the whole Administration and the territory actors;
- Involvement of the local community and the schools into yearly events as a communication link of Sustainability;
- Development of the *Environmental accountancy* and sustainable indicators;
- Management of the thematic *IDEA Centre Library* including texts, documents and books on *Environmental issues*;
- Support of the public Entity cross-cutting activities such as: Certification ISO 14001:04 (revision 2015), the Covenant of Mayors, sustainable events, green public procurements etcetera.

Education, information and training are very linked and so we address to students and citizens directly with the lab activities of our provisional training, offering free courses like "*Teaching Sustainability*" and "*ActivECOlab*". *IDEA Centre* does not neglect the communication campaigns, the conferences, *workshops*, *work-cafés*, *info-points* and events where *gadgets* are given to those present, but also communication by local press and our Website. Some examples: during the National Tree Day, plants are given as present to the town community under the catch-phrase "a tree to reduce CO_2".

Nowadays, we cannot help using the *Social Media* and in fact *IDEA Centre* has a public page on *Facebook* to spread out strategically *Environmental issues*. Using *Social Media* in fact, as everybody knows, means to increase the transparency and the innovative relationship and the transposition of the citizen voice.

The strengths are those linked to the training activities with the youngsters and schools (primary and secondary high school) who are increasingly be interested in the *Environmental matter*. The criticalities, on the contrary, refer mainly to the citizens who show more or less attention according to the theme dealt with. Nevertheless, we must say that the community is more aware and participating. An example is shown by the big participation to the projects "*Ferrara mia*" (My*Ferrara*) promoted by the Urban Centre of Ferrara.

As far as the verification of the satisfaction degree is concerned, on the part of the recipients it is monitored through various indicators according to the number of participants, the *Web* or *Facebook* visits, the questionnaires given to the citizens who have attended our Lab or taken part to our activities.

Finally, the coordination and integration of the different educational experiences and programmes are fundamental as well as the relationships among different actors, the *stakeholders* present on the territory who can compare their efforts and commitments, encouraging a fruitful exchange of ideas and views especially by Networking. The latter is undoubtedly an added value for a collaborative and continuous improvement.

IT 11 Translation of Interview No. 12: Giulia Meloncelli

IT 11.1 Premise to Questions

"*Eliminate the concept of waste, not reduce, minimize or avoid it ... but eliminate the concept the same design through*" this is a quotation by the famous American design, *William McDonaugh* and by the German chemical, *Michael Braungart*.

IT 11.2 Questions

▶ Do you share those contents? Which and how many eco-messages do your works transmit? Which feedback do you get from your customers? Do you collect data for statistical purposes? In the affirmative, can you make an assessment?

▶ As already pointed out also at Community level: Eco-design together with energy labelling is one of the most effective strategies to promote the energy efficiency (it is estimated to contribute to 50% of energy saving by 2020) and push consumers and the market to invest in eco-efficient products. Can your creations of Eco-design contribute to reduce energy consumptions and the *Environmental pollution*? Do you use indicators as eco footprint (or similar) "to certify" the eco performance of the object or item of clothing you designed?

▶   How much does the "green" aspect of your design affects the recipient's life style? According to your experience, communicating the consumers-clients the environmental characteristics or inform them on the sustainability of the production chain, based on certification also, can it be an added value for an enterprise or start-up, increasing the sales? Why?

IT 11.3   Answers in a Nutshell

My career in the world of *design* **started soon**, before the end of my academic studies, in 1995. One of the most positive aspects of the Study Plan was the possibility to range in design at 360°, starting from the planning. Being able to experiment in the academic Lab the feasibility of your own projects, "getting your hands dirty", opens your eyes on the production issues and on the advantages and limitations of the materials to be chosen to make a piece. We can compare those years to the "re-History" of Green Economy and a very few businesses were involved in eco-sustainable productions. Having prepared a thesis on *"The waste as a resource"* in 1999, my commitment was to convert everybody to a production respectful of the *Environment*. Actually only a few clients allowed me to use *secondary raw materials* for the designed lines. Each time I was visiting the production department of a new factory, I could not help peeking into waste bins.

I used to ask: "But, how much waste a day do you produce?" And then: "Can I take this for some creative input?".

I was satisfied because, in my mind, the multi-coloured spurges from injections have already become trivets in *Gaetano Pesce* style but, at the same time, I felt sad for the enormous quantity of waste produced every day and unexploited.

Eco-design is an *avant-garde design* that breaks the grounds in favour of *Environment* without taking anything away from function and aesthetics, and on the contrary adding a very important *quid*: the respect for nature choosing to work with waste materials and the lowest polluting *Environmental impact*. Furthermore, an *eco-design* product is *the bearer of a story, an ideal, a lifestyle*, part of itself and I am firmly convinced that the dissemination of all that happens in the "backstage" is essential.

We must communicate the product in its totality. In the case of an *eco-design*, the information addressed to an audience that does not know is fundamental.

Communication is part of the product itself. Some examples:

A. "Eco-design bag in eco-leather with pocket realized from a sleeve jacket".
B. Eco leather bag realized with a swatch from the manufacturer that would have it weigh on stock or sent it to the incinerator. Using inhomogeneous material means that each piece is unique in the world.

Each of my products informs the consumer about the factory's philosophy and, in the meantime, he is involved in the contribution to the welfare of the Planet.

I fully share the quotation by *McDonaugh* **and** *Michael Braungart* that by eliminating the waste is as if we eliminated the creativeness of the design. A waste for somebody, may become a treasure for another.

My products are bearers of different eco-messages:

1. *do not waste*
   – the 98% of our materials are from obsolete stocks or used;
2. *critical and conscious consumption*
   – we incentivize the customer to the recovery of their own objects;
3. *production with low Environmental impact*
   – waste is then recycled; the cloth washing is with ecologic detergents; the promotion uses FSC (Forest Stewardship Council) recycled paper;
4. *healthy and ecological workplace*
   – we chose the green building for our location; a non-poisonous vegetable garden and a garden full of colours; everywhere low consumption lighting indoor and especially outdoor in order not to be harmful to small animals and nocturnal insects.

All these elements are liked by our customers that show their appreciation through messages on the *social network or by mail.*

Our clients appreciate the craftsmanship and the "tailored". So, I would like to finish with Mark Victor Hansen's quotation which I share:

> *"Garbage is a great resource in the wrong place that lacks the imagination of someone to be recycled for the benefit of all"*

My target is *"making with what has already been made"*. Consequently, the eco footprint is very low even though transforming the materials made by others it is not always possible to certify the E.C (Ecological Footprint) since we do not know the geographical origin or the productive processing.

RICICLI is a brand **appreciated by two categories of people**: the former is fascinated by the design and appreciate the functionality and aesthetics. The latter, on the contrary, appreciates the fact that unused materials have been recycled by Italian labour.

I am fond of my job, and I have been in love with Nature since I was a child.

Therefore, the eco-sustainable aspect of *RICICLI* products wants to make clients, without *Environmental impact priorities*, reflect on "another way of consuming" products. Those who are already aware of this are gratified by the contribution to reduce the eco-footprint.

Last but not least, I think it is essential to communicate the consumers the aspects of the production chain and the origin of the project, to witness that the products are the result of a study and recycle and not a mere commercial operation that sometimes may hide a camouflaging of a cheap fake produced by industries abroad into a *"vintage"* object.

IT 12   Translation of Interview No. 13: Paola Poggipollini

IT 12.1 Questions

▶ "The good organization and the modernisation of a public administration require an efficient communication plan",[1] do you agree with this statement?

▶ In your managerial experience in the Public Administration, focused on Sustainability, which communication tools do you consider more efficient to communicate the *Environment* to the public? Can you give us some practical examples?

▶ The public opinion is increasingly more careful of *Environmental issues*. This has pushed Local Public Administrations to set up *Environmental management tools*, the so called *Environmental Management Systems*—EMS (ISO 14001 Certification/ EMAS Registration). New informative documents were born to "communicate the *Environment*" inside and outside the Municipal body; you were the witness of the *green* path that allowed the Municipality of Ferrara to be certified ISO 14001, in May 2010: one of the first example of Italian medium-large-sized urban centre to achieve this target, what can you tell us about that? Do you consider the EMS as useful tools to disseminate the *Environmental message*?

IT 12.2   Answers in a Nutshell

Yes, I agree with the drawing up of an efficient Communication Plan. An organised and modern public administration implies a demanding professional commitment aimed at defining a strategy at Municipal level, including medium and long-term objectives on territorial management. It is essential, then, to establish a due communication plan to disseminate the agreed *goal settings*, addressed both to the employees and to local stakeholders and citizens.

It really depends on what you want to communicate. If we are referring to simple pieces of information, *Media* like: Internet, e-mails, SMSs, newsletters, press releases, local television or radio programmes are sufficient to provide information.

If we would like to interact with citizens, pushing them, for example, to separate their waste correctly, we need to use more direct communication tools such as: informative meetings, distribution of leaflets in commercial centres, *focus groups*, house-to-house brochure dissemination, carried out by staff in charge to provide accurate information.

In order to involve citizens in the Municipal decision making, it is useful to organise participatory meetings, focus groups and workshops, implementing new methodologies like the open space technology facing a large audience.

---

[1] Rizzo/Bordi, "*La comunicazione istituzionale sul web*", IlSole24Ore, 2009, pages: 106.

I am referring, in particular, to the *Local Agenda 21* workshops involving local *stakeholders* and citizens in the settlement of agreed objectives to encourage a *Sustainable Urban Development* at *Environmental*, economic and social level.

I am thinking also to the participated District Programmes project which acted as a best practice in testing the participative budget. Last but not least, the Communication Plan linked to the dissemination of the Municipal *Environmental Management System* and the Energy Plan.

Agenda 21 **(A21) made citizens reflect on *Environmental matters* and form their eco-conscience in town management**. Nevertheless, A21 risks being forgotten if it is not supported by an effective communication plan and best practices promoted by the Municipality.

Furthermore, the reporting tools, such as *Environmental Budget Balance* or Eco-budget are efficient methodologies to give an overview on the results achieved by the stakeholders at *Environmental level*.

The *Environmental Management Systems* (EMS) involving the Local Public Administration are useful *Media* to promote the process of simplifying administrative procedures aimed at legalising and adjusting the Municipal structures to safety and *Environmental standards*, energy efficiency and green procurement, boosting a more transversal analysis in problem solving solutions in terms of Sustainable Management. Tools as *Environmental* or *Sustainable Budget Balance* should be implemented and integrated into the economic Governmental and European balances, since they include important result indicators on the Programmes in the implementation of sustainable policies, notably with regard to the Italian level.

The *peer review* is a very valuable free of charge tool to improve the *Environmental Reporting Techniques*. Unfortunately, it is not often used by the Local Public Administrations.

The adoption of *Environmental Management Systems* helps Local Public Administrations behave in a *greener* way.

This commitment testifies the real willingness of practicing and promoting *Environmental best practices*, an essential condition to lead citizens and stakeholders to put into practice virtuous sustainable actions.

IT 13   Translation of Interview No. 14: Carlo Ratti

IT 13.1   Premise to Questions

Carlo Ratti Associates, International Design & Innovation Lab, based in Turin with branches in London and Boston where he is the manager of the MIT *Senseable* City Lab representing an excellence with an Italian "heart" for high-tech projects in the field of architecture, urban regeneration and design.

IT 13.2   Questions

▶   On the basis of your multiannual experience how much the new technologies have revolutionized the *Environmental language*? And how much can they affect it?
▶   Should you make a priority list to make communication more efficient and suitable to the modern language "codes", what did you range in the first five place and why?
▶   What *Environmental messages* did you want to convey to the public at Milan EXPO 2015 through the "Future Supermarket" project realized for the Future Food District by your team? What was your source of inspiration? Do you think that the "future supermarket" can soon become the "present supermarket"?
▶   The development techniques of Internet Things allowed you and your team to create *Environments* and objects able to interact and often "understand" the medium-large sized community's needs. We are thinking of the "Office 3.0" project for *Agnelli Foundation* 2016, or *HubCab* or *MONiTOUR* projects presented at the States-General of *Ecomondo*, the Green Technology Expo 2016, Rimini, Italy. *Environmental Communication* is for its nature an interactive circular process which implies an answer by the recipient, how do you verify that it is correctly understood? How much does a suitable education-information to the addressees weigh in the design phase?

IT 13.3   Answers in a Nutshell

In recent decades, the digital technologies have already become a part of our lives hand in hand with a steady eco-conscience, at social and political level. I find this coincidence very interesting. Thus, I am convinced that the digital technologies can help us in implementing a better relationship with the *Environment* we live in, especially if we will be able to use and manage in a proper manner the data provided by these devices.

Few months ago, we launched *Treepedia* project (http://senseable.mit.edu/treepedia), a digital platform to map the tree canopies, that extraordinary "green curtain" whose role is fundamental to guarantee the urban wellbeing.

I would use a proactive approach open to the future, avoiding any menace or catastrophic scenarios, so far as the reality can be serious. I am a great admirer of *Edward O. Wilson*, the American biologist who proposed, in the Eighties of the last Century, for the first time, the so called "*Biophilia Hypothesis*", under which human beings are "programmed" to feel happy when they are plunged into natural elements.

To conceive the *Future Food District*, we were inspired by the image of the Signor *Palomar*, a character of the Italian writer *Italo Calvino* who, immersed in a Parisian *fromagerie*, feels like being in a museum or inside an encyclopaedia: "Behind each cheese there is a different pasture, a different nuance of green under a different sky (…). This shop is a museum: *Signor Palomar,* while visiting it, feels at the *Louvre*, sensing in every object the presence of the civilisation that gave its shape and from which it takes shape." On this basis, our project took shape: trying to use new tools to make products being able to tell their story. A greater traceability of products which allows you to build up new relationships among people.

A suitable education-information is essential. Nowadays, a new "open source" paradigm is born, as stated in a recent book published by Einaudi [Ratti Carlo, Claudel Matthew, *Architettura open source—Verso una progettazione aperta* (Open Source Architecture—towards an open design), Torino, Einaudi, 2014, pages: 142]. We like to think the projects are an open code, to be carried on thanks to the active contribution of a multidisciplinary team which includes the final users. We wish we could convert the "*archi-star*" idea into a "choral architect", able to harmonize different voices in one consonant chord. Consequently, it is very important to communicate citizens a direct line going in two different directions.

IT 14   Translation of Interview No. 15: Niccolò Ronchi

IT 14.1 Premise to Questions

The "green market pressure" is involving the music world.

Sustainable music concerts and festivals are spreading at European and International level, confirming this trend. Being "green" means not only the eco-planning (e.g.: energy efficiency, waste eco-management, etc.) and a suitable eco-location (e.g.: urban gardens or a protected natural area) but also a "green" music event, such as the Festival Øya at Tøyenparken (Oslo, Norway).

IT 14.2   Questions

▶   Maestro, which past, present or future pieces of music (including yours) would you choose to communicate effectively the *Environmental message* on the main issues like: the safeguarding of the biodiversity, Climate Change and the protection of natural vital resources?

IT 14.3   Premise to Questions

Music has been more than inspired by the natural world, translating the natural sounds into the "music language" or even introducing some original natural tunes within the piece of music, thanks to the new technologies. Some pop-stars became an icon of the eco-message (e.g.: the Icelandic songwriter *Björk* or the American Group, *Maroon 5*). Nevertheless, the "green" trend in music is rooted in the classical music. Let us think of the baroque "Four Seasons" by Antonio Vivaldi, you have already mentioned, or by *Ludwig van Beethoven* (e.g.: Simphony No.6), by *Claude Debussy* (e.g.: "Deux Arabesques" and "Moonlight"), by *Frédéric François Chopin* (e.g.: "Prelude in D flat", known as the *Raindrop Prelude*) or more recently by *Edvard Greig* (e.g.: "Morning Mood" composed for the musical work Peer Gynt) and by *Nikolai Rimsky-Korsakov* (e.g.: "Flight of the Bumble Bee").

IT 14.4   Questions

▶   Do you agree with this point of view? Do you think these pieces of music could act as *Media* being used in awareness campaign on *Environmental* and *sustainable issues*? How did you place them in order to "modernise" their significance for the next generations?

▶   If you were asked to compose a piano piece of music (an instrumental solo or music and lyrics) in order to promote the preservation of the ecosystem, of which Man is a part, inspired by the principles of Sustainability as defined by the World Commission on *Environment* and Development, named "Our Common Future" (*Bruntland Report*, 1987), which kind of music would you suggest?

IT 14.5   Answers in a Nutshell

First of all, I believe that the protection of the biodiversity and of essential resources such as Water, just like Climate Change, **are extremely important issues** which derive from the macro-concept of "Respect". Respect for other people, respect for the different, respect for something that does not belong to us and, last but not least, self-respect.

As time goes by, after a 25-year devotion to music (I am turning 30), I realise that studying music is a great education for human life and an instrument through which to raise awareness on the concept of "Respect". Learning polyphony is essential, from the outset. Within the same musical chord, played at piano by the same player, we can rarely find the same intensity in every sound. And, if it happens, we have to respect the specific feature of every single sound within that chord in order to find the right balance able to create harmony, metaphorically and acoustically speaking. Examples from music are countless. Let us think of a musical form like the "fugue" where the polyphony is decisive, highlighting or weakening any single voice to achieve a clear and harmonious result. Let us think of what happens in the orchestra where not only "fugues" but also sections of orchestra have to respect each other, with specific peculiarities and characteristics widely differing and often diametrically opposed. Being less philosophical and more "practical", many musical compositions are able to raise the public awareness on *Environmental issues*. As far as Water is concerned, for example, some of the greater composers from the past have composed descriptive "programme music", in an effort to create suggestions attributable to that precious liquid element.

I am referring to "*Jeux d'eau*" by *Maurice Ravel*, to "*Jeux d'eau à la villa d'Este*" taken from "*Years of Pilgrimage*" by *Franz Liszt*, to "*Reflets dans l'eau*" by *Claude Debussy* or to the well-known prelude "The drop of water" taken from "*Preludes op. 28*" by *Frédéric Chopin*. Musical works whose title is directly linked to Water. Not to mention the work for orchestra "*La Mer*" by *Claude Debussy*. Or the symphonic poem "*The Moldova*" by *Antonin Leopold Dvořák*, composed to celebrate the beauty of the River Moldova and its saga, from its sources in the Bohemian woods to the entering on the River Elbe to the mouth, in the North Sea.

Furthermore, a large number of pieces of music are inspired by "Nature". The great **Ludwig van Beethoven** was positively obsessed by the natural world since he considered Nature an element of "goodness", metaphor of that God who was able to inspire Human Beings and Humanity pushing them to an inner exploration aiming at reinforcing the strength of the human soul.

Moving on our music journey inspired by Nature, we should mention some more or less familiar pieces of music: the famous "*Four Seasons*" by **Antonio Vivaldi**; "*Peter and the Wolf*" by the Russian composer **Sergej Prokofiev** where the animal sound is associated with some of the orchestra music instruments for their timbre; the suite "In the Open Air" by **Béla Viktor János Bartók**, the "*Sacre du Printemps*" by **Igor Strawinskij**, "*The Chant of the Sun*" by the composer **Sofija Gubajdulina**, The "*Seasons*" op. 37a by the Russian composer **Pyotr Ilyich Tchaikovsky**. I am referring, then, to the Northern Europe music tradition such as **Edvard Grieg,** its leading representative who was deeply inspired by the natural, halfway between the magical and the real. Let us think then of "*The morning of Grieg*" or other fantastic compositions like the "*Lyric pieces of music*". Last but not least, a real anthem in honour to the biodiversity : the "*Catalogue des oiseaux*" (The Catalogue of Birds) by **Olivier Messiaen**. His passion for birds was so strong that he used to consider himself an ornithologist more than a composer.Music and Nature. Nature and Music. Moreover, Man is Nature and it is difficult to imagine the artistic human creativity without a direct link, more or less consciously, to his "Divinity" to which he belongs.

How to involve the whole of humanity to make it clear the essential role of every individual commitment in respect of the *Environmental* and *sustainable* issues? Nowadays, fortunately, some pop musicians, maybe closer to the general public, have become icons of the eco-message. I think this is a good sign, unless I would like to make a clarification. According to public opinion, classical music is anachronistic, or even "old-fashioned". I personally think that people confuse the term "old-fashioned" with "immortal". Each of us comes into contact to a *music cell*: the heartbeat. The Heart. That rhythmic element represents Life. The Life is strictly linked to the Planet Earth we live in.

Consequently, an awareness campaign expressed by music should be set up as an Ode to life. *Beethoven* can be related, then, to "*Giardini di Marzo*" (Gardens of March) by the famous Italian songwriter *Lucio Battisti* without distorting both initial musical structure. We live in a world where *chaos*, noise and visual pollution which we are committed have reduced our ability to perceive with our senses. Other forms of art, then, should be involved: photography, painting, filmmaking and writing.

We should create something that bring together all these Beauties, creating a union of communicative strengths capable of influencing the audience and stopping the general insensitivity of this chaotic world.

The music composition, as the outcome of human creativity and sensitivity, belongs to that circle of human expressions that are able to influence human beings. Therefore, I would think to a balanced piece of voice and instrumental music. That is because the human voice is the tool to which anyone else took inspiration and tried to reproduce its endless nuances.

I wouldn't choose any specific language but the mere reproduction of sounds. Finally, I would choose a kind of music which would be a classic music "form" but an *emotional box*. For this reason, I would characterise the piece of music with strong contrasts: high-pitched and deep sounds; and then dynamics from the *pianissimo* to the *fortissimo*. This is important to tell everybody that diversity is not a problem but it adds value. At last, I would conclude the composition with few sounds able to put human beings at their easy, making them focus on soul vibrations, and nothing else. We might forget ourselves to be part of the whole thing, returning to the origins. In my opinion, there is a piece of music that elicits a similar emotion. That is *"Spiegel Im Spiegel"* by **Arvo Pärt**. Touching beauty and anything else: this is what everyone should feel contemplating the wonders of the World.

IT 15   Translation of Interview No. 16: Antonio Salinari

IT 15.1 Questions

▶ "There is need of formal beauty and emotional involvement if we want to be connected to the citizen, consumer of culture, performance and narration" reminded us *Erik Balzaretti*, and *Benedetta Gargiulo*. When the artist deals with the *Environment* theme he cannot help limiting to his "personal feeling" but he must necessarily arouse a positive emotion to be understood at a global level. Do you identify in this statement?

▶ What are you inspired by for realizing your works? Do you use symbolic elements or other forms of communication?

▶ Which kind of emotions do you feel when creating a new work?

▶ Why the choice of wood to make your works?

▶ Which emotions would you like to arouse in the audience?

▶ How does the artist influence the matter?

▶ Parallel to you own activity, you have set up a Laboratory addressed to young people that allows many youngsters of the *Susa* Valley to experiment the art of shaping wood and create artworks, which *Environmental messages* are you seeking to convey through this artistic and educational activity? What about the feedbacks from the students?

▶ How can we convey an *Environmental message* through an artwork, the result of the artist's creativity? Can you give us some examples?

IT 15.2   Answers in a Nutshell

Generally, I share the necessity that art must arouse a positive emotion and that it must be understood at a global level.

I think that in order to move in a positive way, Art should be a narration and have a strength creating a sense, involving the audience in the identification and in the interaction with a value, whatever it may be.

I also think that, to be understood at a global level, Art should rely upon a representation: all the narration has *a meaning* and *a form*, if the latter is *visual art* become more intelligible.

According to André Derain's thought "we *should have deeply penetrated the life of things that we paint*". For me the main sources of inspiration are my experience, the technological background and the passion for nature. More specifically I am inspired by my personal experience; the necessity to rebuild some values of reference; the beauty in general and, in particular, that of Nature. Here I mean the concept of the Greek beauty: *kalogakathia*, literally "the beautiful and the good", as a value that creates emotions.

When I create my works, I try to tell a story through those extraordinary differentiators that are the symbols, the metaphor and the matter.

Furthermore, I think that also the working process, based on manual techniques and on meaningful timing of completion of the art piece, may represent an expressive gesture in communication.

Giving shape to a thought, above all because it was born from an inner urgency to express ourselves, it is always an extremely rewarding experience.

Equally gratifying it is to use your hands to build something that before it did not exist, smelling the wood and feel the tactile sensation one gets when caressing the wooden surface of the works.

I would lie to make people reflect on the value for Man to find again his collocation in the Nature, being a part of it.

Wood is a living, warm, renewable material and I think it is highly symbolic in representing Nature: it expresses the strength of building beauty. Particularly in the project "*search for stability*" it is the harmony of precious woods coming from different countries of the world, assembled together and sculptured, predominating on the form, confirming Marshall McLuhan's thought: "*medium is message*".

I would say that between wood and artist there is a mutual influence, an interaction: on one hand the sculptor gives form to the matter, on the other hand the matter influences the craft, with the designs created on the surface of its grains, essential for the beauty of the work itself. Grains that, sometimes, lead the artist to "revise" the form to follow them.

The development of creativity, the use of manual techniques of processing, the search for natural materials, the necessary observation of Nature promotes a healthy *modus vivendi*, that encourages to spend more time in contact with the *Environment*.

All that with the purpose to generate an empathy with the Nature itself and an active role by the young people. Thanks to these activities and to the context I can appreciate, on the part of the young, an increasing respect for *Environment* and an attitude to valorise the available resources.

Therefore, I prefer to convey messages on the need to be part of Nature. For example, the artwork "*1938—Metamorphosis*" depicts the charm and the wonder of Nature, through the story of a hypothetic dream of *Escher*, which then would result in his masterpiece "*Heaven and Water*" of 1938. It represents the concept that not everything is in harmony with Nature.

In "*Awareness*" the man's ability to identify himself with the other living beings, an attitude becoming a fundamental aspect to help the change of mentality on *Environment*, is tested.

The last work refers to the Greek philosopher *Anaximander*; he thought that at the beginning everything was harmoniously together in the "*Apeiron*", which for him represented the universe, until, on account of Man, this cohesion broke.

In the culture the precious woods from different continents, symbolising also the cultures and the peoples, merge in one classic form.

IT 16   Translation of Interview No. 17: Omero Soliman

IT 16.1 Questions

▶ How important is the *Environmental Communication* in your job? To whom is it addressed?

▶ Do you think that the recipients of the Communication Plan are adequately prepared to receive it? What should be done to improve the understanding of the *Environmental messages*. What can "disturb" the correct interpretation of such a kind of messages?

▶ Do you think the stakeholders' role is important in disseminating the *Environmental message*?

▶ Command and control (legislative instruments) or voluntary instruments like Eco-design, Eco-label, Energy-label, and EMS (*Environmental Management System*), which, according to you, make the approach more effective to communicate the "green" issues?

▶ What are your future perspectives? How do you want to develop your Communication Plan?

▶ Do you think that the *Social Media* as known as social network like Facebook, Twitter, Instagram, etcetera could be reliable instruments of *Environmental Communication*?

▶ What about the advert campaigns conveying *Environmental messages*? Are they effective enough? What do you think should be an essential restraint to protect the addresses from the greenwashing?

IT 16.2   Answers in a Nutshell

Nowadays, more than ever, the *Environmental Communication* in my job of architect is extremely important, an absolute priority. This is also because the holistic vision has influenced my "being an architect". A deep path whose roots are to be found in respect and love for our Mother Earth. For this reason, my *Environmental Communication* is addressed to the community. Unfortunately to deal with scientific topics linked to the *Environment* is not always easy so the expected targets sometimes are not so immediate.

Nevertheless, it is a steady commitment to inform the citizen with a simple, clear and understandable language but in the meantime very effective to stimulate the recipients' innermost sensibility toward the *Environment*. An awareness that is often neglected because of our chaotic everyday life that distracts us. It is then fundamental that some guidelines are established. In the Communication Plan, there should be a clear identification of the targets; a strategic planning and above all a fluent network of relationship with the stakeholders. Finally, it is important not to underestimate the verification of the results and their impacts on the recipient.

Luckily, we are noticing that the attention to *Environmental themes* is "awakening" despite or because of social problems. I think the recipient who I am addressing to is gaining the right maturity.

As far as I am concerned, I got more Responsibility, trying to make the recipient more involved in my *Environmental commitment*.

For example, we are used to sending a questionnaire to collect information about his degree of satisfaction, about the quality of our services and asking suggestions to improve. We also update our customers by our newsletters on new regulations came into force or about programmes our Holistic Firm is going to promote.

Inside a well-established net of relationships, the stakeholders become strategically very important. They are the "glue", the bridge between the links and the necessary feedback to understand the right policy to adopt. All of them are fundamental: from the traditional suppliers and clients to the modern investors, partners, the social community itself, the press and the *Media*.

If we are to choose between legal instruments (*command & control*) and voluntary actions (Eco-label, Eco-design etc.), first of all we must say there should be a balance both sides. They are two different approaches that compensate each other. Of course, we are more linked to the Eco-design which is our favourite.

As far as our future planning is concerned, our first example is to give a constant commitment on the field. As an architect, I am always concerned with the best way to use strategies and find eco-sustainable solutions in all my designs. Consequently, our Holistic Architectural firm has declared his "mission" publicly. Tangible examples of our *Environmental commitment* are: to lower the costs of corporate travels using video-conferences instead, thus reducing the CO_2 emissions; to choose those restaurants with "climate dish" that is the restaurants using regional and seasonal ingredients from "0 Km" bio-farms.

In our Holistic Architectural firm "reborn" in 2015, thanks to the holistic vision we have implemented the Sustainable research even more with new methodologies. One of our goal is the "welfare housing", a high-level comfort for the health of its inhabitants using all the most advanced safety and hygiene-sanitary devices. Our holistic protocol accompanies then the final subject from the domestic *Environment* where he lives to the materials used for the building, the place it has been built and its orientation. But also, the way the inhabitant nourishes himself and the essential use of water. Last but not least the "thin energies" are to be considered, which are still difficult to measure but they are *Media* able to communicate the *Environment*.

Furthermore, we think that the *Social Networks* are very powerful communicative tools but in the same time extremely delicate and fragile. Their strength is "to speak" with people. This aspect implies the choice of the communicative language that must be very weighed, measured, designed, targeted and effective. But its capacity of interacting directly with the addressee is very valuable, being able to monitor his opinion and his needs over the time. The other side of the coin, of course, is to avoid mistakes in communication: only one inaccuracy can compromise the results forever.

Therefore, the *Social Media* must be used wisely and with the correct preparation.

For example, it is not so easy to realize a publicity campaign on *Environmental issues*. First of all, we think that the *Environmental message* should take into consideration long term interests linked to the Sustainability at global level. Secondly, they should be meant as lifestyle towards progress without damaging or exploiting the precious resources of the territory to be maintained for the future generations in the short term. Thirdly, the quality of the services able to save time, to offer advantages and to satisfy the client's requirements. This communicative approach usually inspires confidence and quiet in the addressee. Finally, we can say that the dual vision of the *Environmental sustainability* for the future generations and the satisfaction of an immediate service make the advertising campaign always very effective.

IT 17 Translation of Interview No. 19: Joaquim Tarrasó Climent

IT 17.1 Premise to Questions

"The world will not evolve its current state of crisis by using the same thinking that created the situation" stated *Albert Einstein* highlighting a key aspect of communicating and informing on *Environmental issues*: the continuing need to develop in order to adapt to new "linguistic codes", those belonging to the *Circular Economy*, the *Social Media*, the *Internet of Things*, etcetera. In this perspective, *Environmental Communication* is a multimedia process that aims at presenting new environmentally-friendly values.

An informative, educational and creative "strategy" which allows to face most of *Environmental matters* consciously and responsibly. It is also an ever-changing "tool" able to express both orally and through non-verbal *Media*: shapes, symbols, materials and any other natural element which affects human beings can "revolutionise" the citizens or countryside lifestyle changing the public opinion's point of view on "ecosystems".

IT 17.2   Questions

▶   Mr. Tarrasó, according to your multiannual experience in realizing architectural projects at an international level, which are the driving forces of a bio-architecture based on coherent principles of Sustainability? How can you communicate efficaciously the stakeholders the benefits of an urban eco-system, being designed, starting from a business oriented target? Can the design of a public building, of a square, or of a public garden "send green messages" to the recipients and users? In your opinion how much is the written and oral communication relevant in transmitting the *Environmental performances* of an architectural works as far as the visual appearance is concerned?

IT 17.3   Premise to Questions

Sustainability is a concept which influences many aspects, including the everyday object design, although of decorative nature. Thus, the eco-design goes further than being "eco-friendly" since it affects the whole production chain such as: the manufacturing process, the future scenarios of the "product life cycle" (more and more circular), the innovation and the creativity. All that with the view to act more responsibly.

IT 17.4   Questions

▶   What does the correct interpretation of the eco-messages linked to the architectural projects and eco-designs depend on? How can the expenses affecting the price of ecological products be reduced?

IT 17.5   Answers in a Nutshell

With reference to your questions, I thought to divide my answers in three different parts: (1) Designing; (2) Graphic Communication; (3) Project carried out.
   Project design and its development: There are different ways to communicate at this stage. According to the experience acquired in all phases of the project cycle,

I can tell you that it is convenient and interesting to involve the *stakeholders* from the earliest stages. This target can be achieved at various levels. First of all, through the participatory process. This could be a great opportunity to analyse, or better to gather information, to understand the project conditions and even the parameters that may help developing solutions to meet requirements, questions and other aspects. A participative process could represent also an opportunity for an *Environmental design project*.

It serves as a tool to involve and share the same process. You cannot consider an architect as an external subject, making individual solutions. Being an architect means, on the contrary, to involve all the stakeholders, not just clients, and this approach is beneficial. The participatory phase is an essential concept for design development. A further element to be considered at this stage is the context. This is a necessity which is always linked to specific aspects in all respects. At user level, for example, it would be useful asking: which subjects use or will use a specific architecture work under construction?

Moreover, there is an additional element of the greatest significance from the communication point of view to implement high *Environmental* and landscape value architectural projects: the *added value* embodied in the *Environmental elements* at each step. Architectural impacts on the ecosystem and on the social background has been focusing attention for many years. In my opinion, this is a way to raise the eco-awareness and to encourage eco-communication.

Graphic Communication: during this construction phase both at technical and graphic level, we must take into account all the implications in terms of communication. In the field of architecture, there are different *Media*. I am not referring to verbal communication tools, only, but to the graphic design. The typical way to communicate for an architect, his "language" is the graphic code which could be a "powerful weapon". Instead of different kind of verbal languages, architects use architectural design, diagrams and graphs.

In the project implementation phase, graphic designs are conceived and managed for a purpose other than respecting technical and legal requirements or seeking the Municipal authorisation which remain key objectives, anyway. They play a crucial role to transfer that knowledge able to convert the draft design into a real architectural work.

When we were talking about various systems overlapping the same context, we were not referring only to diagrams depicting some *Environmental cycles* but to real *Media* capable of interacting with users making them aware of the surrounding natural *Environment*.

Project carried out: At design stage, it is always important to reflect on the direct correlation which joins the final user to the surroundings. We are not only looking at the audience but also at the users. I think this is something crucial to take into account, as many projects testify (see Case Study).

Certainly, decoding an *Environmental message* depends on several factors such as: the pollution level reported both in urban and industrial areas. In this latter case, the metals concentrations dispersed in the *Environment* may be relevant which requires specific technical solutions. Furthermore, even in the home *Environment* there are substantial sources of pollution as the noise. Thus, it is difficult to establish a relationship between an architectural project and the required costs.

I am not able to quantify the *Environmental expenses*.

Apart from that, it is essential to do a complete analysis on all the architectural conditions in a given context, examining what we need, which are the priorities and the key elements helping us to achieve workable and sustainable solutions. When I am working on a project, I consider communicating with colleagues and other *stakeholders* a crucial part to better understand any *Environmental matter*.

Architects, even if they are not eco-experts, are in direct contact with the *Environment*. Consequently, we always cooperate with environmental communication advisers and many other professionals like: hydraulic engineers, traffic management experts, biologists, anthropologists or sociologists when cultural aspects are involved.

IT 18   Translation of Interview No. 20: Paolo Taticchi

IT 18.1 Questions

▶   Professor Taticchi, you have been gaining a multiannual experience on the business-economic sector, education and sustainable strategies, the goods and services supplying chain at international level. Which role does the *Environmental Communication* play on the Green Economy? How much can it influence the consumers'—citizens'—users' choices?

▶   How much the know-how to communicate the *Environment* weigh on the market? How can we make the information and the *Environmental data* of technical nature underlying the sustainable commitment in the chain production and service delivery be accessible to the big public? Can you give some examples on the matter?

▶   The voluntary tools for the *Environmental Process Certification* such as: *Environmental Management System* (EMS), ISO Norms, etcetera and of product such as: eco-label, energy label, energy star, Sustainable Product Certification etcetera, implying *Environmental Communication* actions, can they really guarantee the consumers on the quality of information when buying a product or a service?

▶   The setting up of an *Environmental Communication Plan*, can it have positive effects of selling prices and turnover?

▶   To know the stakeholders a priori that is: suppliers, transporters, consumers, outsourcing operators, competitors, companies, etcetera, can influence the success of an *Environmental Communication Plan*?

▶   Which are the new frontiers in the corporate communication strategies?

IT 18.2   Answers in a Nutshell

Surely the *Environmental Communication* has an important role in the Green Economy and in the companies' strategies of Sustainability and Communication. The question opens basically two debates: on one hand the importance of the e-*corporate* (everything concerned with *brand* and *brand identity* of a company using the Net) and on the other hand how much it is used by the citizens. From the e-*corporate* point of view the Communication is essential since the communicative chain is by now part of the *marketing*, which derives from the strategic-operational actions promoted by the companies in order to put their goods in the markets and create their own *brand* to increase the sales of certain products and services. This means that if a company is sustainable it will develop a trademark to be communicated adequately to show the companies' environmental and social commitment that is their eco-sustainable transformation.

Further studies point out that citizens–consumers are nowadays increasingly more informed and careful of the criteria of choice for their purchases. To prefer a product or a service instead of another depends on a multiplicity of factors round the idea of *brand* itself, identity element of excellence. Having a brand with *Environmental sustainability* components it is demonstrated that it leads to the increase of the most important *market shares* (percentage controlled by a company). To inform the consumers-citizens is then necessary to draw up a suitable Communication Plan able to convey data of technical nature.

To communicate the *Environment* in a very effective and credible way is something complex. It deals with the education of the consumers on the *Environmental Sustainability issues*. This includes a certain level of cooperation by the industry.

I think that the consumers are still subject to the risk of different types of *greenwashing* and so also victims of a communication conveying messages that claim an image of the products and services with positive impacts on the *Environment*. In reality, there are not codified key-words to identify the truth of the communicated information. Most depends on the ecological education of each consumer. What can help achieving the task is to rely upon Certification forms or eco-labels, already well-known officially on the market and they are authoritative from the communicative point of view.

Unfortunately, they do not protect totally the consumer in respect to the *Environmental performances*. The markets of the "green" products and services have become a great business so many companies that develop *Environmental certifications* linked to the processes, in reality it is a kind of theoretical and experimental exercise more than practical.

The positive element of these Certifications is to raise awareness in the national governments and in the various industrial sectors to adopt actions meant to protect the *Environment* and to respect the values of the Sustainable Development. And like a *domino effect* it has triggered a virtuous process at industrial level.

In many cases, more than the *Environmental Communication* it is just communication linked to the *brand* so nowadays practically all the business strategies are connected to the creation of a *brand*. To create it "sustainable" can increase the *market shares*, and a greater competitiveness or a greater marginality.

What is called in the industry jargon the *stakeholder analysis* is an important knowledge for the development of any business strategy also because the communication is effective when we know the listener and his needs. It is necessary then to draw up Communication Plans aimed at each group of addressees.

To talk about "green" and *"Environment"* at a company level is not suitable. Nowadays we talk about *corporate sustainability* that is of business models that develop it considering the *Environmental* and *social dimension* (*triple bottom line*) so developing *business* more with the turnover than with the income, minimizing those *Environmental* and *social impacts*. What has taken hold is the *corporate sustainability* or *sustainability business* which means to manage the *trade-off* among Economy, *Environment* and Social, thus developing models of *business* that are more sustainable. Until 10 years ago the decisional process in the business world was based on the economic variables. When to the economic variables were added the *Environmental* and social, the *trade-off* achieved were different to manage. Hence the decisional processes themselves are today different.: a positive revolution that brought a lot of companies to follow routes of *rollover* linked to being more sustainable also thanks to the support from the Research.

We can state that nowadays being more sustainable from the economic-environmental-social point of view made the companies be more competitive on the market. The new frontier is to explore in an active manner what technically is defined *business triple bottom line* in which to build up new industrial models reflecting the Sustainability.

IT 19   Translation of Interview No. 22: Eleonora Vallone

IT 19.1 Premise to Questions

You have been committed for a long time in projects that enhance the importance of the sustainable use of Water, as a natural resource, but also as an element linked to the whole eco-system and Man's survival. To reinforce and make this *Environmental message* more effective you have conceived and directed the first *Aqua Film Festival* in Rome at the Cinema House—Villa Borghese from the 6th to the 9th October 2016: a review completely devoted to short films and "very short films" (of max 3 mins) whose main theme is Water. An event that had a big success in terms of participation at international level, also thanks to the word of mouth conveyed through the Social Media.

IT 19.2   Questions

▶ What has inspired you to conceive this original cinematographic *kermesse*? What can the Water element communicate? Which is your bond with the world of Cinema?

IT 19.3   Premise to Questions

In order to make the *Environmental communication* really effective it is necessary that the contents of the message are listened to, understood and remembered by the recipients. This is what most of the *Environmental scholars* think. Internet Network offers curious attempts to make people be aware of the *Environmental* and social issues but often they tend to lose their original communicative function and become the target of the "current fashion". Last but not least, the viral effect of *"The Ice Bucket Challenge"*, the campaign launched by ALS, American Association for SLA (Amyotrophic lateral sclerosis) patients and their families. A charity message—challenge spread out so much as to produce 24 millions of videos and as many buckets of cold water, coming from all over the world, shot both by amateurs and VIPs.

IT 19.4   Questions

▶ According to your experience, do you think that something similar could happen to an *Environmental campaign*? Which gesture or gestures could be more effective to make the public opinion be aware of protecting an ecosystem like the marine *Environment*?

▶ How much can a film festival like *Aqua Film Festival* influence the audience to increase the general awareness on the *Environment* pushing them to evolve towards a more sustainable lifestyle?

▶ If you should make a short video yourself to make the public opinion more aware to highlight the marine *Environment* and its biodiversity, which words, sounds and images would you use?

IT 19.5   Answers in a Nutshell

The international project *"Aqua Film Festival"* is the goal of my life! I could successfully realize this target with much commitment and struggling for it at the forefront. For me the element "Water" is practically a second "Mother".

I then wanted to associate Water and Cinema because this precious natural resource helped me in a very delicate moment of my life. It has been a key element to recover my motor skills after a serious road accident in 1984.

But Water is much more than this! It has always had a deep symbolic value. On the other hand, the image sequences of a film can tell everything.

The festival is inspired by four thematic areas: (1) sea water; (2) water of lakes, rivers and sources of fresh water; (3) thermal water; (4) water of our *Environment*.

It is subdivided into two categories, very relevant for me: the "short films" (25 mins) and the "very, very short films" (3 mins).

Water is also the metaphor of our life that flows like a river or a sea, as many writers, philosophers and thinkers remind us (e.g.: *Buddha, Eraclito, Lev Tolstoj,*

Giuseppe Ungaretti, Herman Hesse, Jorge Louis Borges, Fabrizio Caramagna, Alessandro Baricco).

Even in the United Nations Universal Declaration of Human Rights Water is considered an instrument to ensure "a tenor of life sufficient enough to guarantee the health and the personal and the family wellness" (art.25).

Through a suitable Communication Campaign to be spread by the Net, the public opinion should be made aware of the waste of the resource Water. Starting from everyday simple gestures as brush-washing our teeth or having a shower or taking a bath.

The modern household appliances help to save electric energy and water thanks to quick washing or eco-programmes.

In this respect, *Legambiente* (Italian Association of *Environmental interest*) has already introduced a special classification called "*Eco-tourism Award*" giving a prize, every year, to the best tourist facilities.

Elba Island (where the Aqua Film Festival second edition took place in June 2017) is one of the first island holiday resort to promote the Sustainable Tourism.

As far as Cinema is concerned, it is an extremely effective tool to convey environmental messages and, particularly, to convey the love for the sea and its eco-system.

The realization of *video-curricula* (video telling about themselves) against a natural and aquatic background has notably increased the ecological awareness.

The idea is, in fact, to raise public awareness and "teaching it" for the protection of the sea so much essential to our existence.

If I was asked how to communicate effectively the Aquatic *Environment,* I would concentrate myself on the choice of the best *medium* to make everybody understands how much joy I feel when I plunge into the water. It is a boundless feeling that puts me at the centre of the world, if not of the entire universe. Everything becomes perfect.

And then, I would find the way to make the public experiment the beauties hidden in the seabed and make them reflect on the importance of the daily eco-gestures to preserve them.

A commitment to raise awareness in the public opinion on the *Environmental issues* particularly dear to the Government of the Principality of Monaco and specifically to Prince *Albert II of Monaco Foundation,* our official partner, that since its birth in 2006 has been promoting high value international projects addressed to safeguard the Ecosystem Earth, thanks also to the constant commitment of S.A.S. Prince Albert II of Monaco. It is not by chance that the Third Edition of the *Aqua Film Festival* 2018 will be taking place both at the Elba Island and in the Principality of Monaco, one of the most densely populated territory in the world (more than 17,000 inhabitants per square kilometre), but environmentally friendly.

Most of my students, at the end of my courses, confessed me about their real consciousness raising to preserve the marine biodiversity and about their change of lifestyle in order to avoid its damage and destruction.

PART III

CASE STUDIES (CS)

Case Study 1: The Buddhas of Bâmiyân—An Unsustainable Loss

Fig. CS 1.1 Statue of Smiling Buddha—Exhibition: *La Cité Interdite à Monaco—Vie de cour des empereurs et impératrices de Chine* (The Forbidden City in Monaco—Court Life of Chinese emperors and empresses)—Grimaldi Forum, Principality of Monaco, 14th July–10th September 2017—Photo by Maurizio Abbati (Photographer and Copyright Holder) © 2017. Picture choice by the author because of the unavailability of the official pictures of the Buddhas of Bâmiyân

Source: Mounir Bouchenaki, Archaeologist, former Director of ICCROM (International Centre for the Study of the Preservation and Restoration of Cultural Property), former UNESCO Assistant Director-General for Culture and Special Adviser to the Director-General of UNESCO.

© Springer Nature Switzerland AG 2019
M. Abbati, *Communicating the Environment to Save the Planet*,
https://doi.org/10.1007/978-3-319-76017-9_31

In between the 9th and 11th March 2002, in Afghanistan an unprecedented attack against the World Heritage took place. The explosion of two huge *Buddha* Statues carved in the rocky walls of **Bâmiyân valley**, 230 km far from the capital Kabul, ordered by the Taliban religious chief *Mullah Omar.* The sculptures, strictly linked to the surrounding *Environment*, dated back to the Fifth Century A.D and they represented an authentic cultural treasure for the Country, heritage of a glorious past thanks to the Silk Road that allowed the peaceful meetings among the Persian, Greek, Hindu, Buddhist and Islamic cultures.

Unfortunately, the wars that have been devastating the territory for a quarter of a century have damaged seriously the numerous archaeological sites as Kabul Museum.

UNESCO (United Nations Educational Scientific and Cultural Organization) has tried, until the very last minute, to prevent this terrible event with a communication strategy aimed at involving all interested parties.

Mounir Bouchenaki was informed of this threat by the Greek ambassador in Pakistan who was at the moment in Afghanistan dealing with the issue with the Taliban, a political and military movement to protect the Country which turned soon into guerrilla warfare to set up a strictly conservative Islamic regime, very repressive against the opponents.

That was not the first time that such a kind of threats were spread out. In 1998 and in 1999, there were some attempts of destroying the statues but, thanks to a delicate work of diplomacy based on the awareness campaign to save them, heritage of a thousand–year-culture of the Country and integral part of the plateau ecosystem (2500 m) at the foot of the *Baba Mountains*.

Then, the UNESCO Director-General *Kōichirō Matsuura,* defined this new threat as "a real cultural disaster" and invited all the interested parties to take all necessary measures to avoid the destruction of what represented an irreplaceable piece of History of Afghanistan.

A series of meetings between the UNESCO Permanent Delegation in the Asiatic area started. The General Director formed a crisis unit and appointed *Mounir Bouchenaki* as negotiations coordinator who privileged the communication aspects creating an international cooperation *Net* in record time.

On the 1st March 2001, *Bouchenaki* contacted the French ambassador *Pierre Lafrance*, former ambassador in Pakistan and an expert of the Afghan territory, as well as a speaker of one of the languages spoken in that Country. In less than 24 h the diplomat arrived in *Kandahar,* stronghold of the Taliban.

On the 2nd March 2001, at the opening of the UNESCO International Colloquia on Central Asian Heritage, the Director-General informed all the presents about the situation evolution, while *Bouchenaki* was setting up quickly a high qualified working group among them *Paola Leoncini-Bartoli*, Chief Executive Office of the General Manager Support for Culture and Christian Manhart, Programme Expert who played a key-role in contacting the most important non-governmental organizations committed for the protection of the Afghan Cultural Heritage (ICOMOS and ICOM) and the Society for the Protection of Afghan Cultural Heritage (SPACH).

On the 6th March 2001, during an official press conference, *Bouchenaki* underlined the multi-year commitment by UNESCO in the interested area to protect the cultural heritage, strictly linked to the natural eco-system and confirmed the unanimous condemnation expressed by the international community against the Taliban's decision to destroy the *Buddhas Statues*. The Hindus and Islamic communities included, and they distanced themselves from the attempt to justify with religious motivations such a kind of destruction.

On the 8th March 2001, the UNESCO Director-General, thanks to *Mufid Shihab's* negotiations, future Minister of Education in Egypt and President of the Egyptian National UNESCO Commission was able to speak on the phone to *Hosni Mubarak*, Afghanistan President, to facilitate an Egyptian mission to Kandahar to make the Taliban reflect on the wrong interpretation of the holy texts.

In the following days, many actions were implemented and promoted by political and religious leaders to make the Taliban be aware and withdraw their decisions. Also, the Arab and Islamic ambassadors at UNESCO were involved, in particular those of Morocco, Qatar and Syria, contacted by *Bouchenaki* himself.

Mounir Bouchenaki with the diplomatic support, succeeded in convincing *Sheikh Youssef Kardaoui*, among the most respected personalities in the Arab religious world, to go to Afghanistan. *"The statues made by elders who came before Islam are part of a historic patrimony [...] I advised our brothers of the Taliban movement to reconsider their decision in light of the danger of its negative impact"*. That was the strong message by the Islamic theologian. He was supported by other authoritative voices among which *Sabri Abdel-Raouf*, Head of the Division of Islamic Studies at *Al-Azhar University in Cairo (Egypt)*.

The delegation on a mission was able to meet the Minister of Religious Affairs and the Minister of Foreign Affairs of Afghanistan. He did not succeed to seat at the negotiating table either with *Mullah Omar* or the religious Authorities in order to convince them to undo the Decree which contained the demolition order of the two statues.

From the 9th to the 11th March 2001, the *Bamiyan Buddhas* were completely destroyed in coincidence with *Aid el Adha* celebrations. The fact was officially announced some days after on the 15th March 2001. It was defined *"A crime against Culture"* to underline its seriousness.

This very sad epilogue shows how in war or post-war situations the cultural heritage is often the target of tensions between those who preach a "punitive" approach and those who try the way of "reconciliation".

A race against time in which the efficacy of communication clashes against particularly difficult ideological preconceptions to change and that can generate very impacting effects as this is the case. They are worsened by the increasingly frequent forms of non-conventional wars led by mercenary armies unrelated to their national identity and the creation of independent states not recognized by the International Law.

From the 21st to the 31st December 2001, during the annual meeting of the Ministers of the Arab World Culture gathered at *Doha (Qatar),* the most specialized leaders in Islamic Law (27 professors and experts coming from twenty-five

Countries) condemned with strength the destruction *of Bamiyan Buddhas* and expressed themselves against any false legal interpretation spoiling the cultural heritage.

On the 17th October 2013, following a prep-work lasted 2 years supported by UNESCO to give a strong answer to what happened at *Bamiyan*, a *"Declaration concerning the intentional destruction of Cultural Heritage"* was made.

A program of action that assumes a concrete contribution, on the international scene, to encourage the post-destruction reconstruction, to help the dialogue between the parties and to prevent the recurrence of any attack against the cultural heritage.

In the light of the recurrence of the terrorist attacks as the ones in the archaeological site of Palmyra, in Syria, the protection of what represents the people identity and its territory will be always a priority in UNESCO programs hoping in favorable conditions to set up a strict cooperation.

Case Study 2: An Adriatic Sea Project That Communicates the Idea of Sustainability

CS 2

Fig. CS 2.1 Studio Camerini © (Copyright Holder)—project of the historic centre redevelopment, *Isola di San Nicola (Tremiti)* [*Saint Nicolas Island, Tremiti Archipelagos, Italy*]

Source: Mariaelena Camerini, Professional Architect and Designer focused on culture and Environment.

I would like to talk about a study we carried out in 2008, for the redevelopment of the town centre of Saint Nicholas Island in the *Tremiti* Archipelagos in the Adriatic Sea, Italy.

Twelve miles North far away from *Gargano* promontory, *San Nicola* island is the historical, religious and administrative centre of *Tremiti* Archipelagos.

With its historic fortified structure starting from one thousand and one hundred century, the abandoned vast abbey complex and the village, whose origins date back to a penal colony wanted by the royal *Bourbons* in 1792, is the destination of short daily excursions by tourists coming from the mainland or from other inhabited islands of *San Donnino*.

In *San Nicola,* there are no vehicle accessible roads, all foodstuff is imported by the mainland and the locals, devoted to tourism with restaurants, souvenir shops, and boat tours, live in the city centre with a few infrastructures.

The island succumbs regularly to the seasonal laws and to the commuting phenomenon: during wintertime, it is depopulated to revive again in the springtime when the long touristic season starts.

On account of a great historical and cultural identity, Saint Nicholas is affected by its inhabitants' and visitors' approach; the **phenomenon of commuting tourism** and of the depopulation are the most serious dangers: on one hand the lack of resources, on the other hand the progressive weakness of tradition and the local communities.

The structures of the Benedictine Monastery, integrated into the landscape of extraordinary *Environmental value*, are living the degradation and decay typical of these forgotten places and for this reason they should be the object of valorisation projects of the spaces and of new models of touristic reception, for example the **"scattered" hotel**.

An idea of reuse born after a landscape study analysing its historical and formal characteristics, with a particular attention to its building blocks over time, that could define a model of hospitality reviving the places by considering the preservation and durability; the latter meant as an identification and pursuing of social, *Environmental* and economic targets without reducing hopelessly the resources of the place but, on the contrary, it could sustain it in a dynamic way.

Hence, the idea of the "scattered" hotel, a horizontal hotel, situated in the historical centre with rooms and services dislocated in different existing buildings but close to each other.

This hotel typology should be a receptive and shared structure addressed to the demand of tourists interested in staying in a quality urban context in contact with the inhabitants and to know them and respect their lives, using normal hotel services.

This *win-win formula* has proved to be particularly suitable to the villages and small towns of artistic and architectural interest that in this way could recuperate and enhance old closed and useless buildings, avoiding to build new ones.

The realization of this case has pointed out the aspect of the "*cultural and economic animator*" developed by the "*scattered hotel*" towards the local communities of which it has increased the income and occupation, without altering the

Environment, the spaces and the identity; but encouraging the presence of a motivated, durable and responsible tourism.

The value of the existent is implemented through a requalification which starting from the design intent goes to the use of techniques and materials employed reminding our responsibility toward the *Environment. If one of the main principle of Sustainability is the reuse, why isn't it always applied in a large scale?*

Case Study 3: When Communicating the *Environment*: The Common Sense Makes the Difference

CS 3

Fig. CS 3.1 *GreenMinded* © Project (Copyright Holder)—500 cigarette ends found in 30 min by one person within a radius of 60 m

---

@GreenMinders (Twitter)
www.facebook.com/GreenMinders (Facebook)
www.greenminded.fr (Web Site)

Source: Alice Comble, Engineer of Telecommunication, Project Manager and Founder of the start-up GreenMinded first one classified at the European Competion "METHA Europe 2016" (EVER Exhibition, Principality of Monaco).

© Springer Nature Switzerland AG 2019
M. Abbati, *Communicating the Environment to Save the Planet*,
https://doi.org/10.1007/978-3-319-76017-9_33

We organize a waste collection in the parks and in the city centre also to make questions about the quantity of waste recovered and to liven up our "*GreenMinders*" community. Each collection is also the occasion to meet together and share a picnick, an aperitif *etcetera*. This kind of action enables us to realize how much positive the impact of a group may be. "Alone you go fast, in a group you go further".

When we do not organize big collection, we demonstrate inside our team that we are involved in our daily routine. And not only realizing this project but also being real citizens in our free time.

For example, when on Sunday we go walking or during the way home from the office, we pick up cigarette ends. It takes some time but it allows us to know the huge amounts we find throughout the town.

Each of us is able to state that there are "a lot of" cigarette ends in town. Nevertheless, nobody is really capable of counting them, or giving their weight, their degree of toxicity or density. Acting in our everyday routine with a view to increase the awareness on separate collection, allows us to appeal to the community, to make people talk about us and to underline our goal in favour to the Planet.

You will find hereafter an example of the visual support we publish through the Social Network at the end of a separate collection:

In short, our daily leitmotiv is based on the well-known saying: "The world has changed for your example, not for your opinion". We try then to behave as well as we would like others to do.

By the creation of "*Borne to Recycle*" we effectively help the access to responsible and respectful *Environmental behaviours*.

Milan, a town in the middle of one of the biggest metropolitan areas in Italy (3,176,180 inhabitants) has been developing shared actions for a long time in order to face *Environmental issues* particularly serious, considering also its geographical position: in the heart of the *Po* Valley surrounded by the Alpine range acting as a barrier to the outflow of the atmospheric pollutants. According to 2014 data from the *European Climate Agency* the concentration of PM10 (polluting particles with a diameter inferior to 10 μm) reaches, in the Milan area, alarming values from 50 to 75 mg *per* metric cube (normal values should be inferior to 20 mg per m^3). These worrying peaks of pollution are the consequence of the high level of the urban traffic congestion both in the historical city centre and on the roads in and out Milan and its Metropolitan Area.

To overcome this delicate situation from 2006 to 2009, the City Administration, then led by the Major *Letizia Moratti*, decided to adopt a series of sustainable actions to make the urban fabric and public transportation a better place to live. For example, a Sustainable Mobility Plan was adopted to increase the use of the underground network, the "clean" technological innovation and a system of *road pricing* for all the polluting vehicles having access to the urban area. A Covenant of Mayors was ratified, a voluntary commitment on Climate and Energy issues involving thousands of local and regional authorities of any size and levels in the European Community.

To promote the energetic efficiency and the use of renewable energetic resources at local level is a very important step towards the climate mitigation change. Last but not least, in 2015 Milan hosted the world exhibition, EXPO 2015: *"Feeding the*

@**MilanoSiMuove** (Twitter)
www.facebook.com/MilanoSiMuove/ (Facebook)
www.milanosimuove.it (Web Site)

Source: Edoardo Croci, Professor, Researcher and Director IEFE Center for Research on Energy and Environmental Economics and Policy – Bocconi University – Milan (Italy).

© Springer Nature Switzerland AG 2019
M. Abbati, *Communicating the Environment to Save the Planet*,
https://doi.org/10.1007/978-3-319-76017-9_34

Planet, Energy for Life", a good opportunity to deal with the most relevant questions on Sustainable Development and food safety at the international level.

To show the increasing awareness for *Environmental issues* in 2010 *MilanoSiMuove* (Milan-moves), a Committee for the *Environment* and Quality of Life in Milan, was born. It was composed of representatives of the scientific world (Polytechnic University of Milan, "Bocconi"), *Environmental organizations*: *WWF Lombardia, Genitori Antismog* (Anti-smog Parents), *FIAT Ciclobby*, politicians and VIPs. The Committee planned to strengthen the sustainable management of Milan metropolitan areas through the presentation of a series of consultative *ad hoc referendum* (five questions on the lifestyle in Milan).

The five questions of the consultative *referendum* were about the following issues:

- traffic and smog reduction to the enhancement of the public transportation system, extension of the subject to toll urban area and the city centre pedestrianization;
- doubling of the trees and green areas and new housing with no additional land use for building;
- protection of the green area used for Expo 2015;
- Energetic consumption and reductions of greenhouse gas emissions through the implementation of the *Sustainable Energy Action Plan*—SEAP—a key-document promoted by the Covenant of Mayors to allow each member signatory to clarify the necessary actions to achieve the targets of CO_2 reduction by 2020.
- reopening of the Milan river system of "*Navigli*" Canals.

The fixed goal was achieved through a capillary action of promotion meant to raise funds and reaching the number of signatures to validate the referendum (more than 15,000 signatures authenticated by a public official). All this thanks to a communication campaign working on many levels, to a plurality of forms and mediums according to the recipients' targets: citizens, schools or volunteers. Leaflets, posters, Web Networks, Social Networks like Facebook, blog and online forum, testimonial, press releases, events, concerts, advert messages on the main newspapers or on TV broadcasting. Just to make some examples.

Fig. CS 4.1   Communication Campaign © MilanoSiMuove (Copyright Holder)

Fig. CS 4.2   Communication Campaign © MilanoSiMuove (Copyright Holder)

In June 2011, despite the legislature transition of Milan Municipal Administration and the new Mayor *Giuliano Pisapia*, all the *referendum* five questions were approved, binding the political choices of the newly established Council. The Municipal Administration approved officially the referendum and established a Commission and a Consultative Body in the Municipal Council with the specific target to manage the process of execution of the referendum issues. The promoters, representing the citizens, are still committed to monitor and promote the referendum execution of the objectives, some of them being defined.

The route of a participated communication experimented in Milan shows numerous strengths. The involvement of the stakeholders and the big variety of information about the Administration Action Plans made available have increased the citizens' *Environmental awareness*. The communicators' professionalism and competence have enabled a big success thanks also to innovative forms of communication.

All that made the project *MilanoSiMuove* (Milano moves) a best practice in the public communication sector which is keeping on its commitment to promote sustainable actions through the democratic tool of referendum supported by a suitable Communication Plan. The promoter *MilanoSiMuove* and its secretary, *Marco Cappato*, have just proposed other four *referendums* pro-active (non-advisory) questions and therefore constraining for the Municipal Administration, on the basis of the new Statute of the City including the previous issues. Social housing: availability of 25,000 buildings without land use through the urban regeneration of old buildings; the "Navigli" Canals navigability; mobility: an integrated transport plan and the Green Public areas: with the realization of the project "*Green Rays*" (a project promoted by Milan Municipality—Sector Development), Land and Urban Furniture and Decor for a new network of pedestrian routes and cycling tracks meant to enrich the urban fabric. These are the issues whose *Environmental* and *social targets*, although ambitious, are to be achieved within 2020.

Fig. CS 5.1 Project Logo #skateTOsostenibility (Copyright Holder) © 2016

In the school year 2015–2016 among the projects proposed to the city schools, "*The unbearable lightness of consuming*" sets its main goals for giving the students tools to investigate the numbers of consumerism, to extrapolate data of the *Environmental impact* and above all to build *Media interactive* channels to take practical action, facilitating inside the young community of peers and not only (school system, family system, young community system) good practices of innovative sustainable consumerism. The young people themselves had the possibility to propose and above all to carry out these practices independently or together experts.

@**skatetosostenibility (Facebook)**
www.padovanet.it (Web Site)

Source: Daniela, Local Agenda 21 Coordinator and Responsible for Informambiente, the official Communication Service offered by the Municipality of Padua – Italy.

© Springer Nature Switzerland AG 2019
M. Abbati, *Communicating the Environment to Save the Planet*,
https://doi.org/10.1007/978-3-319-76017-9_35

The project was realized with nine first-year classes of the Technical Institute Technology Sector "*G. Marconi*" of Padua.

During the school year the young students chose to focus their course of study on the innovative concept of using *the skate-board as a sustainable means* substituting or alternating the bike: **"skating towards sustainability because walking is not enough anymore! We must move, time is short!".**

The **#skateTOsostenibility** campaign, done and managed by youngsters on the main social network (Facebook, Snapchat, Twitter e Instagram) and translated in all the original languages of the students (Moldavian, English, French, Portuguese, Arab, Spanish, Chinese, Rumanian and Italian) contains all the proposals conceived to promote through music, videos, theatre, graphic arts and technology the concepts of Sustainability.

Fig. CS 5.2 Source: video rap made by *Marconi* Institute first-year class B 1 in Padua—Project of Environmental Education: "Consumo Sostenibile" (*Sustainable Consumption*). Title: This world is rotten. Filmaker and musician: Alessandro Cosentino—Theatrical Lab: Lucia Schierano. © 2015–2016. *Questo mondo è marcio* (YouTube)

The rap song ***IMPATTOZERO***, written and sung by the youngsters after a musical lab with *Seven Tacks* (a famous rapper on the Padua scene) whose **video clip** has been shot by the **Padua** filmaker and musician *Alessandro Cosentino*, also author of **a spot on *Environmental impact* during our daily routine**, in the framework of the project. The **play** performed by the students about *Environmental impact* and social network, was structured during the theatrical Lab with *Lucia Schierano* (filmmaker and Theatre actress). **Graffiti and permanent panels** *promoting the practices of moveable sustainability* focused on skate-board, as means of transport to move in a eco-sustainable way and not only as an amusing tool; a message conveyed also by **T-shirt** #skatetosostenibility – SKATE = $CO_2 \times 0'$ made to promote the project. Among the students' proposals there has been also space to technology with the presentation of the **e-cycle**, a bicycle connected to an electric engine that recharge an *ipod*, a smartphone or a tablet. The message the young people want to send is: "We don't want to give up to comfort but we want to make it sustainable. Which is the accessible sustainable energy? The energy of our muscles!"

The end of the journey coincided with the networking of the two videos chosen by the students among many others at the beginning in participated modality.

In the framework of this project, it has been possible to make an experience exchange with other schools of the *Este* city in Padua province, through a skype connection made at the *Marconi* Institute *auditorium* with the involvement of three first-year classes by the teacher of reference of the same Institute for an exchange of good and sustainable practices in occasion of the Good Practice Festival: **"FROM THE SUSTAINABLE TO SAY TO THE SUSTAINABLE TO DO"**, organized by the Municipality of *Este* (near Padua).

In September, for the ecological Sunday, realized in the framework of the *European week for sustainable mobility*, a delegation of students of this project and a delegation of students who realized the **"I like Green"** met together and compared their experiences. All this in the Square so that the big public was informed.

Fig. CS 5.3 Source: video rap made by *Marconi* Institute first-year class B 1 in Padua—Project of Environmental Education: "Consumo Sostenibile" (*Sustainable Consumption*). Title: This world is rotten. Filmaker and musician: Alessandro Cosentino Theatrical Lab: Lucia Schierano. © 2015–2016. *Questo mondo è marcio* (YouTube)

Case Study 6: The Tools of the Urban Planning Give Voice to *Environment*

CS 6

CS 6.1 *Metrominuto* Ferrara/Ferrara Metre-Minute

http://servizi.comune.fe.it (Web Site)

A project to promote the soft mobility: **'*metrominuto Ferrara*'** is a map whose information is shown on short distances and journey times on foot between the main places of interest. *Ferrara* is one of the most important Italian art towns in the North, near *Po* valley.

Since 1995, it has been included as a heritage site by UNESCO and it has both a Medieval and Renaissance historical centre enclosed by well-preserved ancient walls. The town of Ferrara has a remarkable tradition lined to soft mobility: for instance, the 27% of its citizens chooses the bicycle to move around every day.

Metrominuto Ferrara looks like the underground map of big cities but the information cover walking and cycling distances only. The project whose recipients are students, tourists and citizens, aims at removing the obstacles of uncertainty as far as the distance and the journey times are concerned which may prevent people from walking: to know the distance and the foreseen time facilitate the moving walk and the exploration of the beauties of Ferrara historical centre.

Metrominuto Ferrara reproduces a local urban good practice realized already by *Pontevedra*, the Spanish town that activated a strong political criticism to the use of cars. Ferrara project has received sincere appreciation both from the citizens and the institutions, schools and associations. The map is advertised by road signage in five places in the city and can be downloaded from the website.

@**centroideaferrara** (Facebook)
www.comune.fe.it (Web Site)

Source: Elisabetta Martinelli, responsible coordinator of the Sustainable Education Centre IDEA - Municipality of Ferrara – Italy.

© Springer Nature Switzerland AG 2019
M. Abbati, *Communicating the Environment to Save the Planet*,
https://doi.org/10.1007/978-3-319-76017-9_36

Fig. CS 6.1.1 The *Metrominuto* map of Ferrara © Centro Idea Comune di Ferrara

CS 6.2 A Social Wood for the City of Ferrara (Italy)

http://servizi.comune.fe.it (Web Site)

A **community Food Forest** (or **Food Garden**), is a wood designed and managed by the community that looks after it and can use the fruit, thanks to an active synergy between the local community and the Administration for the sustainable management of the common goods. Ferrara Municipality has started the project of a "community food wood" called *Food Forest* to answer the requests of a group of citizens, supported by the Social Coop Integration Work of Ferrara. IDEA Centre (Educational and Sustainability of Ferrara Municipality Centre) coordinated all the actions of the project through participatory workshops. Thanks to a financial contribution by the *Emilia-Romagna* Region they have realized:

- the **educational labs/workshops** on the job-field "*Let us create a forest garden*" intended for the project designers;
- six **labs of participatory planning** for the creation, study and design of the Food Forest with the Permaculture technique for the design,
- conservation and ethic awareness of **productive eco-systems** taking into account *Environmental*, economic and social aspects;
- **planting of identified species**.

All the path planning has been accompanied by facilitation techniques and integration of the working group, with the purpose to harmonize the differences and enhance the talents of everyone. At the end of the first module of the educational workshop, the participants, through a path of participatory planning, have designed the wood, with a map to fill in virtually with trees, bushes and herbs going to make it up. The path is supervised and monitored by teachers, in all the phases. The second module of the educational workshop provides the realization on the job-field of the design and the planting of plants and not only a workshop for listening but also practical place for learning while doing, preparing the soil and planting the plant species that will be going to give birth to the first *Social Food Wood* in Ferrara. The project started in 2013 and ended in December 2015.

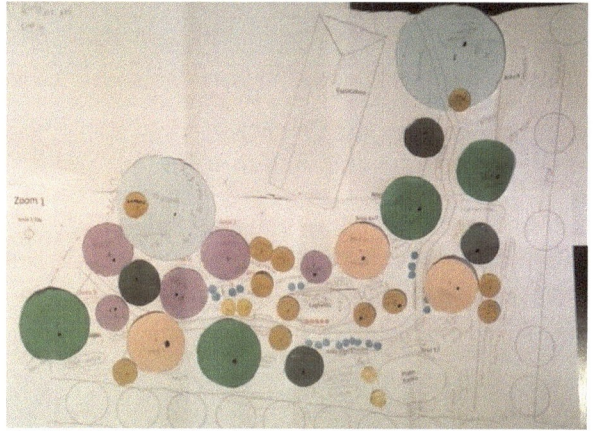

Fig. CS 6.2.1 The Map of the "Food Forest" in Ferrara © Centro Idea, Municipality of Ferrara

---

CS 6.3 ECOWASTE4FOOD Project ECOWASTE 4 FOOD Interreg Europe

@ecowaste4food (Twitter/Facebook)

ECOWASTE4FOOD project was financed by the programme of territory cooperation "*Interreg Europe*", and wants to improve the planning tools of the local policies integrating them with support actions to the use of effective resources through the promotion of eco-innovation for the reduction of *food waste*. The Municipality of Ferrara has participated to the ECOWASTE4FOOD as a partner with *Environment Service CEAS–IDEA* centre.

ECOWASTE4FOOD is given coherence to the local policies to the effective use of natural resources and the reduction of food waste. This objective will be achieved through an Action Plan resulting from the learning interregional process. ECOWASTE4FOOD is then also a project of *capacity building* oriented to increase the knowledge of the policy makers and administrators, officials, experts, technicians and the civil society to enhance the adoption of good practices with the main policy tools.

The objective is to set up a shared Action Plan with an organic framework of promotion and support to further eco-innovative solutions, services and products for the effective use of the resources and reduction of the waste food, through new strategic objectives declined in many new operational objectives, redefinition of priority of investment, forecasting and assessment and also financial means and sources.

In doing so Ferrara will be supported by a group of local stakeholders to share the local needs as a result of the activities carried out at transnational level and held up during the course of transfer of good practices and in the development of a Local Action Plan impacting on the local policy tools.

The main activities to be carried out in Ferrara:

1. **setting up of a group of stakeholders** to meet regularly to discuss issues and solutions; they stakeholders will take part to local events and projects and they will cooperate with Ferrara Municipality to draw up a local Action Plan which the results of the whole course will flow into;
2. **mapping of all the activities and initiatives** for the reduction of local food waste;
3. **participation to all activities of communication** at project level and to the communication output and drafting: biannual newsletters, brochures etc. and the established communication Platform;
4. **preparation of a document, together with the Provence-Côte d'Azur region, to share** at the second meeting of the project on the theme of services for the reduction of food waste;
5. **participation to four interregional workshops** involving the stakeholders and local communities;
6. **seven study visits** (one in Ferrara) involving also the stakeholders;
7. **drawing up of an Action Plan** and a **peer review** meeting at the end of the project.

Case Study 7: When Objects Communicate the *Environment*

CS 7

Fig. CS 7.1.1 © RICICLI by Giulia Meloncelli (Copyright Holder)

www.riciclidesign.it (Web Site)

Source: Giulia Meloncelli, *industrial designer* **and eclectic designer of everyday objects and fashion accessories made of sustainable materials.**

CS 7.1 **Computer Bag/Rucksack** (Fig. CS 7.1.1)

Unisex *computer bag/rucksack* useful for technology up to 19′, for working ad free time. It is made with eco leather, once used for the upholstery of the *yachts* interior, and technical fabric.

The front has a jacket *renvers* that forms a pocket closed by a zipper. In the pocket, a striped, flowery or other fantasies *pochette* is embedded, to play with the contrasts. The zipper is opened by a special puller, that is an old recycled coin.

This external pocket is capacious and very practical to contain all those objects of frequent use such as keys and mobile for example.

Also, the main zipper is opened by an old coin and the inside is lined with a very resistant fabric and it is structured and rich in compartments. In fact, there are two small pockets made from pieces of jackets or trousers and a side pocket can carry a *tablet* or A4 documents.

It is padded so that the technology is well protected.

The ergonomic shoulder strap is adjustable and removable thanks to two strong carabiners.

By changing the anchor of the two carabiners it is transformed then from a shoulder strap into a rucksack in a few minutes. Therefore, it is very practical to use when you are moving by bicycle or by motorcycle.

The technical materials and the high performances make it light and flexible so as to be able to fold on itself to put it inside. Washable of course.

CS 7.2 **Billiard Cloth Hanger** (Fig. CS 7.2.1)

We recover used balls of incomplete *sets* and then we screw them in used wooden boards. The result is a hanger. Each piece is unique and the colours may be different according to the availability in stock. On demand, it is possible to personalize the hanger with the client's lucky numbers or dates to remember such as anniversaries, birthdays, etcetera.

Fig. CS 7.2.1 © RICICLI
by Giulia Meloncelli
(Copyright Holder)

CS 7.3 **Half-Light Lamp** (Fig. CS 7.3.1)

The interior lamp is made of recovery of car headlights or vintage motorcars.

The wooden front is also recovered, in particular it was part of old doors or wooden boards. It can be hung on the wall, or put on a table or in a shelf or on the floor projecting the light to the ceiling. It works with electric current 220 V and the light bulb given is at low consumption and easy to be found at the hardware shop. The maintenance is carried out from the back and it is very easy to substitute a light bulb. Each piece is unique and the Lamp is accompanied by its special *pedigree* stating the originality of the period.

Fig. CS 7.3.1 © RICICLI by Giulia Meloncelli (Copyright Holder)

Case Study 8: *Treepedia, the Project That Makes Trees the Eco-communicators*

The green areas in the urban centres have always been small ecosystems for the citizens where to stay in contact with nature and oxygenate. The management of the public green is therefore a major element in the urban planning. But how to communicate the green perception quarter by quarter? Street by street? The *software Treepedia*, conceived by a working team of *Senseable City Lab—Massachusetts Institute of Technology (MIT) of Boston*, wants to give an answer to this question. Thanks to Hi-tech a series of detailed maps can visualize by a click the green "spots" in any corner of the main capitals and world megalopolis, by a click. Its objective is to improve the ecological rate of the big towns.

The project, led by *Carlo Ratti*, aims to enhance an indicator of vital importance: the mapping of the trees. The dangerous increase of atmospheric pollutants and in particular *smog* are affecting large areas of our Planet.

Hence, the importance to reassess the capacity of trees, and so of the green areas, to act as a "natural filter" of the air, thanks to the process of photosynthesis. But they also act as "sponges" keeping the excess water, in case of exceptional atmospheric phenomena thus avoiding flood events and creating a "barrier" sound absorbing the noise; as well as a "regulator" of the global temperature which always tends increasingly more to achieve extremes of heat and cold.

Accordingly, the desire of MIT to create an information and communication tool, based on the trees Census planted in big cities. Starting from an algorithm used by Google Street View, able to process data from the point of view of those walking in the street. The software *Green View Index*, is the parameter measuring the visual

---

http://senseable.mit.edu/treepedia (Web Site)
@SenseableCity (Facebook/Twitter)

Source: *Carlo Ratti,* **Engineer, Architect and Director of the MIT Senseable City Lab, Massachusetts Institute of Technology – Boston – USA – founder of the Carlo Ratti design and consulting firm.**

perception of the green around us. The project then has started to take a census of the trees, as a first experimental phase in ten sample-cities.

At present the urban territories recorded are more than ten. Among them we remember: Boston, New York, Los Angeles, Seattle, Sacramento in the United States; Vancouver and Toronto in Canada; London, Paris, Amsterdam, Geneva and Turin in Europe; Tel Aviv in Middle East. Anyway, beside the graphic visualization and cadastral map there is much more.

In fact, the software allows "to interview" the green "spots" that give information on the vegetal species, even in the modality of "*street view*". It is, then, possible to know the tree degree of concentration, area by area, with the possibility of comparing the data, stating which part of the city is "greener" and which town has more public green. But the project, as already said, aims to improve the management of the urban centres.

Hence, last but not least, ***Treepedia*** objective is to transform itself in a *medium* of open source communication increasingly more detailed with the implementation of a number of video cameras which detect in real time the state of health and the emergency situations or the state of decline that need of prompt interventions.

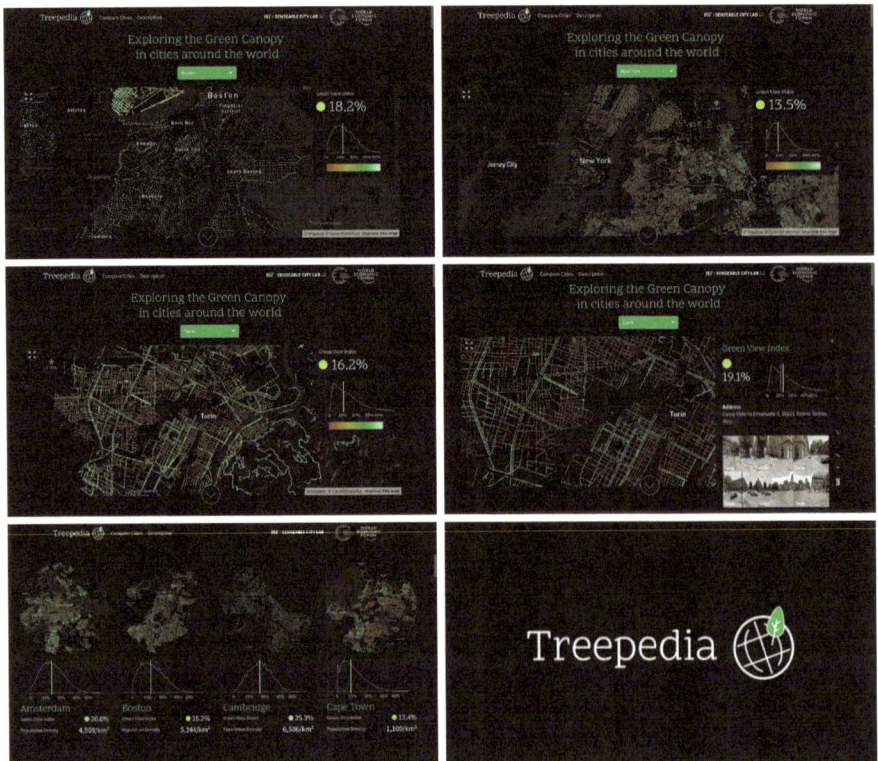

Fig. CS 8.1 © Treepedia application—extracts from the official website referring to: Boston, New York, Turin (general map and detailed map per district), Multi-city comparison—© MIT Senseable © Mapbix © OpenStreetMap. http://senseable.mit.edu/treepedia (Web Site)

The high technology is a good allied of the *Environment* and is available to the citizens who will be able then to report promptly *Environmental criticalities* or deficiencies related to the management of the green; and to the decision makers or the technical public administration that will manage responsibly the green areas, taking shared decisions with the stakeholders, facilitating that dialogue at the base of the communication chain.

In this way, awareness campaign to defend the urban green, are to be encouraged, promoted by public institutions, private companies or citizens, individual or group.

Case Study 9: The "Soul" of Architecture Communicates the Respect of the *Environment* Through the Five Senses

CS 9

Fig. CS 9.1.1 Residence Hotel—© Studio Olistico (Copyright Holder) Sirmione (Lake of Garda)

---

www.studio-olistico.com (Web Site)
@StudiOlistico (Facebook)
@Studio_Olistico (Twitter)
Studio Olistico (LinkedIn)

Source: Omero Soliman, Architect and Expert in bio-architecture, eco-design and Feng Shui – co-founder of the Studio Olistico – Legnago (Verona, Italy).

CS 9.1 The Case Study of the *Hotel Residence Rossi* of *Sirmione* (Garda Lake)

The *Environment* on which we live both in a house or in a tourist hotel where we stay overnight can they contribute to make more harmonious the relationship between Man and Nature? In other words, can they contribute to our welfare? "To stay well inside", in fact, does not identify a physical condition but a psychological and spiritual one strictly connected to the *Environment* surrounding us which we interact with through various forms of communication. An element not to be neglected also to preserve our health. If the three aspects: body, psyche and spirit are not harmonically tuned among themselves the probability to fall ill will increase.

To achieve the wellness is not absolutely clear-cut but it depends on a complex series of actions and circumstances, besides a personal commitment, as a result of a soul searching and of a careful critical sense of things and great powers of observation. But not only this. The *Environment* surrounding us must be "suitable to our needs". This is what the Chinese Taoist philosophy teaches us since they considered it an essential aspect of living in harmony with ourselves and the nature. A harmony which became an ancient philosophy of Life, a perfect equilibrium among the energies of Heaven, Man and Earth. *Feng Shui* (Feng = wind; Shui = water), is a discipline applied to architecture, based on the observation of natural elements that interact with us shaping the surrounding landscape. Therefore, each single detail becomes a *medium* of communication. An example is the water element which becomes a communicative flux of ideas and positive energies of light (Yang) and the negative element of the darkness (Yin) [*Isabella Puliafito*).

On the basis of these principles, the architect *Soliman* has realized his *Feng Shui* design whose target is to transform the simple *Hotel Residence Rossi* hall in *Sirmione*, famous tourist resort on the *Garda Lake* (Italy) into a harmonic *equilibrium*. A balance among shapes, colours, materials, able to recreate a well-being atmosphere for the guests staying or working in the hotel hall. A project studied in detail in which nothing is left to chance especially as far as the "shape" is concerned. *Omero Soliman* thinks that one of the most important role of the architect is to create the "shape" from the "unshaped".

A very characteristic element that belongs the human beings, and it puts us in relationship with everything around us, the Ecosystem Earth included. "To understand who we are, what is our life about, is giving shape to architecture. The starting point is not the pencil, the paper-sheet or the drawn line, anymore. What really matters is our soul and our heart!" *Soliman* underlines. The synergies between the interior and the external human dimension is the best way to convey ideas, prepared to listen to, to create a debate with other men and any other living or non-living being we get in touch with. This is the straightforward way to communicate.

In *Feng Shui* project, many aspects have been considered by *Soliman*: the orientation of the lounge and the other rooms for the guests and the hotel personnel; the relationship between the hall and the other entrances; the light; the edges of the walls or cornerstones; the corridors; the furniture; the materials; the scents and the colours. All the elements that influence the wellness which means also health and hygiene. In this way, the interchange of necessary energies is free to breathe the *Environment* and make the Nature to be preeminent and it enters the spaces blending into a unique "eco-system". In fact, *Feng Shui* project reproduces, in the hall of the *Hotel Residence Rossi*, the shapes associated to the "Five elements of the Oriental tradition" (Wood, Fire, Earth, Metal, Water).

The communicative function of the colour is also emphasized. The chromatic nuances are among the most immediate and simplest forms of communication and they enable to make complex concepts to be easily understood through their "symbolic meaning". The colour is considered by the scientific community as a "visual sensation", the result of the stimulation of the *retina* of the eye by electromagnetic waves to differentiate yellow from red, green from blue and so on. It is not therefore the light only to determine the visual perception of the colour but also the capacity of the coloured material to transmit a particular wavelength of light. There is then the inclination of the light source, the colour of the light itself, the visual condition of the eye, in that very moment we are looking at.

All these factors may influence the perception of what the colours are communicating, both primary or secondary, "warm" or "cold"; or those known as "chromatic colours" that is the white" (Yang) or the black (Yin), in fact all chromatic nuance is associated to a specific energy.

Fig. CS 9.1.2   Residence Hotel Rossi © Studio Olistico (Copyright Holder). Sirmione (Lake of Garda)

To sum up: the type of plaster of a wall (e.g.: smooth or spatula), the furnishing fabrics, the furniture, the paintings and the photographs on the wall, the lamps and any other source of light contribute to the chromatic perception that everyone has of the surrounding *Environment*, thus giving origin to different types of communication that influence our way to interpret the eco-system, increasing or decreasing our wellness sensation.

To enter into communication with the natural *Environment* means also to "decode" the thinnest and most impalpable existing element: the scent. It is not by chance that we are often talking about *olfactory architecture*. A new way to perceive the space where the "sense of smell" is the protagonist. Here the element "air" becomes the *medium* to convey messages through smells and fragrances to our brain. An intense scent as the jasmine, orange blossom, or some varieties of roses, for instance, allow the human brain to identify the plant species and transmit emotions, "acting on the physical, psycho-emotional and spiritual plane instilling well-being", the architect *Soliman* reminds us. A primordial form of communication becoming a science with the birth of the modern *aromatherapy* associated to architecture.

"The sense of smell is one of the five senses to which less importance is given even though it is the oldest sense. The interest toward the "olfactory innovations" is always from people who are studying deeply the subject for a "more holistic knowledge" than the architectural design itself. In fact, this is because we are expected to compare ourselves to the "thinnest aspects of the reality surrounding us", underlines *Soliman* and he adds: "the benefits of the essential oils may offer are numerous. For example, they may give energy, attention, relax and physical wellness. For this reason, it is useful to select them carefully. There is really a "hidden language" of the olfactory communication whose psychological reaction between scent and context should exert a considerable influence on our mind and memory with indescribable "effects".

It is not by chance that some well-known brands at international level like: *Abercrombie & Fitch, Hollister, Kartell* and big hotel chains are using aromatic products such as *Moooi*, and they have invested and bet on this new frontier of communication that can go far beyond the commercial purpose. *Sissel Tolaas*, founder of the *frangrance Re_Search Lab*, claims more than seven thousand synthetized smells or inspired by objects and historical periods, created an exhibition devoted to the First World War, in 2014, at the Museum of Military History in *Dresden* (Germany). Such a "shocking" olfactory experience so as to prevent visitors to enter in one of the exhibition halls; the sense of a violent repulsion to war felt by the soldiers while approaching the battlefield.

"We must not forget therefore, the "evocative power" that some fragrances have on our psyche, also to make the public opinion to be aware on *Environmental issues*. Such a kind of experiments have already been used by big production companies like *Disney*, with the development of the sensory cinema, during the years Nineties. You can smell all scents while sitting comfortably in a film production hall able to recreate the olfactory sensations of a Northern boreal forest or the Mediterranean fragrances of South France".

Case Study 10: A New Visual Communication Tool to Facilitate Any *Decision-Maker* Task—The *State of Green Transition Index*

CS 10

Communicating green competencies through an online innovative data platform could be a major asset in both raising awareness among citizens and promoting the national *Circular Economy* performances worldwide. Communication becomes even more effective when it is promoted by the national government, relying on a team of qualified experts.

A high tech "green revolution" is involving **Denmark** whose Government has decided to invest in a major project addressed to convert the Country into the first fossil-fuel-free State by 2050. In order to achieve this target, a public-private partnership, named **State of Green** (www.stateofgreen.com), was established in partnership with the Confederation of Danish Industry, the Danish Energy Association, the Danish Wind Industry Association, and by appointment of H.R.H. *Prince Frederik of Denmark*.

The key idea is to gather the leading national stakeholders, representing different fields (e.g.: energy, climate, water, *Environment*) strongly linked to the concept of Circular Economy, with the aim to create an interactive tool showing, in one *click*, the Danish green growth economy evolution, compared with other Countries, worldwide.

Dual Citizen LLC conceived a tailor-made/custom-made Index named *State of Green Transition Index* which enables to compare 69 Danish cities to the rest of the world (59 Countries). The core idea was to offer the audience a visual communication tool showing the social and economic changes towards a sustainable lifestyle. The result is a multifunctional platform which allows, first of all, commercial and political decision makers, as well as international *Media*, to interact, triggering a communication process able to reorient global growth on more sustainable pathways;

@stateofgreendk (Facebook / Twitter)
www.stateofgreen.com (Web Site)

Source: Jeremy Tamanini, Strategic Communication Advisor in Circular Economy, Environmental policies and Sustainable Development – Founder and Lead Consultant at Dual Citizen Inc. – Washington DC, U.S.A.

© Springer Nature Switzerland AG 2019
M. Abbati, *Communicating the Environment to Save the Planet*,
https://doi.org/10.1007/978-3-319-76017-9_40

and, secondly, to inform Web surfers on *Environmental issues* boosting a spontaneous debate among them.

A radar histogram preview, based upon the 4th edition of the Global Green Economy Index (GGEI), is changed in real time according to the Selected Country, using quantitative and qualitative indicators (18 indicators, in total) to measure green performances and perception. A further in-depth option allows you to get easily a detailed percentage overview of each of four main areas of interest and their sub-categories with a detailed description of any indicator:

- Leadership and Climate Change: Head of State, *Media* Coverage, International Forums, Climate Change Performance.
- Efficiency Sectors: Buildings, Energy, Tourism, Transport.
- Markets Investments: Investment Attractiveness, Cleantech Technologies, Cleantech Commercialization, Green Investment Facilitation.
- *Environment* & Natural Capital: Agriculture, Air Quality, Water, Biodiversity & Habitat, Fisheries, Forests

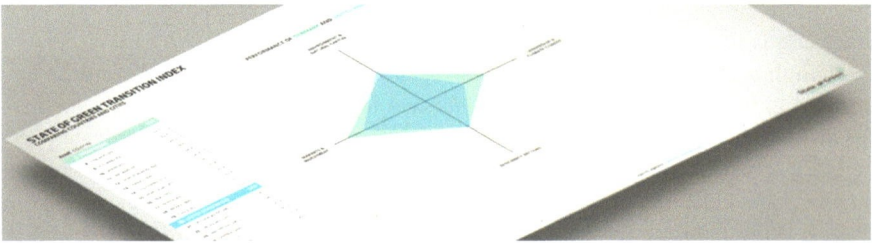

Fig. CS 10.1 Source: State of Green Transition Index—© Dual Citizen LLC (Copyright Holder)

An additional online virtual "journey", discovering Denmark's solutions for a greener future, is providing updated multimedia reports on a wide range of best practices and projects which are being carried on locally and a national level. A sustainable virtual "puzzle" where any piece corresponds to real actions pursued by some specific stakeholders in ten main sectors: Energy Efficiency, Heating & Cooling, Intelligent Energy, Wind Energy, Solar and other Renewables, Bioenergy, Water, Climate Adaptation, Resources & *Environment* and Sustainable Transportation.

A unique reference point to explore solutions, learn about products and connect with stakeholders, providing the visitors key information on the Danish ambitious *State of Green* projects, on sustainable performances and incentives, at national and international level. But also, a clever *arena* aimed at strengthening the synergy among companies, institutions and organisations as a driving force to disseminate the official *Danish* "green" brand beyond its borders in order to attract investments, business opportunities and partnerships.

The recent Curating Cities: *Sydney-Copenhagen* cultural and business cooperation regarding a special exchange of sustainable experiences is particularly significant. Danish furniture design, eye-opening business seminars and art exhibitions promoted by *State of Green* have been held in *Sydney* and *Melbourne* (Australia), in November 2016. An innovative way to exchange ideas and increase eco-awareness around the World, symbolized by a perception-installation which allows citizens to realise the urban fluctuating levels of carbon emissions through colour shades and sound melodies.

A new dimension in eco-communicating that can be summed up in the catch phrase "State of Green. Join the Future. Think Denmark" accompanying the promo video which explore the action programme to power "our common future".

Fig. CS 10.2 Source: State of Green Transition Index—© Dual Citizen LLC (Copyright Holder). "State of Green. Join the Future. Think Denmark" [Video]. StateofGreenDK (YouTube)

CS 11.1 Architecture and Design Are Evolving into Living Organisms

Architectural works and design objects are able not only to communicate the *Environment* but also to interact directly with the ecosystem as if they were "*living beings*". Hence the birth of **Théâtre Evolutif**—Evolutive Theatre (Fig. 41.1) created *in Bordeaux* (*France*) on the project by the **Architecture firm *OOZE*** with the purpose to "placemaking" an ideal site to exchange ideas, experiences, sensations, emotions and memories. The inspiration takes it origins from an existing urban space: the city square *André Meunieur*. A historical place that had lost in time its "social" value of a meeting point, transformed into an anonymous pedestrian area to walk through quickly, in the frantic daily routine.

What could be done to integrate this area into the city "ecosystem"? Coming back to the natural world. It was in fact considered the best *medium* to facilitate the dialogue and the communication. In the city heart of Bordeaux, then, a natural area, continuously evolving was born made up of a wild bush, cultivated fields and even an open-air farm to be eco-managed. This is the result of a decisional participated project involving directly the residents of *Saint-Michel* Quarter particularly sensitive to innovation at the service of *Environment*.

Théâtre evolutif, a pilot project that takes on a communicative value. The square, place of meeting and cultural exchange, heir of Classic Greek *agorà* is then turned into a "green stage" in direct contact with natural elements: sky, air, water, flora and fauna.

A place where everybody can change himself into an "actor" making room to his creativity to discover new forms of coexistence between Man and Nature.

---

http://espinasitarraso.com/index.html (Web Site)
Source: *Joaquim Tarrasó Climent*, architect and senior lecturer at the *Urban Design Architecture*, *Chalmers University of Technology*, Goteborg (Sweden).

Fig. CS 11.1.1 *Théâtre Evolutif—OOZE* (Copyright Holder) ©. www.ooze.eu.com

A laboratory of ideas, based on the bottom-up strategies, to rethink in an ecological sense the city lifestyle with eco-gestures able to shape continuously the *Environment* following the laws of Nature.

But also, a place where to establish the highest level of communication:

1. among the citizens, directly involved in the natural restyling of the urban place;
2. among the citizens and the natural elements, specifically Water, vital resource for any living being;
3. among the citizens and biodynamic cycle of biodiversity which is being acted upon a direct management of the green areas and the beehives by Man, for example. A strong *Environmental message* for the present and future generations.

CS 11.2   The Nature "Embraces" the Architecture to Eco-communicate

Communicating the respect of the *Environment* is strictly linked to the educational and informational activity. An increasingly close symbiosis that, as already mentioned, brought to coin the neologism "*edu*-communication". Can the *Environment* of our education and information reinforce the *Environmental message* received? The project **Primary School of Sciences and Biodiversity**, realized in A4 Est Quarter, as known as (aka) "*Seguin-Rives de Seine*" ZAC (Urban Developing Area) in *Boulogne-Billancourt*, near Paris, by the **Architecture firm "Chartier Dalix"** (www.chartier-dalix.com) has proved it is possible.

The structure hosting an educational centre composed of eighteen classrooms and a gym, opened to public, is surrounded by a "*Natural Environment*" created *ad hoc.* This building is fully-fledged part of the most innovative works from the *Environmental point of view.* The basic concept is to create a "personal" eco-system, above the urban context where it is built up and from which it wants to "defend itself" by using the strength of Nature.

The project mission reflects a trend increasingly followed on architecture: to create favourable conditions to the development of bio-diversity in the heart of the urban centres, united to the strong pedagogical value and to its aesthetics. The natural element, symbolically recalled by the lawn, the trees and its inhabitants, the birds and the insects, "embraces" all the building floors and it assumes a more effective "green" value if we consider that the area on which it has been built is part of a bigger project of *industrial reconversion area* where once the Renault car factory plants were situated. From an architectural point of view the building is composed of two structures linked by a "mineral wall", that is the masonry sinuous façade is strictly linked to a green linear surface to avoid contrasts between volumes, and it becomes metaphorically a "green cornerstone" able to join the world of a Flora to Fauna and the human beings.

It is not by chance that it has been considered an example of "*landscape as vital space*". The recreational park used by the School recreates deliberately a *Natural Environment* free from expressing itself according to the natural laws. This is what determines the same external aspects of the building in continuous evolution. An authentic "green heart" beating in a strongly urbanized context.

CS 11.3   The Square Communicates the Importance of the Water Resource

Water is a natural resource of vital importance in our eco-system. For this reason, it is necessary not to forget its role also where it is abundant. How can we communicate its importance to make the citizens aware of its rational use? The ***Benthem***

Fig. CS 11.3.1   Benthem Square of Rotterdam (Netherlands) in dry and wet conditions. © De Urbanisten (Copyright Holder). www.urbanisten.nl

Square **of** *Rotterdam*, Netherlands (Fig. 41.2), realized by the **Architecture Firm De Urbanisten**, represents a perfect example of architectural work that communicates a strong *Environmental message*. This is the first urban *"fountain of water"* ever made. It is available to the whole community that participates to its management and enhancement that is: the students and the teachers of *College Zadkine* and the Graphic *Lyceum*; the young theatre and the *David Lloyd* gym, the members of the adjacent Church; and all the inhabitants of *Agniese* district.

A very clever system of rainwater drainage shall ensure that the spaces usually animated by students, children, sportsmen or simple passers-by when the climate is dry turn into water reservoirs thanks to the rainwater collection that after a series of treatments and filtering is introduced in the city water system. A dynamic *Environment* suitable to the needs of the moment depending on the atmospheric conditions in full respect of the natural Environment and without water waste. The same project idea was born on the basis of a participative process among the stakeholders of reference who accepted enthusiastically the innovative "square of water".

Let us try to understand how it works: two non-deep "small-square reservoirs" act as water reservoirs when it rains. A third larger square collect the water only during significant downpours. On dry climate conditions the same spaces are used for skateboarding and rollerblading and as a dance floor and the bigger square into field sports like football, basketball, volleyball etcetera.

The rainwater is directed to the three reservoirs through special stainless-steel gutters whose diameter is superior to the average so that the skaters can go practicing during the dry days. The rainwaters are then collected through a wall and a *rain well* realized for this purpose.

The visual result is very agreeable. The water in its run forms a series of small falls and fountains that animate the whole area. The water is therefore a vital and precious resource reused to "water-feed" trees, lawns and field-flowers surrounding the square. Water is seen also as a "source" of amusement and wellness for Man and any other living being of the square.

To reinforce this valuable *Environmental message*, also the colours chosen to redecorate the small squares and the routes of this "square-garden" play an important role. The prevailing blue in all its nuances and the brightness of the stainless-steel gutters are a visual metaphor of the element "Water".

Case Study 12: Communicating the Sustainability as a *Driver* of Innovation: Three Italian Eco-Virtuous Examples

CS 12

Fig. CS 12.1 "Efficient Idea", sketch by Maurizio Abbati (Copyright Holder)

Source: Paolo Taticchi, Professor of Business Administration, Supply Chain Management and Sustainability – Director of MSc, MBAs and Global Initiatives at the Imperial College Business School - London, UK.

To meet the great *Environmental challenges* that our Planet is going to face in a view of the increasing growing of the population, the traditional model "*business as usual*" oblivious to the negative effects following the choices in the economic world are not sustainable, anymore. The availability of the natural capital by Man, that is the threshold of annual Sustainability calculated on the basis of the capacity of *Environmental resources* and services to regenerate, is always constantly at the limits and it tends to be overcome since the very beginning of each year (*Earth Overshoot Day*).

We must therefore rethink completely the way to do *business* addressing simultaneously to the *Environmental*, economic and social issues as "sides" of the same model, following a *holistic approach*. An authentic revolution of thought and action in the *company management* that is not limited to introduce *Environmental characteristics* and *products* and *services* but to revolutionize the whole *modus operandi* (the way of working). Through models that minimize the energy consumption; to create a "close circuit system" in which the waste of materials is reset to zero preferring the reuse of the same material in respect to its recycling; to create a value from the "waste"; to reduce in the consumers the sense of unjustified "need", to avoid that the demand of goods and services exceed the real need; to enhance the competence, the creativity and the forms of professional cooperation directed to an eco-sustainable management; focus on functionality before the profit. The final target is therefore to integrate really the values of Sustainability in the *business management*. Goals that can be achieved also through the investment in innovation in terms of *eco-design and other eco-efficient* methods able to reduce the energetic consumptions and each form of pollution, in each phase of production.

Hence the *Sustainable Business Models* can work as effective *medium* of communication for their capacity to pursue interests of a large group of *stakeholders* which includes also the *Environment* and the Society.

An aspect, that of *Environmental sustainability*, that boosts already significant *case studies* showing how the business management can pursue the highest *Environmental* and *sustainable values* without losing its competitive and productive *performance*.

This has allowed industrial realities to play a key role in the application of innovative responsible models that give a business propriety to the *Environmental* and *social targets*. This is the case of **Cucinelli S.p.A.** (www.brunellocucinelli.com/it/), a world market leader in the processing of cashmere and leather and crown jewel of the Italian fashion division. "A firm must not damage the *Environment* and the human beings" this is the philosophy of *Bruno Cucinelli*, proud founder and CEO (Chief Executive Officer) of his "humanistic enterprise" as he calls it. And just right from the Medieval village *of Solomeo*, a few kilometres far from Perugia (Center Italy), that was born the sustainable "revolution" of the firm model. Founded in 1978, *Cucinelli S.p.A.* encloses 360° the concept of Sustainability. The enterprise has been able to create 700 work places and a net of more than 300 suppliers, all Italians and from local areas (80% *Umbria* region). A quality *made in Italy* that allows the economic increase of the territory in full respect of its social and *Environmental values*. The financial income, in fact, enables optimal working conditions for the employees and a part of it is reinvested to improve the old village and the adjoining castle, useful stages for cultural and sports events. The company policy has promoted also the cultivation of olive trees and autochthone fruit trees thus creating new business opportunities respecting the *Zero-kilometre bio agriculture*. A productive model *socially oriented* that has been giving very good results in terms of turnover and has been catching the world's attention for its capacity of managing successfully more productive factors integrating competitiveness, innovation, evolution, cooperation, perseverance with an *Environmental* and *Sustainable Management* of the territory and its actors.

In other cases, it is the technological innovation to convey the *Environmental message*, instead. It is shown by the example **Black&Decker Italia** [*brand*: *DeWalt Industrial Tools* S.p.a.] (www.blackanddecker.it), with headquarters in the area of *Perugia*, a regional branch of the well-known multinational specialized in bricolage tools for professional and amateur users, mechanical and patented electronics solutions, fixing systems and much more. With its 200 employees, this entrepreneurial reality devotes much of production, equal to 90%, to the European market, reserving only 4% to the Italian one. So far nothing special but the investments in innovation of processing and of products. Under a careful analysis we realize, nevertheless, that the whole chain of production is inspired by key values of Sustainability without mentioning it. Productive efficiency and improvement in the health conditions and safety have always been the company cornerstones that since the beginning of the Years Nineties, has started a processing of organizational streamlining first, to face the needs of being competitive on the market. This involved the introduction of management systems finalized to the reduction of waste while keeping the performance productive quality. As well as the creation of a flexible supply system based on the territory to minimize the transport costs and the *Environmental impacts* and also an improvement of the socio-economic conditions and technological development of the whole area. A sustainable virtuous process generating an upstream return of some phases of production from Asia to Centre Italy with the contextual formation of new *clusters* (supporting the firm professionally in specific areas and for medium-long periods). To show that also a manufacturing *business oriented* organization, if it changes the traditional rules of production, can generate indirect benefits towards the sustainable management of the territory.

Sometimes, the economic crisis of market has inspired the sustainable innovation with a low *Environmental impact*. This is the case of **CSC S.r.l.**, an Italian small-medium enterprise, specialized in concrete supplied mainly to the local business in the region *Umbria*. The financial crisis of 2008 had seriously endangered the production, decreased of 25–30%, a threat also for the 40 employees of the firm. Hence the breakthrough that led to renew the business model by creating two innovative products at high *Environmental value*. Thus *"Plastoc"* was born, a new concrete obtained by mixing a percentage of plastic material with the traditional concrete. The project, supported by the local authorities and by the scientific support of *Terni* University during testing and certification, has made a product particularly performing and with great application potential. The lightness (50% less than the traditional conglomerate), the impact resistance and the capacity to absorb noise have made *Plastoc* very suitable for areas at risk from earthquake, opening the enterprise to new business opportunities far beyond the initial expectations. A second project then sees the birth of a new *joint venture company*, **RMT** (Recycled Materials of *Terni*) **S.r.l.** (limited Company), set up to allow the experimentation of a new building materials as a result of the union of some types of municipal solid waste with the concrete in view to decrease the waste quantity destined for landfill, thus reducing the disposal costs. The research in fact, has allowed to use dust, gravel and other debris collected in the street of medium-large sized cities (mainly *Terni* and *Perugia*) as inert components of the concrete for building. All this enables the reuse of 60–70% of the collected material, as well as the re-entering the market of all the selected scrap metals. An entrepreneurial process, based on the cross-sectorial cooperation that enhances, in a totally voluntary way, the principles of Sustainability without downgrading the quality of the final product, thanks to a technology able to transform "waste" into a new raw material. A perfect example of industrial symbiosis, open to innovation, able to create *Environmental*, social and economic benefits, simultaneously.

About the Author

Maurizio Abbati Born in Bologna in 1977, he first attended *Liceo Classico* (Classical High School) and in 2000 he graduated in *Law* at the University ***Alma Mater Studiorum of Bologna***, with an experimental *thesis* in *International Commercial Law and Maritime Navigation* prepared at the *Institute of Advanced Legal Studies in **London*** and the *University of **Southampton***.

He got two Second Level *Master's Degrees* respectively in *International Relations* and *Environment and Sustainable Development.* Between 2002 and 2006, he attended *workshops* in some of the most famous academic centers in Europe: *The International Institute for Industrial Environmental Economics IIIEE—**Lund*** (Sweden); *Université de Versailles-Saint-Quentin-en-Yvelines—**Paris** (France); Sykli School of Environment—**Helsinki** (Finland); Centro de Estudios Ambientales del Mediterráneo (CEAM)—Universidad Autónoma de **Barcelona*** (Spain).

This educational root led him to his professional career as *European Manager of the Environmental and Social Sustainability*, *Project Manager and Auditor ISO 14001.*

In the academic domain he published several essays, among them a specific chapter *in Luca Mezzetti*: *"La Costituzione delle Autonomie"*/The Constitution of Autonomies, *Edizioni Giuridiche Simone,* Naples, *2004.* He held conferences and courses among which the seminar *"Urban Sustainability and Local Agenda 21: the Ferrara Case Study"* at the ***International Venice University***, in July 2009 [in English]; the conference *Environmental Management Systems Boost Green* at the ***Italian Cultural Institute of Krakow***, May 2012; and the *lecture* on the same topic at ***Jagiellonian University of Krakow,*** *Economy* Faculty, May 2012 [in English].

In the professional field he has successfully completed European projects, held conferences, and has published some articles on *best practice* and on certified *Renewable Energy* and *Joint green public procurement* (for Ferrara Municipality), *Project Intelligent Energy pro-EE, 2009–2010* [in English]; *"5 Passi nella Green Way*/Five steps in the Green Way—*Green Management at Zero km: Sustainability*

© Springer Nature Switzerland AG 2019
M. Abbati, *Communicating the Environment to Save the Planet*,
https://doi.org/10.1007/978-3-319-76017-9

Systems in comparison", *Giants' Club* Sanremo (Italian Flowers' Riviera), May 2015 [in Italian-English]; Article: "The Environmental Message sets up Networks" in the educational magazine: "La Ricerca" by Loescher Publisher, Turin (November 2017).

In particular, the author has been experiencing, for almost 10 years, a multitasking role with advanced communication skills. Consequently, he was able to create a network of international experts operating in different sectors: a selected community whose contributions are included in this manual in "*The interviews with professionals*" part.

After a long training and a careful examination, he got the journalist's press card and since then he has been working as a freelance *qualified editorial journalist* in the Principality of Monaco, in France, and in Italy.

He has been involved in press projects in the different known languages (English, French, Spanish, and Italian as a mother tongue) writing articles specialized in legal, environmental, sustainable, and communication topics and he is the author of this book on "***Environmental Communication***" also in the Italian version.